Advances in Solar Thermal Energy Harvesting, Storage and Conversion

Guest Editors

Yu Qiu

Qiliang Wang

Chao Xu

Evangelos Bellos

MDPI

Basel • Beijing • Wuhan • Barcelona • Belgrade • Novi Sad • Cluj • Manchester

Guest Editors

Yu Qiu
School of Energy Science and Engineering
Central South University
Changsha
China

Qiliang Wang
Department of Architecture and Built Environment
University of Nottingham
Nottingham
UK

Chao Xu
School of Energy Power and Mechanical Engineering
North China Electric Power University
Beijing
China

Evangelos Bellos
Department of Mechanical Engineering
University of West Attica
Athens
Greece

Editorial Office

MDPI AG
Grosspeteranlage 5
4052 Basel, Switzerland

This is a reprint of the Special Issue, published open access by the journal *Energies* (ISSN 1996-1073), freely accessible at: https://www.mdpi.com/journal/energies/special_issues/L7428MG65A.

For citation purposes, cite each article independently as indicated on the article page online and as indicated below:

Lastname, A.A.; Lastname, B.B. Article Title. *Journal Name* **Year**, *Volume Number*, Page Range.

ISBN 978-3-7258-4253-7 (Hbk)
ISBN 978-3-7258-4254-4 (PDF)
https://doi.org/10.3390/books978-3-7258-4254-4

Cover image courtesy of Yu Qiu

Contents

energies

MDPI

Article

Modern SCADA for CSP Systems Based on OPC UA, Wi-Fi Mesh Networks, and Open-Source Software

Jose Antonio Carballo [1,2,*], Javier Bonilla [1,2,*], Jesús Fernández-Reche [1], Antonio Luis Avila-Marin [1] and Blas Díaz [1]

[1] Centro de Investigaciones Energéticas, Medioambientales y Tecnológicas-Plataforma Solar de Almería (CIEMAT-PSA), Point Focus Solar Thermal Technologies, P.O. Box 22, E-04200 Tabernas, Spain; jesus.fernandez@psa.es (J.F.-R.); antonio.avila@ciemat.es (A.L.A.-M.); bdiaz.serviciosexternos@psa.es (B.D.)

[2] Centro Investigaciones Energía SOLar (CIESOL), Joint Institute University of Almería—CIEMAT, P.O. Box 22, E-04120 La Cañada, Spain

* Correspondence: jcarballo@psa.es (J.A.C.); javier.bonilla@psa.es (J.B.)

Abstract: This study presents a methodology for the development of modern Supervisory Control and Data Acquisition (SCADA) systems aimed at improving the operation and management of concentrated solar power (CSP) plants, leveraging the tools provided by industrial digitization. This approach is exemplified by its application to the CESA-I central tower heliostat field at the Plataforma Solar de Almería (PSA), one of the oldest CSP facilities in the world. The goal was to upgrade the control and monitoring capabilities of the heliostat field by integrating modern technologies such as OPC (Open Platform Communications)) Unified Architecture (UA), a Wi-Fi mesh communication network, and a custom Python-based gateway for interfacing with legacy MODBUS systems. Performance tests demonstrated stable, scalable communication, efficient real-time control, and seamless integration of new developments (smart heliostat) into the existing infrastructure. The SCADA system also introduced a user-friendly Python-based interface developed with PySide6, significantly enhancing operational efficiency and reducing task complexity for system operators. The results show that this low-cost methodology based on open-source software provides a flexible and robust SCADA architecture, suitable for future CSP applications, with potential for further optimization through the incorporation of artificial intelligence (AI) and machine learning.

Keywords: SCADA; heliostat; OPC UA; Wi-Fi mesh network

1. Introduction

1.1. Concentrated Solar Power

Concentrated solar technologies (CSTs) are renewable energy technologies focused on concentrating sunlight onto small areas using mainly lenses or mirrors. Inside CST, concentrated solar power (CSP) concentrates solar energy and transforms it into thermal energy at high temperatures. Then, the thermal energy is used to produce steam that drives a turbine connected to an electrical power generator, unlike, for example, photovoltaic (PV) technology, which converts sunlight directly into electricity. In the case of central tower systems, which are the most promising CSP technology, the heliostat is responsible for reflecting and concentrating solar energy on the receiver. Due to the constant Sun movement, the heliostat has to modify its position to reflect and concentrate the solar energy into the receiver and transform it into thermal energy. Central tower systems are composed of several miles of heliostat that have to be controlled individually.

CSP plays a crucial role in the global energy transition for several reasons. First, CSP plants can generate large amounts of clean electricity, making them suitable for utility-scale power generation. The concentrated heat can reach temperatures high enough to produce steam for traditional steam turbine generators, which are a proven and efficient method of generating electricity. Secondly, one of the significant advantages of CSP over other renewable energy technologies is its ability to store a large amount of thermal energy at a low cost. This storage capability allows CSP plants to generate electricity even when the sun is not shining, such as at night or during cloudy periods. This feature provides a stable and reliable power supply. The ability to store and dispatch power makes CSP an excellent complement to other intermittent renewable energy sources like PV or wind power, helping to stabilize the grid and ensure a continuous power supply.

Despite its advantages, concentrated solar power (CSP) faces significant challenges compared to other major renewable energy technologies. CSP plants generally require higher initial capital investment due to the complexity and scale of the technology, which in turn leads to a relatively high levelized cost of electricity (LCOE) [1]. Several factors contribute to this issue. First, the technological maturity of CSP lags behind that of photovoltaics (PV) and wind power technologies, which have benefited from rapid advancements, economies of scale, and technological innovation. As a result, CSP has not seen the same level of widespread deployment and investment, limiting potential cost reductions through learning curves and mass production. Furthermore, the operation and maintenance (O&M) of CSP plants are more complex than that of PV systems, requiring specialized skills and infrastructure, thus increasing operational costs. This challenge is exacerbated by the relatively limited innovation and experience within the CSP sector. Additionally, CSP technology has been slower to adopt digitalization and automation compared to PV and wind power, missing opportunities for significant reductions in costs and improvements in operational efficiency. The integration of advanced digital tools and automated systems could substantially lower operational and maintenance costs, enhancing overall efficiency and reducing LCOE, as seen in other sectors, for example, power grid management [2]. For example, efficient data processing and the extraction of hidden patterns can dramatically improve knowledge discovery, offering valuable solutions for optimizing system operations. By combining data mining techniques with knowledge discovery, CSP can transition from traditional SCADA systems to intelligent, data-driven frameworks capable of optimal, adaptive, real-time decision-making [3]. Also, modern SCADA systems, leveraging data-driven approaches, can significantly reduce damages through effective alarm management. These systems facilitate the identification of root causes for major failure modes, helping pinpoint critical subsystems. By doing so, they not only optimize maintenance costs but also contribute to design improvements, enhancing their reliability and efficiency over time [4].

1.2. *Plataforma Solar de Almería*

The Plataforma Solar de Almería (PSA) is a leading global research center focused on solar technology, particularly concentrated solar power (CSP). Located in Spain's Tabernas Desert, PSA benefits from high solar radiation, making it ideal for solar research and innovation. The center plays a key role in CSP development, testing solar thermal components under real-world conditions, and advancing sustainable solar power systems. PSA's research sets industry benchmarks and influences energy policy decisions, with collaborations involving universities, research institutions, and industry partners worldwide [5].

PSA hosts various CSP technologies, including parabolic troughs, solar towers, and Fresnel reflectors. The center is pioneering the integration of artificial intelligence (AI) and the Internet of Things (IoT) to optimize solar power plant operations. A major focus is the CESA-I central receiver system, see Figure 1, which uses 300 heliostats with a surface area of 39.6 m^2 each. Despite over 20 years of operation, the heliostat field remains in optimal condition due to ongoing maintenance. The system has recently incorporated an AI-based solar tracking system (HelIoT), allowing it to operate alongside traditional tracking systems.

(a) CESA-I central receiver system.

(b) Heliostat and mirror image of the tower.

Figure 1. CESA-I system captions

Also for this work, a new Wi-Fi mesh communication system has been successfully deployed for the first time for use in a CSP installation with excellent results; firstly, it allows removing costs in civil works and material due to the elimination of wiring, and secondly, it allows great flexibility in terms of communications. This communication system is used by the newly developed smart heliostat (HelIoT).

1.3. Smart Heliostat

The HelIoT smart heliostat is an innovative technology designed to optimize solar energy collection through precise and intelligent tracking of the Sun. This system employs Internet of Things (IoT) technology, low-cost hardware, computer vision, and machine learning algorithms to reduce the need for manual interventions and costs, and improve efficiency, reliability, and automation of solar power plants [6]. Furthermore, its decentralized control architecture enables scalability, making it suitable for both small and large-scale solar fields. HelIoT also provides highly valuable information to the main control system, further enhancing the performance of the entire plant and reducing costs. This includes predictions on cloud movements and solar energy transients, which enable more efficient management and operation of the solar power plant [7].

The PSA is the main actor developing and testing the HelIoT smart heliostat [6]. The research of PSA in this field focuses on the design and implementation of wireless communication technologies, sensor networks, data acquisition platforms, and control algorithms based on neural networks to create a robust and flexible heliostat field capable of autonomous operation under varying environmental conditions. Performance evaluation tests have shown significant improvements in tracking precision and system reliability for the HelIoT system [6].

1.4. Supervisory Control and Data Acquisition System in Energy Sector

Supervisory Control and Data Acquisition (SCADA) systems are a key part of plants in the energy sector [2], providing essential monitoring and control processes that are crucial for efficient energy management [8]. SCADA systems—widely used in various energy applications, including power generation, transmission, and distribution [9]—enable real-time data acquisition, system supervision, and automated control. These capabilities are vital for maintaining the reliability of power grids, as they allow for continuous monitoring and fast identification and correction of issues. SCADA systems also support data collection and analysis, which can be used for predictive maintenance and refinement of operational strategies, ultimately enhancing system performance and reducing downtime. However, increasing connectivity and integration of SCADA systems expose them to cyber attacks, which requires the implementation of robust security measures to protect critical infrastructure [10–12].

The recent appearance of open-source SCADA systems and IoT technologies has democratized access to advanced monitoring and control systems, for example, open-source platforms combined with low-cost hardware and microcontrollers enable the development of low-cost SCADA systems [13]. These systems leverage IoT to provide enhanced connectivity and data acquisition capabilities. These types of solutions not only represent a cost reduction because they are open source, but they also present a continuous updating and evolution by large communities of developers that allow SCADA systems developed with these tools to be kept up to date.

Specifically, CSP plants require the coordination of multiple processes to achieve efficient operation due to the complexity of the system. To manage this complexity, SCADA systems play an essential role in ensuring that each process is synchronized and operated at optimal performance levels. As commented before and despite their relevance, CSP plants have been relatively slow to adopt advanced digitalization tools, and SCADA systems, as a key component of CSP infrastructure, can benefit particularly from digital advancements. Enhanced digitalization can significantly improve the functionality and effectiveness of SCADA systems, offering better integration, more precise control, and improved decision-making capabilities. Embracing these digital tools could greatly enhance the efficiency and reliability of CSP operations. For example, machine learning algorithms can forecast equipment failures in solar power plants, allowing timely maintenance and minimizing downtime. This predictive capability is essential for maintaining high operational efficiency and reducing maintenance costs. In addition, machine learning can offer valuable insight into key operational aspects, such as transients in solar energy. By analyzing these transient variations, machine learning algorithms can improve overall plant performance, mitigate critical situations, extend plant operational lifespan, and further reduce operational costs [7].

In conclusion, CSP has the potential to significantly increase the penetration of renewable energy in the global energy mix, particularly due to its ability to manage power production through thermal energy storage. To fully realize this potential, the CSP must focus on reducing costs and enhancing efficiency. Despite its current challenges, CSP has substantial opportunities for improvement, mainly because it has yet to widely adopt advanced tools from energy digitalization, such as artificial intelligence (AI) and the Internet of Things (IoT). SCADA systems are crucial in the control and operation of CSP plants, particularly given the complexity of these systems. Note that SCADA systems could benefit particularly from digital advancements. Research centers like PSA, which is dedicated to being at the forefront of solar technology, should prioritize the study and development of SCADA systems that integrate the latest digital tools. Incorporating technologies such as HelIoT into SCADA systems can drive technological advances, improve operational efficiency, and reduce costs. This approach will make CSP a more competitive and reliable source of renewable energy.

For all that, this work presents the development of a new SCADA methodology designed to incorporate cutting-edge digitalization technologies such as AI, IoT, and Wi-Fi mesh networks. Tailored for a research environment, the system is highly flexible, enabling seamless adaptation to new developments such as HelIoT, experimental setups, new prototypes, storage solutions, and advanced aiming strategies. Using open-source platforms, the system reduces costs. In addition, the proposed SCADA enhances plant monitoring, control, and overall efficiency. This approach significantly contributes to the competitiveness and reliability of CSP technologies in the evolving renewable energy landscape.

The following work is structured as follows: The Materials and Methods section describes the architecture of the proposed SCADA, which includes a server managing information traffic between different components. The Results and Discussion section presents the test and the performance evaluation of the new SCADA implementation for CESA-I. Finally, the Conclusion section summarizes the key findings and emphasizes the potential of the new SCADA system to enhance the competitiveness and reliability of CSP technologies.

2. Materials and Methods

The architecture of the proposed control system is based on a server that is responsible for managing the information traffic between the different actors (see Figure 2).

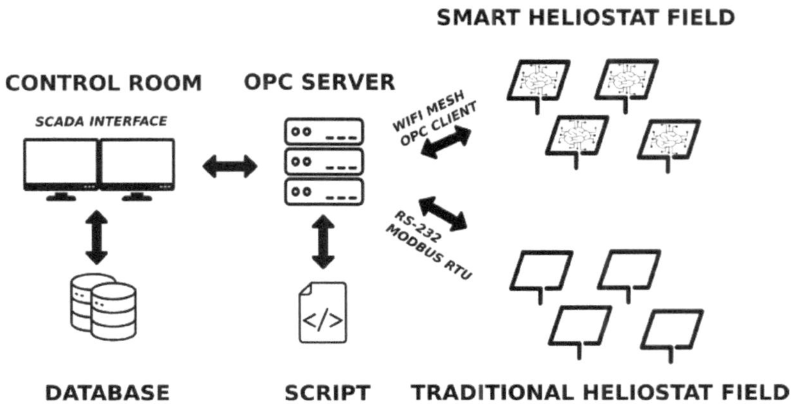

Figure 2. SCADA CESA-I architecture.

In this work, three major developments have been carried out, firstly, the work was conducted to develop the OPC server, which is the key element of the whole project. Secondly, the gateway between the traditional communication system and the server is used to integrate the new developments with the traditional local control. Finally, the SCADA interface allows operators to interact with the CESA-I field.

2.1. OPC Server

As illustrated in Figure 2, the core component of the system is an Open Platform Communications Unified Architecture (OPC UA) server, which acts as a central hub, facilitating communication between the different subsystems.

The OPC UA protocol is a widely adopted client-server communication standard in industrial automation, designed for secure, reliable, and platform-independent data exchange. As the successor to OPC Classic, OPC UA addresses the limitations of platform dependency and enhances interoperability across different systems and networks. Key improvements include platform independence, enhanced security features, and support for complex data types, making OPC UA suitable for both small-scale and large-scale industrial applications [14]. Its service-oriented architecture (SOA) allows modularity and facilitates the integration of additional functionalities. The robust security model incorporates encryption, authentication, and auditing, ensuring data integrity and confidentiality in sensitive industrial environments such as CSP systems.

The present work employs the open-source Python [15] library opcua-asyncio [16], an implementation of OPC UA standards that provides asynchronous support for non-blocking operations. Such functionality is crucial for high-performance Industrial IoT (IIoT) systems, where real-time data processing is essential. The server implementation also includes a SubHandler-Class that manages subscription-based notifications, enabling efficient handling of events and data changes to provide real-time feedback to the SCADA system. The server leverages Python's asyncio library to concurrently manage multiple tasks, a critical feature for maintaining real-time performance in industrial environments. This implementation represents a robust approach to managing a large-scale heliostat field using OPC UA technology.

Using this library, the OPC UA server constructs a hierarchical structure of nodes, each representing one of the 300 heliostats in the CESA-I field. Each device node is an instance of a heliostat object type, encapsulating attributes and behaviors specific to both the traditional and the smart heliostat. Table 1 shows the structure of the heliostat node (nested variables and their description).

Table 1. Heliostat object type.

Node/Variable	Description
Command_node	A child node of the heliostat node used for sending commands to the heliostat's local control system.
Command	Writable variable that holds the current command to be executed (e.g., sun tracking, stop, cleaning).
Command_op	Writable variable containing additional operational parameters for the command.
Diag	Diagnostic variable covering aspects of the system's diagnostics, such as failure states and local control conditions.
Event	Logs event information related to the heliostat (e.g., setpoint achievements, reference point detections).
HEl_IoT_id	Represents the unique identifier of the heliostat device.
Smart_node	A child node of the heliostat node for advanced operations within the heliostat system.
State	Writable variable storing the current state of the smart node; used to trigger smart operations.
HelIoT	Writable variable storing complex data related to the heliostat operations, formatted as a JSON-like string.
State	Holds the operational state information of the heliostat (e.g., sun tracking, stop).
pos_az and pos_el	Store the actual azimuth and elevation positions of the heliostat.
setpoint_az and setpoint_el	Indicate the setpoint positions for the azimuth and elevation axes.

As depicted in Figure 2, the OPC server interacts with various OPC clients. Firstly, the OPC server communicates with the SCADA interface through the PSA communications network, allowing operators to provide or receive information. Additionally, external operators can automate external processes through an application or script through another OPC client. The OPC server also exchanges information with the smart heliostats deployed at CESA-I, which are equipped with OPC clients and communicate over a Wi-Fi mesh network. Finally, the OPC server is connected to traditional heliostats through an OPC-Modbus gateway developed for this work. All data exchanges are handled by the server's event subscription system. This modular and flexible design facilitates the seamless integration of new technologies, such as advanced data analytics or machine learning algorithms, making it highly adaptable to future innovations at PSA.

Furthermore, the architecture incorporates log storage systems, which are essential for maintaining historical data, tracking operator actions, helping in system analysis, and optimizing performance over time. The tracking of server errors is implemented through logging mechanisms, ensuring efficient troubleshooting and system reliability.

2.2. Gateway Modbus-OPC

As commented before, with the main objective of maintaining the existing Modbus-based wired communication and local heliostat controls of the CESA-I field, a gateway has been developed to interface the traditional architecture with the new OPC server. This gateway functions as an application with an integrated OPC client that periodically (less than 2 s) queries the values of the node variables corresponding to each heliostat's local control. On the one hand, if modifications are detected in the local control data, the gateway updates the variable in the corresponding heliostat node in the OPC server by writing the new values. On the other hand, if a Command child node on the server is altered by another client, the gateway is responsible for transmitting this command to the corresponding local heliostat control via Modbus.

Given the design of the communication network in the traditional heliostat field, where the 300 heliostats are distributed across 16 independent RS-232 communication lines, this communication process consists of 16 parallel threads. Each thread operates continuously, sequentially querying (polling function, see Figure 3) each heliostat on its assigned communication line about the variables defined in the heliostat object type. If a change in any variable is detected, its updated value is sent to the OPC server through the OPC client.

```
def polling(self, reg_name_list = ['ESTADO','EVENTOS','POSICION_AZ',
 'POSICION_EL','CONSIGNA_AZ','CONSIGNA_EL','DISTANCIA_CENITAL_SOLAR',
 'AZIMUT_SOLAR','CORRECCION_REFRACCION','HORA_CALCULO_SOLAR'], print_poll = True):
    ''' Pregunta al heliostato informacion sobre los registros listados (reg_name_list)'''
    polling = {}
    for name in reg_name_list:
        polling.update({name:0})
    for reg in reg_name_list:
        #Posicion y consigna elevacion y acimut
        if self.READABLE_REG[reg] in [20,21,22,23]:
            #value = self.heliostat_modbus.read_register(self.READABLE_REG[reg], number_of_decimals=0,functioncode=3, signed=True)
            value = self.heliostat_modbus.read_register(self.READABLE_REG[reg],signed= True,functioncode=3)
            polling[reg] = value
        #Distancia cenital, azimut solar, correccion refraccion, hora calculo solar
        elif self.READABLE_REG[reg] in [24,26,28,30]:
            value = self.heliostat_modbus.read_float(self.READABLE_REG[reg])
            polling[reg] = value
        #diagnostico az, diagnostico ele
        elif self.READABLE_REG[reg] in [18,19]:
            value = self.diagn(self.READABLE_REG[reg])
            polling[reg] = value
        #estado
        elif self.READABLE_REG[reg] == 16:
            value = self.state()
            polling[reg] = value
        #eventos
        elif self.READABLE_REG[reg] == 17:
            value = self.event()
            polling[reg] = value
        elif self.READABLE_REG[reg] == 9999:
            value = self.heliostat_modbus.read_registers(16,4)
            polling[reg] = value
        else:
            print('Error: Registro no disponible')

    if print_poll:
        print('Polling:')
        print('======================')
        for reg in reg_name_list:
            print(str(reg)+'--->  '+str(polling[reg]))
        print('======================')

    return polling
```

Figure 3. CESA-Modbus library polling function.

As mentioned earlier, these 16 processes run cyclically and indefinitely, unless the OPC client is notified of a modification in a server node via a subscription event, such as a change in the command node. In such a case, the corresponding thread stops its routine querying process and sends the updated command to the local control of the specified heliostat. Once the command is transmitted, the querying process resumes.

This loop process is fully implemented in Python and has necessitated the development of a custom Python library, called CESA-Modbus, for communication with the local heliostat controls. This library helps us to abstract from the lower-level programming needed for Modbus communication and is mainly based on the Python library MinimalModbus. Note that, as commented on earlier, the gateway also is equipped with an OPC client, developed using the opcua-asyncio library, which handles events and data exchanges with the OPC server.

2.3. SCADA Interface

A new interface for the SCADA system was developed in this work. Like the previous developments, this was entirely implemented in Python, primarily utilizing the open-source library PySide6. The initial phase of the development involved gathering all the requirements from the operators of the CESA-I system, some of whom have over 30 years of experience operating these systems. Afterward, a visual style was defined, and all graphical components, windows, and user options were designed. This process was iterated several times in collaboration with the operators, incorporating their suggestions to optimize usability.

Subsequently, all the necessary code to provide functionality to the components was developed. The interface was created as a standalone Python application capable of exchanging information with the OPC server through an integrated OPC client and running on any device within the PSA network, ensuring flexibility and accessibility across the entire infrastructure.

The main window is divided into four areas (see Figure 4): heliostat field, information, control, logging, and console.

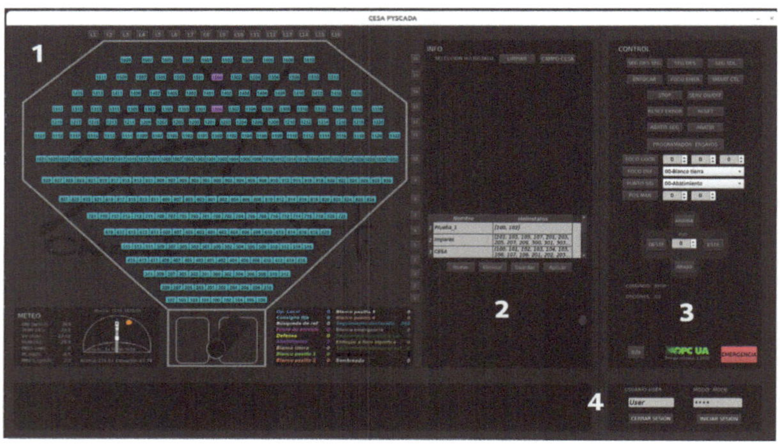

Figure 4. SCADA main window (1—heliostat field, 2—information, 3—control, 4—logging and console).

The primary area, the heliostat field, contains three different sub-areas: meteorological information, a legend, and a central area that represents the CESA-I field.

The meteorological information sub-area (see Figure 5) provides the operator with all the necessary real-time meteorological data, primarily related to solar conditions, required for system operation. This information is obtained from another OPC server at PSA that publishes meteorological data from the extensive network of sensors deployed across PSA. This server was recently introduced as part of PSA's modernization efforts, improving accessibility to the data.

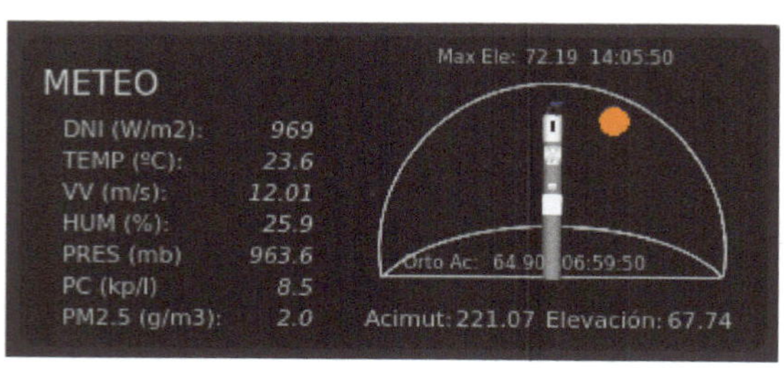

Figure 5. Meteorological data subsection.

As shown in Figure 5, variables such as direct normal irradiation (DNI), temperature, wind speed, humidity, atmospheric pressure, total particles, and particulate matter under 2.5 microns are continuously sampled and updated. Additionally, a dynamic graph provides information on the relative position of the sun (azimuth and elevation), along with the times and values for sunrise and sunset.

The central part of the main area contains a scaled representation of the CESA-I field, where each button corresponds to a heliostat. The button background color indicates the current state of the heliostat, while the color of the text on the button signifies whether the state has been reached (black) or is in transition (green) (see Figure 6).

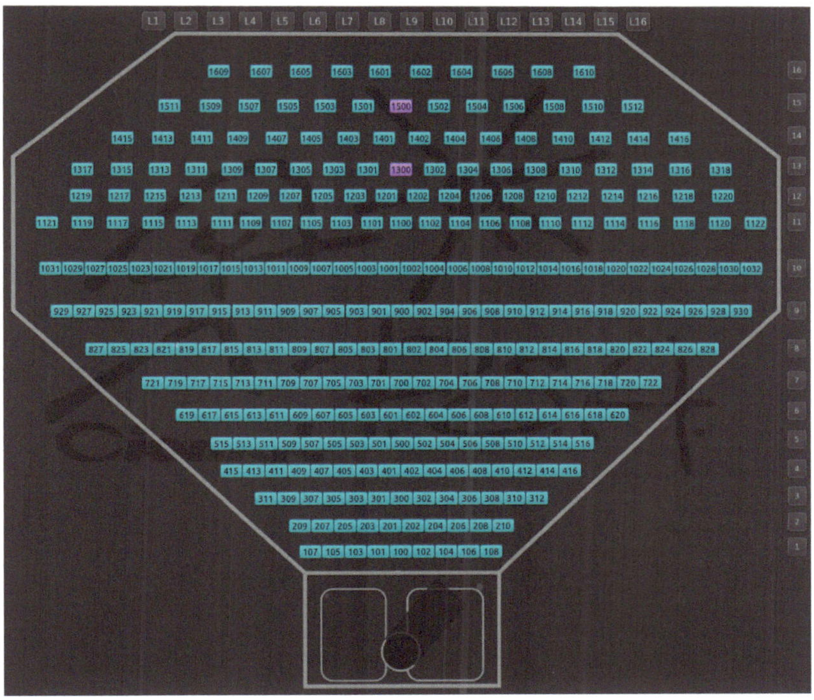

Figure 6. Central subsection of the main area.

In this diagram, individual heliostats can be selected by clicking on the corresponding button, dragging and dropping, or using the top or side rows of buttons to select heliostats by communication lines (L1 . . . L16) or rows (1 . . . 15), respectively.

At the bottom of the diagram, which represents the tower and auxiliary elements, a rectangular polygon shows the shadow projection over the field. The transparency of the polygon's color indicates the shadow's intensity based on DNI, providing operators with valuable information when selecting heliostats for testing.

A dedicated sub-area within the main interface displays the color legend for heliostat states in Spanish (see Figure 7). Each state is represented by a unique background color for the corresponding buttons, which matches the text color in the legend. Additionally, a counter next to each state indicates the number of heliostats currently in that specific state. Operators can easily select all heliostats in a given state by clicking on the respective text in the legend.

The meaning of each state is as follows:

- Operación local: Heliostat operating under local control only.
- Consigna fija: Heliostat positioned at a fixed setpoint for both azimuth and elevation.

- Búsqueda referencia: Heliostat searching for the 'zero reference' in both azimuth and elevation.
- Fuera de servicio: Heliostat that is out of service and not operational.
- Defensa: Heliostat in a defensive position to protect against high winds.
- Abatimiento: Heliostat in its rest or stow position.
- Blanco tierra, pasillo 1, pasillo 2, pasillo 3, pasillo 4: Consecutive safety corridor points used to guide the heliostat toward the receiver.
- Seguimiento desfasado: Heliostat is tracking, but aimed at a standby position near the receiver, not directly at it.
- Blanco emergencia: A designated target point used in emergency situations.
- Seguimiento normal a caldera: Heliostat tracking a specific point at the top of the tower, different from the main receiver.
- Enfoque a foco significativo: Tracking toward a user-defined significant target.
- Seguimiento normal al sol: Heliostat tracking the sun with its aim vector parallel to the solar vector.
- No comunica: Communication with the heliostat has been lost.
- Sombreado: The heliostat is shaded by the tower.

This system ensures that operators can quickly interpret the current state of the heliostats and manage them efficiently.

Figure 7. Legend subsection of the main area.

To the right of the main area is the information section (see Figure 8), which displays the numbers of the selected heliostats and includes a component for managing predefined heliostat groups. This component allows users to create, delete, and load predefined heliostat lists. Additionally, this section features a "CAMPO CESA" (CESA field in Spanish) button that enables the selection of all heliostats in the field at once.

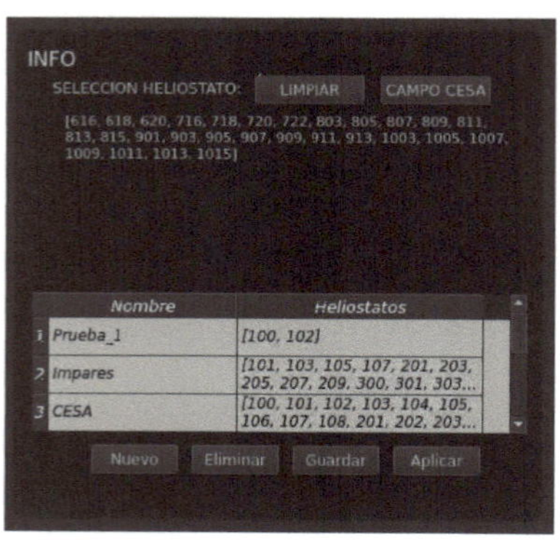

Figure 8. Information area.

The "control" area (Figure 9) is next located, which contains buttons to control commands to either a single heliostat or a group of heliostats, selected using the various methods previously described. Additionally, a text display shows the most recent command and its associated options. At the bottom of the control area, there is an "INFO" button that opens a window providing information about the interface. An "OPC UA" icon is also visible, indicating the status of communication with the server (green if the connection is active and red in case of failure) along with the refresh time in milliseconds. Finally, there is an "EMERGENCIA" (emergency in Spanish) button, which safely sends all heliostats in the field into stow mode.

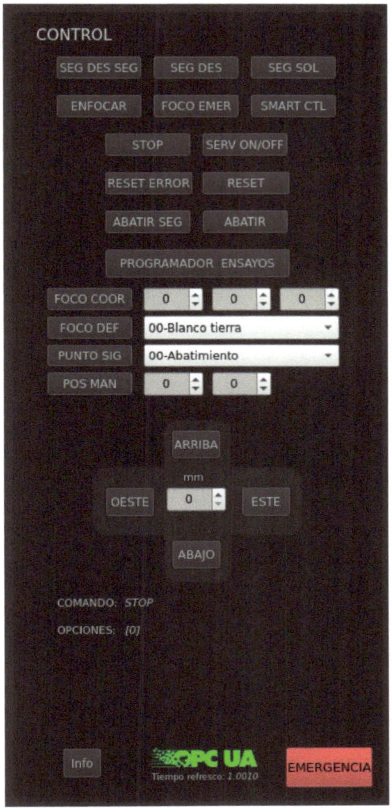

Figure 9. Control area.

In the lower right section, the "Login" area (see Figure 10) is dedicated to user management for the interface. In this area, users can log in, log out, and view the currently logged-in user. A user permission control system has been developed to restrict the actions that each user can perform. There are three permission levels: guest, operation, and configuration. Each user is assigned a permission level that limits their allowed actions. The guest level only allows users to consult data, the operator level enables actions related to the routine operation of the field, such as sending heliostats to tracking mode, and the configuration level grants full access to all interface functionalities, including both routine field operations and heliostat configuration settings.

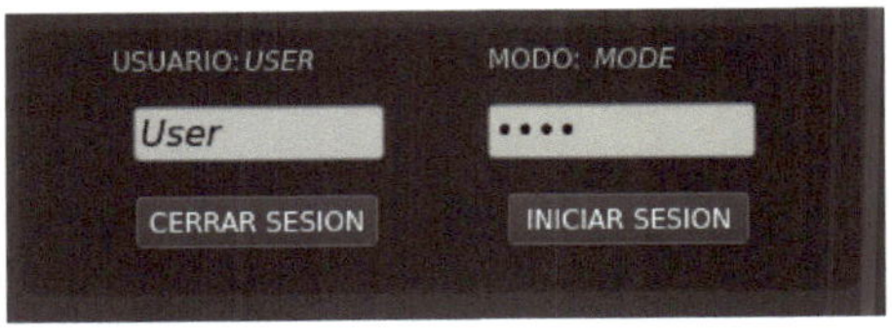

Figure 10. Login area.

Finally, in the lower left section (see Figure 11), the console area displays the most recent actions performed on the interface by the operators. The actions are shown in the following format: date, time, user, permission level, command/action, and options. These logs are also automatically saved to a log file for future reference.

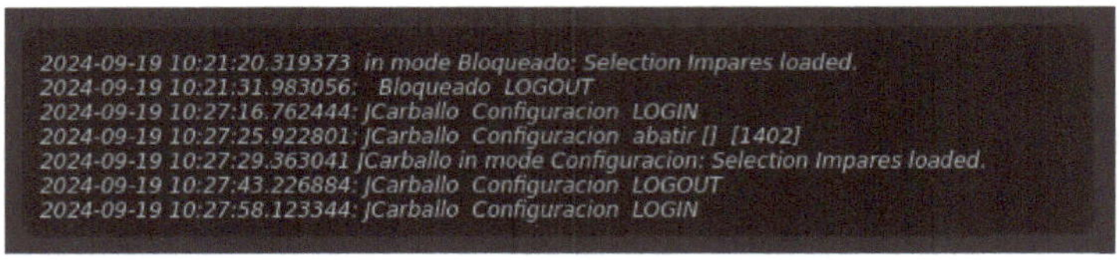

Figure 11. Login area.

Clicking on any button representing a heliostat within the CESA field diagram in the central area opens a new window (see Figure 12), which provides all available information about the selected heliostat, along with operational and control options. These options will be enabled or disabled depending on the permission level assigned to the user currently operating the interface.

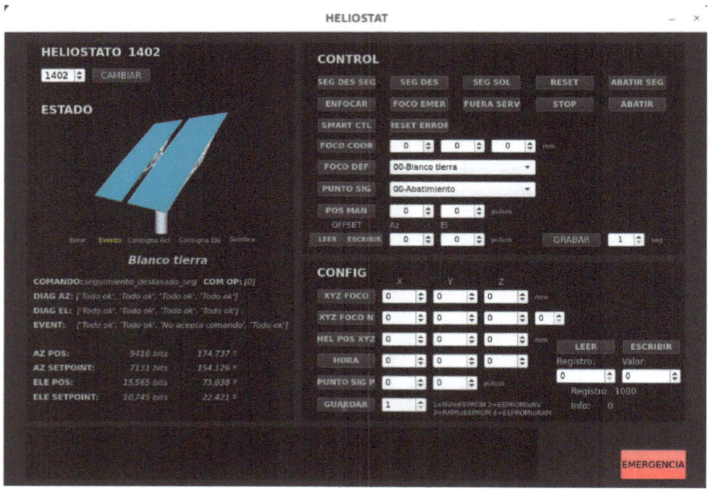

Figure 12. Heliostat window.

The left side of the window is dedicated to displaying relevant heliostat information. Here, an image showing the approximate position of the heliostat is presented, along with five indicator LEDs accompanied by text. These LEDs will remain blank unless a specific condition is met, such as an error, an event, reaching the setpoint in elevation or azimuth, or being in the shadow of the tower, in which case the corresponding text will change its color. Below this section, a larger text field displays the current state of the heliostat, followed by a structured display of all data collected from the local control system during each polling cycle.

The heliostat window contains three additional areas: the control area, which provides all the options related to the normal operation of the heliostat; the configuration area, where parameters of the heliostat's local control can be modified; and a console, similar to the one in the main window, displaying recent actions. Additionally, the window is equipped with an emergency button, identical to the one in the main interface, which sends the heliostat into a safe stow position in case of an emergency.

If, at this point, the "SMART CTRL" button in the control area is clicked, a new window will open (see Figure 13), allowing the initiation of the smart control of the heliostat, provided the heliostat is equipped with the smart control system (HelIoT). In this window, in addition to the previously available heliostat information, data related to smart tracking is displayed. This includes images used by the system for tracking, along with neural network detection results, and two graphs showing the tracking error, the setpoint, and the heliostat's position in both azimuth and elevation.

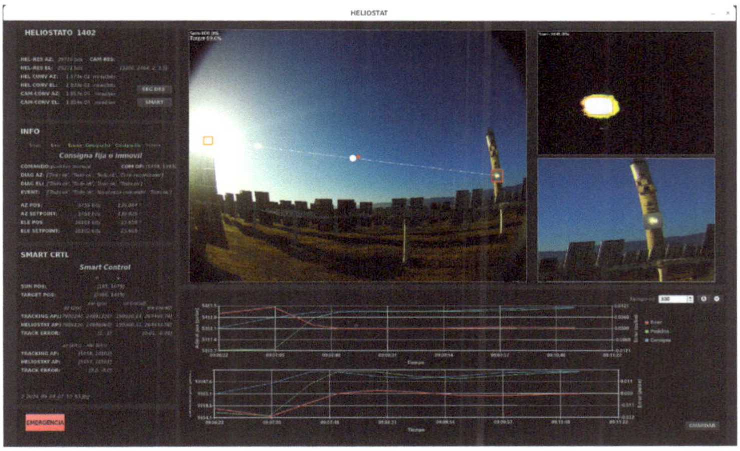

Figure 13. Smart heliostat control window.

3. Results

Various tests, described below, have been conducted on each component of the system to identify and prevent potential operational issues.

3.1. Wi-Fi Mesh Communication Test

The first set of tests focused on the Wi-Fi mesh communication network deployed across the heliostat field. This test aimed to verify the stability and reliability of the wireless communication system under real-time operational conditions. The test involved connecting 10 smart heliostats at different locations within the CESA-I heliostat field and continuously pinging each one during normal operation. The Wi-Fi mesh network successfully maintained stable connections with all smart heliostats distributed across the field, demonstrating reliable communication and network performance under operational conditions. Data transmission rates were consistently high, with minimal packet loss observed (less than 0.1%) even during peak operational periods. Latency remained below 100 ms, ensuring that real-time commands were delivered promptly to the heliostats. The mesh structure also demonstrated robustness against single-point failures, with no significant performance degradation when individual nodes temporarily lost connection. In general, the Wi-Fi mesh system proved to be both resilient and efficient for the intended application. Note that during normal operation, each heliostat is controlled locally and only needs to exchange information if a new setpoint/command is sent or if detailed status information is requested, the latter does not occur frequently during normal operation and is done on an individual and occasional basis.

3.2. Gateway and Modbus-OPC Library Test

The gateway system, responsible for interfacing between the OPC server and the local heliostat control systems via Modbus, was tested for two weeks under full load with all 16 communication lines active before the deployment. The recursive communication process, which polls each of the 300 heliostats sequentially, was run without any problems during the tests. Nowadays, while the system is under normal operation, thanks to the gateway and Modbus-OPC library the state of each heliostat is updated in less than

2 seg, well within the acceptable range for the system's real-time requirements. The handling of control commands and status updates between the local controllers and the OPC server occurred without delays or inconsistencies. If inconsistencies or errors are detected, the system handles them correctly without errors that would slow down the process. This demonstrated that Modbus-OPC integration works effectively under full-scale field conditions.

3.3. OPC Server Test

The OPC server, developed using the Python opcua-asyncio library, was tested for scalability, performance, and reliability. Under normal operational conditions, the server efficiently handled the simultaneous connections of the 300 heliostats, operators, and external clients without noticeable performance degradation. Memory usage remained stable, and CPU load was manageable, even when handling high-frequency event subscriptions and real-time data updates, for example, during field start-up or field shut-down. The latency in the server's responses to client requests averaged 100 ms, which is sufficient for real-time SCADA applications. The event subscription system worked as expected, with notifications regarding heliostat status changes being processed immediately and reflected in the SCADA interface. No critical errors were encountered during prolonged test sessions, confirming the reliability and robustness of the implementation of the OPC server.

3.4. Usability

The new SCADA interface was tested for functionality, usability, and stability by operators with over 30 years of experience with previous systems. After several iterations of feedback and refinement, the final version was rated highly in terms of usability, particularly for its intuitive layout and clear depiction of system states. The integration of predefined heliostat groups and the color-coded status legend proved to be especially useful for managing large numbers of heliostats efficiently. Operators reported that the interface simplified routine tasks, such as sending heliostats into tracking mode or managing emergency situations. User login functionality has been well received, by the head of operation, due to the permission-based control system with appropriately blocked restricted actions based on user roles. The system's emergency response mechanism was also tested, and the "Emergency" button successfully triggered the safe stow procedure for all heliostats without delay. In general, the operators expressed satisfaction with the efficiency and responsiveness of the system. The visual clarity of the interface, the feedback in real-time, and the functionality of the interface were well received. During usability testing, the average task completion times were reduced compared to the legacy system, particularly in heliostat group management and status monitoring tasks.

3.5. Stability

The system's overall stability was evaluated during a 72-h continuous test under normal operating conditions before the final deployment without crashes. The interface successfully exchanged information with the OPC server, and all control commands from the interface were accurately transmitted to the heliostats. The graphical representation of the CESA-I field and heliostat statuses is updated in real-time with minimal latency, providing operators with clear and current information.

The results of the series of tests indicate that the developed SCADA system meets the performance, usability, and stability requirements for real-time heliostat field management. The Wi-Fi mesh communication, gateway, and Modbus-OPC integration proved to be reliable, while the OPC server maintained robust performance under heavy loads. The new SCADA interface improved operational efficiency, with positive feedback from experienced operators, confirming its effectiveness for industrial-scale solar plant control. In addition, the whole system is still under development, open to changes and improvements, either at the suggestion of the operators or because of the need for further development.

Since the tests were completed, the SCADA system was adopted as the main control and supervision system of the CESA-I field by the operators, abandoning the traditional one, and was put into production.

4. Discussion

The results of the tests conducted on the CESA-I SCADA system indicate significant advancements in the operation, control, and monitoring of the heliostat field. These results align with previous studies on the integration of SCADA systems in solar thermal plants but with notable improvements in terms of costs, flexibility, scalability, and real-time performance. The implementation of a Wi-Fi mesh network, combined with the OPC UA server, provides a robust framework for handling the complexities of large-scale heliostat control, which has been a persistent challenge in the CSP sector.

The successful deployment of the Wi-Fi mesh communication system, as highlighted in Section 3.1, confirms its viability to replace traditional wired communications in CSP plants. This innovation reduces infrastructure costs and enhances the flexibility of heliostat deployment and maintenance, a finding consistent with earlier research that demonstrated the advantages of wireless communication in industrial automation. The low latency and minimal packet loss achieved in the tests are particularly encouraging, suggesting that the Wi-Fi mesh can support real-time solar tracking and control operations. This aspect marks a significant improvement over the limitations of wired systems, which have traditionally constrained the adaptability and scalability of heliostat fields.

The gateway system linking Modbus-based local controls with the OPC UA server was extensively tested, with results that demonstrated seamless integration between legacy systems and modern SCADA architecture. This is an important step forward in maintaining the operational continuity of older heliostat fields while benefiting from modern data acquisition and control technologies. The recursive communication process, which efficiently updates heliostat statuses and commands via OPC-UA, showed no signs of delay or failure, validating the effectiveness of the custom CESA-Modbus library developed for this purpose.

The ability of the OPC UA server to manage the simultaneous connections of more than 300 heliostats, external clients, and operator stations without significant performance degradation demonstrates the scalability and robustness of the server. This confirms the viability of OPC UA for CSP plant operations, a key finding that addresses the gaps in previous studies, which often highlighted the difficulties of managing such extensive data streams in real-time (SCADA). The efficient handling of event subscriptions, with latency averaging around 100 ms, further supports the suitability of OPC UA for industrial control systems, ensuring timely updates and operational safety.

The usability tests of the new SCADA interface, conducted with operators who have decades of experience with the CESA-I field, offer strong evidence of its practical efficiency. The positive feedback on the interface's design and real-time response underscores the importance of operator-centered design in industrial SCADA systems. Furthermore, the stability of the system during prolonged testing, with no crashes or failures, confirms its reliability for continuous operation, an essential requirement for solar thermal plants.

The results highlight several areas for future research. First, combining AI and machine learning with SCADA systems presents a promising path for predictive maintenance and operational optimization, as highlighted in previous studies. AI-driven analytics could enable the system to forecast equipment failures and enhance operational efficiency by analyzing heliostat performance and environmental conditions in real-time. Furthermore, integrating cloud-based SCADA systems with edge computing could revolutionize scalability and responsiveness. Edge computing would allow data processing to occur closer to the heliostats, minimizing latency for real-time control, while the cloud would provide scalable resources for centralized data management and advanced analytics, making the system capable of efficiently managing even larger heliostat fields.

5. Conclusions

SCADA methodology presented represents a significant step forward in the control and management of solar thermal power plants. The results demonstrate the system's scalability, reliability, and cost-effectiveness. Key outcomes include stable communications with low latency, robust real-time control, and an intuitive user interface that streamlines operational tasks. The implementation of the Wi-Fi mesh network significantly reduces infrastructure costs and enhances flexibility in managing the heliostat field. Additionally, the SCADA system is designed as an open and adaptable platform, offering the potential for future optimizations through artificial intelligence and machine learning. Overall, this methodology strengthens the competitiveness and feasibility of CSP technologies within the renewable energy sector.

Further research should focus on the integration of advanced AI algorithms and the potential for global deployment in other CSP facilities. Also, future works include receiver control, storage management, and smart aiming point strategy [17] integration in the SCADA.

Author Contributions: Conceptualization, J.A.C., J.B. and J.F.-R.; methodology, J.A.C. and J.B.; software, J.A.C. and J.B.; validation, J.A.C., J.F.-R. and B.D.; formal analysis, J.B., J.F.-R. and A.L.A.-M.; investigation, J.A.C., J.B. and J.F.-R.; resources, A.L.A.-M.; data curation, J.A.C.; writing—original draft preparation, J.A.C.; writing—review and editing, J.A.C., J.B., J.F.-R., B.D. and A.L.A.-M.; visualization, J.B. and J.F.-R.; supervision, J.F.-R. and A.L.A.-M.; project administration, J.F.-R. and A.L.A.-M.; funding acquisition, A.L.A.-M. All authors have read and agreed to the published version of the manuscript.

Funding: The authors would like to acknowledge the EU for the financial support provided through the Horizon Europe Program under the ASTERIx-CAESar project (contract number 10112223) and the Spanish Ministry of Science and Innovation's National R+D+i Plan Projects PID2021-126805OB-I00 (HELIOSUN project).

Data Availability Statement: The original contributions presented in the study are included in the article, further inquiries can be directed to the corresponding authors.

Conflicts of Interest: The authors declare no conflicts of interest.

Abbreviations

The following abbreviations are used in this manuscript:

AI	artificial intelligence
CSP	concentrated solar power
CST	concentrated solar technology
DNI	direct normal irradiance
HelIoT	smart heliostat
IoT	Internet of Things
IIoT	Industrial Internet of Things
LCOE	levelized cost of energy
LED	light-emitting diode
OPC	open platform communications
PSA	Plataforma Solar de Almería
PV	photovoltaics
SCADA	Supervisory Control and Data Acquisition
UA	Unified Architecture

References

1. International Renewable Energy Agency (IRENA). Renewable Power Generation Costs in 2022. 2023. Available online: https://www.irena.org/Publications/2023/Aug/Renewable-power-generation-costs-in-2022 (accessed on 12 July 2024).
2. Kiasari, M.; Ghaffari, M.; Aly, H.H. A Comprehensive Review of the Current Status of Smart Grid Technologies for Renewable Energies Integration and Future Trends: The Role of Machine Learning and Energy Storage Systems. *Energies* **2024**, *17*, 4128. [CrossRef]
3. Grigoras, G.; Gârbea, R.; Neagu, B.-C. Toward Smart SCADA Systems in the Hydropower Plants through Integrating Data Mining-Based Knowledge Discovery Modules. *Appl. Sci.* **2024**, *14*, 8228. [CrossRef]
4. Castillo-Navarro, J.; Kristjanpoller, F.; Mena, R.; Godoy, D.R.; Viveros, P. A Methodological Framework for Managing the Alarms in Wind Turbine Control and Data Acquisition Systems for Failure Analysis. *Machines* **2024**, *12*, 597. [CrossRef]
5. Bonilla, J.; Blanco Galvez, J.; Zarza, E.; Alarcón-Padilla, D.-C. Feasibility and Practical Limits of Full Decarbonization of the Electricity Market with Renewable Energy: Application to the Spanish Power Sector. *Energy* **2021**, *239*, 122437. [CrossRef]
6. Carballo, J.A.; Bonilla, J.; Berenguel, M.; Fernández-Reche, J.; García, G. New approach for solar tracking systems based on computer vision, low cost hardware and deep learning. *Renew. Energy* **2019**, *133*, 1158–1166. [CrossRef]
7. Carballo, J.A.; Bonilla, J.; Fernández-Reche, J.; Nouri, B.; Avila-Marin, A.; Fabel, Y.; Alarcón-Padilla, D.-C. Cloud Detection and Tracking Based on Object Detection with Convolutional Neural Networks. *Algorithms* **2023**, *16*, 487. [CrossRef]
8. Biswal, A.; Bansal, H.O. SCADA and Its Applications to Renewable Energy Systems Integration. In Proceedings of the 2014 9th International Conference on Industrial and Information Systems (ICIIS), Gwalior, India, 15–17 December 2014; pp. 1–6. [CrossRef]
9. Wang, G.; Xie, J.; Wang, S. Application of Artificial Intelligence in Power System Monitoring and Fault Diagnosis. *Energies* **2023**, *16*, 5477. [CrossRef]
10. de Azambuja, A.J.G.; Plesker, C.; Schützer, K.; Anderl, R.; Schleich, B.; Almeida, V.R. Artificial Intelligence-Based Cyber Security in the Context of Industry 4.0—A Survey. *Electronics* **2023**, *12*, 1920. [CrossRef]
11. Nankya, M.; Chataut, R.; Akl, R. Securing Industrial Control Systems: Components, Cyber Threats, and Machine Learning-Driven Defense Strategies. *Sensors* **2023**, *23*, 8840. [CrossRef] [PubMed]
12. Wali, A.; Alshehry, F. A Survey of Security Challenges in Cloud-Based SCADA Systems. *Computers* **2024**, *13*, 97. [CrossRef]
13. Khalid, W.; Jamil, M.; Khan, A.A.; Awais, Q. Open-Source Internet of Things-Based Supervisory Control and Data Acquisition System for Photovoltaic Monitoring and Control Using HTTP and TCP/IP Protocols. *Energies* **2024**, *17*, 4083. [CrossRef]
14. OPC Foundation. OPC Unified Architecture. Available online: https://opcfoundation.org (accessed on 5 September 2024).
15. Van Rossum, G.; Drake, F.L. *Python 3 Reference Manual*; CreateSpace: Scotts Valley, CA, USA, 2009; ISBN 1441412697.
16. FreeOpcUa Contributors. Opcua-Asyncio: Asynchronous Python Client and Server for OPC UA. Available online: https://github.com/FreeOpcUa/opcua-asyncio (accessed on 5 September 2024).
17. Carballo, J.A.; Bonilla, J.; Cruz, N.C.; Fernández-Reche, J.; Álvarez, J.D.; Avila-Marin, A.; Berenguel, M. Reinforcement Learning for Heliostat Aiming: Improving the Performance of Solar Tower Plants. *Appl. Energy* **2025**, *377*, 124574. [CrossRef]

energies

MDPI

Article

Photothermal Conversion Performance of Fe_3O_4/ATO Hybrid Nanofluid for Direct Absorption Solar Collector

Jeonggyun Ham [1], Hyemin Kim [2] and Honghyun Cho [1,*]

[1] Department of Mechanical Engineering, Chosun University, 309 Pilmundaero, Dong-gu, Gwangju 61452, Republic of Korea; jgham@chosun.ac.kr
[2] Graduate School of Chosun University, Chosun University, 309 Pilmundaero, Dong-gu, Gwangju 61452, Republic of Korea
* Correspondence: hhcho@chosun.ac.kr; Tel.: +82-62-230-7050; Fax: +82-62-230-7055

Abstract: In order to enhance the efficiency of direct absorption solar collectors, this study carried out an experimental analysis about the optical and photothermal conversion performance of Fe_3O_4, ATO (Antimony-doped tin oxide), and Fe_3O_4/ATO nanofluids with a total concentration of 0.1 wt%. According to the results of the experiments, Fe_3O_4 nanofluid outperforms ATO nanofluid in terms of optical absorption; nevertheless, at wavelengths shorter than 600 nm, it also shows significant scattering reflection. The solar-weighted absorption coefficient of Fe_3O_4/ATO nanofluid rose from 0.863 (m_{Fe3O4}/m_{Total} = 0.2) to 0.932 (m_{Fe3O4}/m_{Total} = 0.8) when the optical path length increased from 0.01 m to 0.06 m. Moreover, the Fe_3O_4/ATO hybrid nanofluid achieved a photothermal conversion efficiency of 0.932 when the mass ratio of Fe_3O_4 to total mass was 0.2, surpassing the efficiencies of 0.892 and 0.898 recorded for 0.1 wt% ATO and Fe_3O_4 nanofluids, respectively. When present together, the opposing optical characteristics of Fe_3O_4 and ATO boost photothermal conversion performance, which is anticipated to raise the efficiency of direct absorption solar collectors.

Keywords: hybrid nanofluid; optical properties; solar weight absorption coefficient; photothermal conversion coefficient; direct absorption solar collector

1. Introduction

Due to global industrialization and population growth, the demand for energy has continuously increased. This increase has been accompanied by a rise in greenhouse gas emissions from fossil fuel consumption, contributing to serious issues such as global warming and climate change. These challenges underscore the urgent need for sustainable alternative energy sources, with solar energy offering significant potential as an abundant and clean energy source [1].

Solar energy converts radiant energy emitted by the sun into thermal energy, making it applicable across various fields. This conversion facilitates two critical objectives: reducing energy costs and minimizing environmental impact. Solar collectors, devices that convert solar radiation into thermal energy, play a pivotal role in applications such as heating, hot water supply, and industrial processes [2–4]. Solar collectors are generally classified into indirect and direct absorption types. In the indirect absorption method, the heat-transfer fluid absorbs solar energy indirectly from the absorbing surface. In contrast, the direct absorption method involves the heat-transfer fluid directly absorbing solar radiation, potentially achieving higher efficiency [5–7].

In direct absorption solar collectors (DASCs), the heat-transfer fluid directly absorbs solar radiation and converts it into thermal energy, thereby reducing thermal resistance in the energy absorption process and minimizing heat losses. However, the performance of these systems is highly dependent on the photothermal conversion efficiency of the fluid. Conventional heat-transfer fluids, such as water, antifreeze, and oil, exhibit low optical absorption, limiting their ability to absorb sufficient solar radiation and achieve optimal performance in the solar collector [8,9]. Nanofluids are an efficient solution for enhancing thermal conductivity, convective heat transfer, and optical absorption [10–12].

Recent efforts to enhance the photothermal conversion performance of nanofluids have concentrated on systematically examining the composition, concentration, and structure of nanoparticles to identify optimal operating conditions. Studies employing various materials, including metals [13–15], metal oxides [16–19], CNTs [20,21], graphene [22,23], and environmentally friendly materials, have shown significant progress in improving optical absorption performance. These investigations have confirmed notable enhancements in the optical properties and photothermal conversion efficiency of heat-transfer media [7,24–28]. While nanofluids have typically been synthesized with single-component nanoparticles, they face a key limitation in expanding the optical absorption peak band [28]. To address this issue, ongoing research is exploring the photothermal conversion characteristics of nanofluids, incorporating coexisting nanoparticles with distinct optical absorption spectrum peaks. Chen et al. [29] developed nanofluids by combining spherical and rod-shaped Au nanoparticles to extend the absorption wavelength range. They demonstrated that this approach allows for red and blue shifts, leading to a broader optical absorption spectrum. Wang et al. examined the photothermal conversion efficiency of Hedgehong-Oil nanofluids, which incorporate a combination of Au and ZnO nanoparticles, and observed a 240% enhancement compared to the base oil. Shang et al. [30] sought to improve optical absorption by creating Ag@Al_2O_3 nanocomposites with a core-shell structure using atomic layer deposition, aiming to maximize localized surface plasmon resonance. Their findings revealed that Ag nanoparticles coated with Al_2O_3 exhibited a red shift in the absorption band due to localized surface plasmon resonance, resulting in a wider absorption band and enhanced optical absorption performance.

To address the limitations of single-component nanofluids with narrow absorption bands, extensive research has been conducted on multi-component or hybrid nanofluids, which contain two or more types of nanoparticles with different absorption spectra. Joseph et al. [31] used ANOVA analysis to examine the factors that affect the enhancement of photothermal conversion properties in binary nanofluids and to identify the optimal mixing ratio of SiO_2/Ag-CuO. Their findings revealed that a nanofluid consisting of SiO_2/Ag: 206.3 mg/L, CuO: 864.7 mg/L, and SDS (surfactant): 1996.2 mg/L achieved a solar-weighted absorption fraction as high as 82.82%. This indicates that multi-component nanofluids can significantly boost photothermal conversion efficiency. Hazra et al. [32] explored the photothermal conversion characteristics of a BN/CB hybrid nanofluid, composed of carbon black and hexagonal boron nitride (BN), and observed a 34.55% improvement in photothermal performance compared to the base fluid, ethylene glycol (EG). This demonstrates the potential of hybrid nanofluids in enhancing photothermal efficiency. Similarly, Kim et al. [32] investigated the photothermal conversion performance of SiC/ITO nanofluids, aiming to improve the limited infrared absorption of SiC by incorporating ITO. The study showed that a SiC mixing ratio of 8:2 resulted in an increase in photothermal efficiency by 38.7% over that of the SiC nanofluid alone, with a maximum efficiency of 34.1%. Collectively, these studies provide clear evidence that hybrid nanofluids can effectively enhance photothermal conversion performance.

Despite considerable progress in nanofluid research, challenges remain in developing nanofluids that can achieve an optimal balance of absorption across different spectral regions. Current single-component nanofluids, although effective to some extent, often have limited absorption capabilities, especially when broad-spectrum absorption is needed to enhance the photothermal conversion efficiency of direct absorption solar collectors (DASCs). This limitation suggests the need for approaches that combine different nanoparticles to improve overall performance.

Fe_3O_4 and ATO (Antimony-doped tin oxide) nanofluids present an approach to address this issue. Fe_3O_4 nanofluids [33,34] exhibit strong absorption primarily in the visible light spectrum, while ATO nanofluids [35,36] show enhanced absorption in the near-infrared region. Individually, these nanofluids have specific limitations, but when combined, they can potentially utilize the complementary absorption characteristics of both materials. This combination may expand the overall absorption spectrum and improve the photothermal conversion efficiency of DASCs. However, despite this potential, research specifically focusing on Fe_3O_4/ATO hybrid nanofluids is limited, and comprehensive studies evaluating their effectiveness in DASCs are scarce.

This study aims to address this gap by evaluating Fe_3O_4/ATO hybrid nanofluids as heat-transfer fluids for DASCs. A comparative analysis with single-component nanofluids was conducted to assess the potential advantages of the hybrid approach in enhancing photothermal conversion performance. The influence of varying the Fe_3O_4 and ATO mixing ratio on key performance metrics, such as optical absorption, absorption rate, solar-weighted absorption fraction, and photothermal conversion efficiency, was systematically examined. The results provide insights into the feasibility of using Fe_3O_4/ATO hybrid nanofluids for DASCs and their potential to perform more effectively than single-component nanofluids in this application. The contribution of this research lies in its systematic evaluation of Fe_3O_4/ATO hybrid nanofluids and their ability to address some of the limitations associated with existing single-component nanofluids. The findings aim to support the optimization of solar collector systems for improved energy efficiency and offer insights that may inform the development of sustainable energy technologies.

2. Experimental Method

2.1. Preparation of Fe_3O_4/ATO Hybrid Nanofluid

In this study, Fe_3O_4/ATO hybrid nanofluids were synthesized by combining Fe_3O_4 and ATO single-component nanofluids. Fe_3O_4 was synthesized using the coprecipitation method, and surface modification with PAA (Polyacrylic acid) [18,37] was performed to ensure dispersion stability. The ATO nanofluid was prepared by diluting a 50 wt% suspension provided by K&P Nano Co. (Jecheon-si, Republic of Korea) with distilled water to reach the desired concentration. To ensure proper dispersion, both the single-component and hybrid nanofluids underwent stirring at 500 rpm for 2 h, followed by a 5 h ultrasonication process. The characteristics of the Fe_3O_4 and ATO nanoparticles are detailed in Table 1.

Table 1. Specification of nanoparticles.

NP	Fe_3O_4	ATO
Purity		99%
Color	Dark brown	Blue
Outer diameter	5–20 nm	
Thermal conductivity	80 W/m·K	4.4 W/m·K
True density	5.1 g/cm^3	6.8 g/cm^3
Manufacturing method	Coprecipitation	50% aqua solution

Figure 1 illustrates the synthesized Fe_3O_4, ATO, and Fe_3O_4/ATO nanofluids, while Figure 2 shows the TEM images and particle size distributions of the dispersed nanoparticles. The Fe_3O_4 and ATO nanoparticles exhibited distorted spherical shapes, and the Fe_3O_4/ATO nanoparticles in the hybrid nanofluid were found to be agglomerated. The average sizes of the clustered Fe_3O_4 and ATO nanoparticles were 10.9 $\pm$ 4.2 nm and 11.72 $\pm$ 4.2 nm, respectively, with the Fe_3O_4 nanoparticles showing a distorted morphology.

According to the study by Tong et al. [28], the optimal concentration range for achieving maximum photothermal conversion efficiency for Fe_3O_4 and ATO is between 0.075 and 0.1 wt%. Based on this result, Fe_3O_4/ATO hybrid nanofluids were prepared at a total concentration of 0.1 wt%, with mass ratios set at 0.2, 0.4, 0.6, and 0.8. Previous research identified the critical concentration for achieving maximum photothermal conversion efficiency in Fe_3O_4 nanofluids as 0.1 wt%. Guided by this result, Fe_3O_4/ATO hybrid nanofluids were formulated at a total concentration of 0.1 wt%, with mass ratios of m_{Fe3O4}/m_{Total} set at

0.2, 0.4, 0.6, and 0.8. The zeta potential measurements of the Fe_3O_4, ATO, and Fe_3O_4/ATO nanofluids were -40.4 ± 2 mV, -49.9 ± 0.41 mV, and -40.2 ± 1.8 mV, respectively. The zeta potential is a critical parameter for evaluating the stability of colloidal systems. A zeta potential magnitude greater than 30 mV (in absolute value) typically indicates a stable dispersion, as the strong electrostatic repulsion between particles minimizes aggregation. In this study, the pH of Fe_3O_4, ATO, and Fe_3O_4/ATO nanofluids was measured at 7. Therefore, the prepared Fe_3O_4, ATO, and Fe_3O_4/ATO nanofluids were assessed to have adequate dispersion stability based on this criterion. Furthermore, previous research [28] conducted in this study confirmed that Fe_3O_4 and ATO nanofluids exhibited no change in optical transmittance after five cycles of photothermal conversion experiments, as well as no variation in optical transmittance two weeks after preparation. Therefore, the hybrid nanofluid demonstrates reliable dispersion stability.

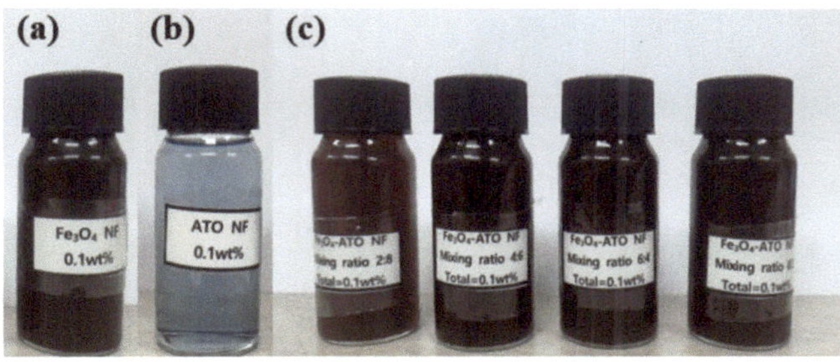

Figure 1. Manufactured nanofluids: (**a**) 0.1 wt% Fe_3O_4 NF, (**b**) 0.1 wt% ATO NF, (**c**) Fe_3O_4/ATO nanofluids.

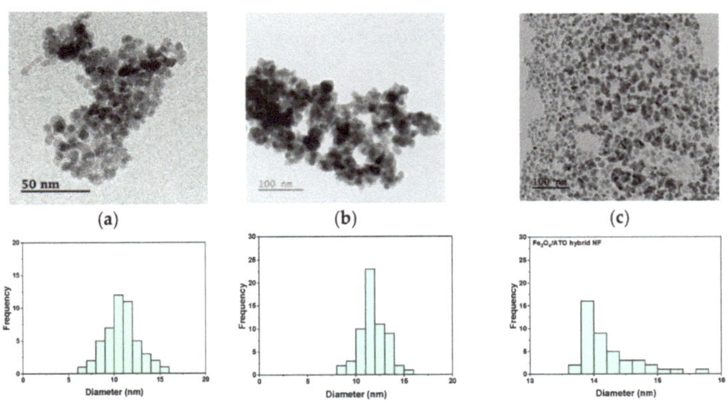

Figure 2. TEM images: (**a**) Fe_3O_4; (**b**) ATO; (**c**) Fe_3O_4/ATO and nanoparticle diameter distribution; (**d**) Fe_3O_4; (**e**) ATO; (**f**) Fe_3O_4/ATO nanofluid.

2.2. Optical Property Measurement Method

When solar energy passes through a nanofluid, some of the energy is reflected at the interface between the air and the nanofluid, while the rest is absorbed as it moves through the fluid. The portion of solar energy that is not absorbed is transmitted through the nanofluid. The relationship describing the transmission of solar energy through the nanofluid is given by Equation (1).

$$A + T + R = 1 \quad (1)$$

where A, R, and T are absorbance, reflectance, and transmittance.

In this research, the optical properties of the nanofluids, including absorption, reflection, and transmission, were analyzed using the Novel Double-Thickness Transmittance Method (NDTTM). This method involves measuring the transmittance of the nanofluid through two cuvettes with different optical path lengths to determine its optical characteristics. By applying the NDTTM, the optical extinction coefficient (κ) and the reflectance at the interface between the cuvette and air (ρ_{op}) are derived. These values are calculated using the transmittance data from cuvettes of different optical lengths and applying Equations (2) and (3).

$$\kappa = -\frac{\lambda \ln\left(\frac{T_1}{T_2}\right)}{4\pi\left(L_{op,1} - L_{op,2}\right)} \quad (2)$$

$$\rho_{op} = \frac{1 - \sqrt{T_1^2 + T_1\left[\exp\left(\frac{4\pi\kappa L_{op,1}}{\lambda}\right) - \exp\left(\frac{-4\pi\kappa L_{op,2}}{\lambda}\right)\right]}}{1 + T_1\left(\frac{-4\pi\kappa L_{op,1}}{\lambda}\right)} \quad (3)$$

where T_1 and T_2 are transmittance at $L_{op,1}$ and $L_{op,2}$, respectively, and λ is the wavelength.

Subsequently, the extinction optical coefficient is calculated by incorporating the reflectance and transmittance values at different optical depths into Equation (4). This calculation is repeated until the condition $|\kappa_{ex,as} - \kappa_{ex}| < 10^{-8}$ is met. During this iterative process, optical properties such as n, κ, $T(\lambda)$, $R(\lambda)$, and $A(\lambda)$ are continuously tracked and refined.

$$\kappa = \frac{\lambda}{4\pi L_{op,2}} \ln\left[\frac{\left(1 + \sqrt{1 + 4\left(\frac{T_2}{(1-\rho_{op})^2}\right)^2 \rho_{op}^2}\right)}{2\left(\frac{T_2}{(1-\rho_{op})^2}\right)}\right] \quad (4)$$

To perform the NDTTM analysis, nanofluids were placed in cuvettes with optical path lengths of 5 mm and 10 mm, and their optical transmittance was measured using a visible-infrared spectrometer (AVANTES-2048, Avlight-DHc, Apeldoorn, The Netherlands), because longer path lengths could lead to low transmittance at the concentrations used, increasing the likelihood of errors in the NDTTM analysis. To comprehensively evaluate the solar-radiation absorption by the nanofluid across different wavelengths, the solar-weighted absorption coefficient $S(L_{op})$ was calculated. This coefficient was determined using Equation (5), with $I_{AM1.5}$ representing the spectral solar irradiance based on ASTM G173-3 AM 1.5 Global [38].

$$S(L_{op}) = \frac{\int I_{AM1.5}(\lambda)(1 - \exp(\alpha(\lambda)L_{op}))d\lambda}{\int I_{AM1.5}(\lambda)d\lambda} \quad (5)$$

2.3. Experimental Setup for Photothermal Conversion

Figure 3 depicts the experimental setup used to assess the photothermal conversion performance. This setup consists of a solar simulator, an acrylic container, and a supporting stand. The solar simulator (Oriel Xenon Arc lamp, LCS-100, Newport Co., Franklin, MA, USA) is equipped with an AM 1.5 filter to replicate the solar spectrum. The solar irradiance intensity at the top of the acrylic container, where the nanofluid is placed, is set at 1000 W/m^2. The container has internal dimensions of ϕ40 mm $\times$ 42 mm, with three T-type thermocouples installed at intervals of approximately 10–11 mm.

The photothermal conversion experiment involves a heating phase, where the nanofluid is exposed to light for 2.5 h, followed by a cooling phase lasting 0.5 h to measure the heat loss coefficient. During the cooling phase, the light source is turned off after the nanofluid reaches its equilibrium temperature under light exposure, allowing the heat dissipation rate from the receiver to be evaluated. The data gathered during the experiment were recorded at 1 s intervals using a data logger (MX-100, Yokogawa Inc., Tokyo, Japan).

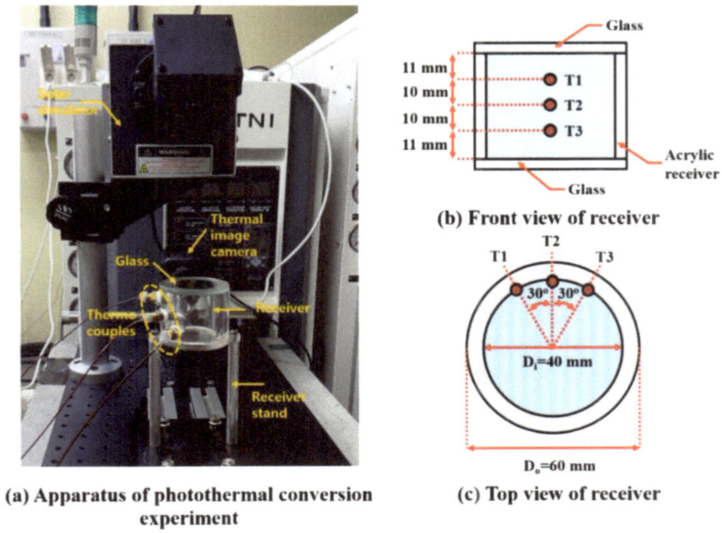

Figure 3. Experimental setup of photothermal conversion experiment.

The photothermal conversion efficiency of the nanofluid, which represents the proportion of solar energy absorbed and converted into thermal energy, is calculated using Equation (6).

$$\eta_{PTC} = \frac{Bm_{nf}c_{p,nf}\left(T_{eq} - T(0)\right)}{IA} \tag{6}$$

In this equation, B denotes the thermal diffusivity to the surroundings, T_{eq} is the equilibrium temperature, $T(0)$ is the initial temperature, $c_{p,nf}$ is the specific heat capacity of the nanofluid, I represents the solar irradiance, and A is the area of the receiver exposed to light. The expression for B is provided in Equation (7) and is derived during the cooling phase. The value of B was measured to be within the range of 4.1×10^{-4} to 4.1×10^{-5} s^{-1}.

$$In\frac{T(t) - T_{sur}}{T_{eq} - T_{sur}} = -\frac{hA_{dis}}{\sum\limits_{i} m_i c_{p,i}}t = -Bt \tag{7}$$

In Equation (7), $T(t)$ is the temperature of the nanofluid at time t, T_{sur} is the ambient temperature, h denotes the heat loss coefficient, m_i refers to the mass of both the receiver and the nanofluid, and $c_{p,i}$ is the specific heat capacity of the receiver and the nanofluid.

The receiving efficiency, η_{rec}, is defined as the efficiency with which the nanofluid captures thermal energy and is expressed by Equation (8).

$$\eta_{rec} = \frac{H\rho_{nf}c_{p,nf}(T(t) - T(0))}{\int_0^t IAdt} \tag{8}$$

In this equation, H indicates the height of the receiver, and ρ_{nf} represents the density of the nanofluid.

The uncertainties in photothermal conversion and receiving efficiency were calculated using Equations (9) and (10).

$$\frac{\delta\eta_{PTC}}{\eta_{PTC}} = \sqrt{\left(\frac{\delta m_{nf}}{m_{nf}}\right)^2 + \sum_{i=1}^{3}\left(\frac{\delta T}{T}\right)^2 + \left(\frac{\delta A}{A}\right)^2 + \left(\frac{\delta I}{I}\right)^2 + \left(\frac{\delta B}{B}\right)^2} \tag{9}$$

$$\frac{\delta\eta_{rec}}{\eta_{rec}} = \sqrt{\left(\frac{\delta m_{nf}}{m_{nf}}\right)^2 + \sum_{i=1}^{3}\left(\frac{\delta T}{T}\right)^2 + \left(\frac{\delta A}{A}\right)^2 + \left(\frac{\delta I}{I}\right)^2 + \left(\frac{\delta t}{t}\right)^2} \quad (10)$$

The uncertainties in measuring the mass of the nanofluid, the temperature of the nanofluid, the light exposure area, the irradiance intensity, time, and the heat dissipation rate to the surroundings were 2.36%, 2.47%, 0.5%, 0.5%, 0.01%, and 2.35%, respectively. The uncertainties in photothermal conversion and receiving efficiency are 4.21% and 3.5%, respectively.

3. Results and Discussion

3.1. Optical Characteristics of Fe_3O_4, ATO, and Fe_3O_4/ATO Hybrid Nanofluid

The optical properties of nanofluids are critical factors influencing their photothermal conversion efficiency. Figure 4 presents the optical absorbance and transmittance spectra for 0.1 wt% Fe_3O_4, ATO, and Fe_3O_4/ATO nanofluids. As depicted in Figure 4a, Fe_3O_4 and ATO nanofluids exhibit distinct absorption characteristics within the wavelength range of 400 to 1200 nm. The ATO nanofluid shows a peak absorbance near 1100 nm, but overall, it demonstrates a lower optical absorbance compared to the Fe_3O_4 nanofluid. In contrast, the Fe_3O_4 nanofluid displays strong absorbance between 800 and 1100 nm, with a marked reduction in absorbance at wavelengths below 600 nm. This reduction can be attributed to the significant scattering reflection that occurs in the ultraviolet region for Fe_3O_4 nanofluids.

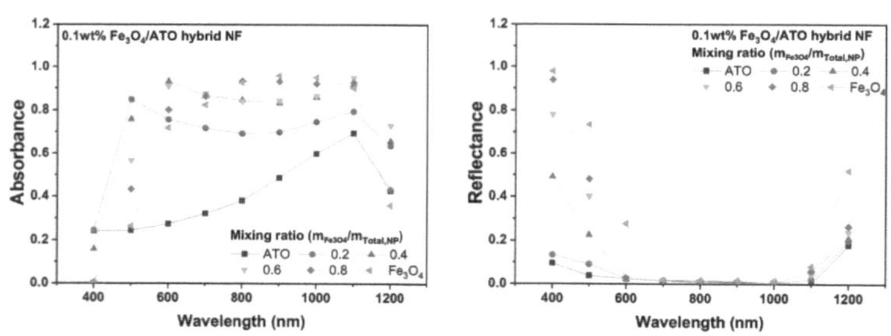

Figure 4. Optical absorbance and reflectance of Fe_3O_4, ATO, and Fe_3O_4/ATO NFs. (**a**) Absorbance of Fe_3O_4/ATO nanofluid. (**b**) Reflectance of Fe_3O_4/ATO nanofluid.

Figure 4b illustrates this scattering phenomenon, where the Fe_3O_4 nanofluid exhibits increased reflectance at wavelengths shorter than 600 nm, with reflectance intensity increasing as the wavelength decreases. On the other hand, the ATO nanofluid maintains a reflectance below 0.1 in the same wavelength range, indicating minimal scattering. These findings suggest that, although Fe_3O_4 nanofluids possess superior optical absorption properties, their high reflectance at shorter wavelengths can hinder their performance. Therefore, the combination of ATO and Fe_3O_4 in a hybrid nanofluid presents an opportunity to enhance overall optical absorption by leveraging the complementary properties of both materials. It is well known that the optical absorbance of hybrid nanofluids, containing multiple types of nanoparticles, tends to follow a linear combination of the optical absorption properties of each individual nanoparticle [39]. Therefore, the combination of ATO and Fe_3O_4 in a hybrid nanofluid presents an opportunity to enhance overall optical absorption by leveraging the complementary properties of both materials.

Figure 5 shows the solar-weighted absorption coefficients of the Fe_3O_4/ATO hybrid nanofluids (NFs). For the Fe_3O_4 nanofluid, the solar-weighted absorption coefficient exhibited a modest increase from 0.854 to 0.883 as the optical path length increased from 0.01 m to 0.06 m. In comparison, the ATO nanofluid displayed a more pronounced increase, with the solar-weighted absorption coefficient rising from 0.529 to 0.876 over the same range. This indicates that, although the Fe_3O_4 nanofluid has superior optical absorbance, the improvement in photothermal conversion performance with increasing optical path length is limited due to scattering reflection.

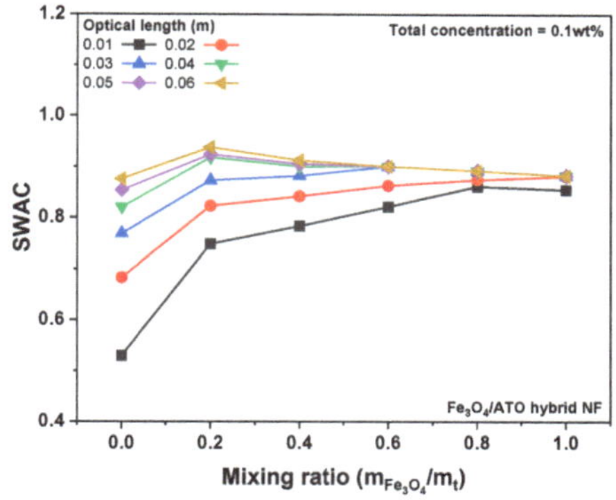

Figure 5. Solar-weighted absorption coefficient according to the optical length.

The Fe_3O_4/ATO hybrid nanofluids demonstrated enhanced solar-weighted absorption coefficients relative to both Fe_3O_4 and ATO nanofluids. At an optical path length of 0.01 m, the solar-weighted absorption coefficient of the Fe_3O_4/ATO hybrid nanofluid (where $m_{Fe3O4}/m_{Total} = 0.8$) was 0.861, indicating a slight improvement over the coefficients of 0.854 for the 0.1 wt% Fe_3O_4 NF and 0.529 for the 0.1 wt% ATO NF. However, at an optical path length of 0.06 m, the solar-weighted absorption coefficient of the Fe_3O_4/ATO hybrid nanofluid (where $m_{Fe3O4}/m_{Total} = 0.2$) increased from 0.748 to 0.938. As the optical path length extended from 0.01 m to 0.06 m, the optimal m_{Fe3O4}/m_{Total} for maximizing the solar-weighted absorption coefficient gradually decreased. These findings suggest that, with an increase in optical path length, enhancing optical absorbance and suppressing scattering reflection are crucial for the performance of Fe_3O_4/ATO hybrid nanofluids.

3.2. *Photothermal Conversion Performance of Fe_3O_4/ATO Hybrid Nanofluid*

The photothermal conversion performance of nanofluids is a critical parameter that directly influences the solar radiation harvesting efficiency of direct absorption solar collectors. This performance is affected by several factors, including the optical absorbance of the nanofluid, the fluid temperature, and the optical path length of the containment vessel. Figure 6 presents the temperature variation of 0.1 wt% Fe_3O_4/ATO nanofluids over time under solar-radiation exposure. The results indicate that the temperature of the 0.1 wt% Fe_3O_4/ATO nanofluids increases progressively with prolonged exposure to solar radiation, demonstrating the effective conversion and storage of solar energy into thermal energy. Over an exposure period of 9000 s, the nanofluid with $m_{Fe3O4}/m_{Total} = 0.2$ exhibited the highest temperature increase, reaching 12.2 °C. In contrast, the nanofluid with $m_{Fe3O4}/m_{Total} = 0.6$ showed the lowest temperature increase of 11.7 °C, a result that is comparable to the temperature increase observed for the ATO nanofluid.

Figure 7 depicts the energy conversion efficiency of water and nanofluids (0.1 wt% Fe_3O_4 NF, 0.1 wt% ATO NF, and 0.1 wt% Fe_3O_4/ATO NFs) during solar exposure. The 0.1 wt% Fe_3O_4/ATO nanofluids exhibited photothermal conversion efficiencies that were either comparable to or greater than those of the 0.1 wt% Fe_3O_4 and 0.1 wt% ATO nanofluids. Among the mixtures, the Fe_3O_4/ATO hybrid nanofluid with $m_{Fe3O4}/m_{Total} = 0.2$ achieved the highest collection efficiency, reaching 76.8%. As the solar exposure duration increased from 300 to 9000 s, a decrease in collection efficiency and an increase in thermal losses were observed across all fluids, with nanofluids showing higher thermal losses compared to water. Notably, the Fe_3O_4/ATO hybrid nanofluid with $m_{Fe3O4}/m_{Total} = 0.2$ not only enhanced collection efficiency but also reduced thermal losses. Its thermal loss rate increased from 13.2% to 64.7% as exposure time extended from 300 to 9000 s, which is lower than the thermal loss rates observed for the single-component nanofluids: 16.2–65.7% for the 0.1 wt% Fe_3O_4 nanofluid and 15–64.9% for the 0.1 wt% ATO nanofluid. However, increasing the proportion of Fe_3O_4 nanoparticles in the Fe_3O_4/ATO nanofluid led to a rise in thermal loss, indicating a potentially adverse effect. This behavior suggests that three-dimensional absorption of solar energy within the receiver contributed to a reduction in heat loss to the surroundings.

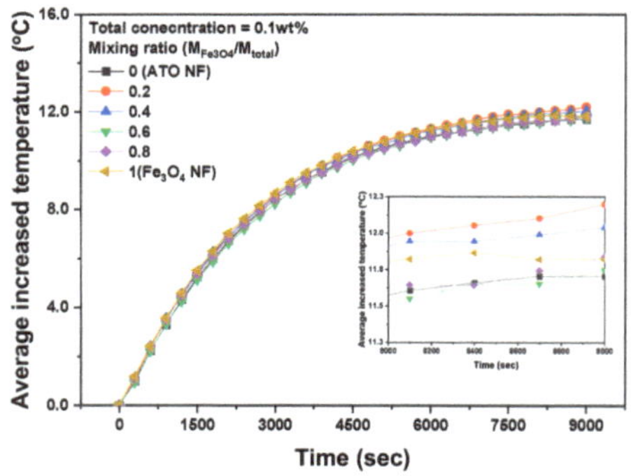

Figure 6. Increased temperature of Fe_3O_4/ATO NFs according to lighting exposure time.

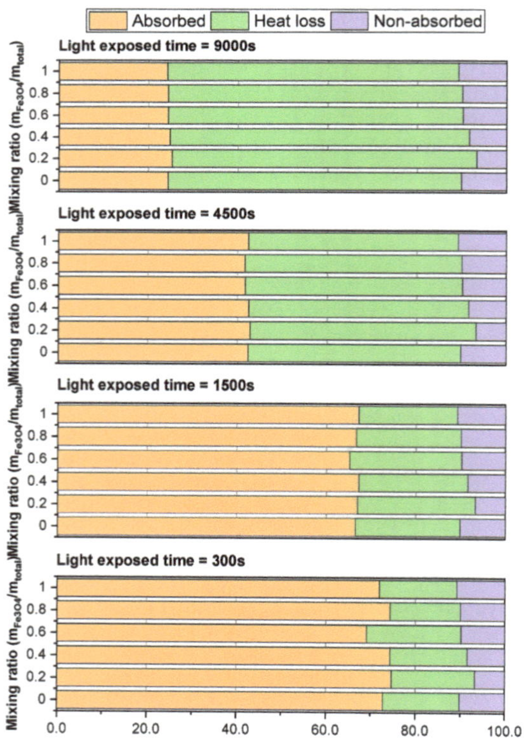

Figure 7. Total energy portion of water, 0.1 wt% Fe_3O_4 NF, 0.1 wt% ATO NF, and 0.1 wt% Fe_3O_4/ATO nanofluid according to light exposure time.

Figure 8 shows the temperature increase and local collection efficiency of the nanofluids after 9000 s of solar exposure. The temperature distribution within the receiver is primarily influenced by the solar energy absorption characteristics of the nanofluids. Both water and nanofluids exhibited higher energy storage in the upper section of the receiver, attributed to the concentration of absorbed thermal energy in this region. The average temperature increases and receiving efficiency for water were observed to be 8.16 °C ($T_{r,tp}$ = 10.3 °C, $T_{r,md}$ = 8.4 °C, $T_{r,bt}$ = 7 °C) and 17.5% (η_{tp} = 7.02%, η_{md} = 5.73%, η_{bt} = 4.77%), respectively. The single-component nanofluids, specifically 0.1 wt% Fe_3O_4 and 0.1 wt% ATO, demonstrated enhanced solar energy absorption capabilities compared to water, resulting in elevated temperature increases and receiving efficiencies. The 0.1 wt% Fe_3O_4 nanofluid achieved an average temperature increase of 11.8 °C ($T_{r,tp}$ = 15.8 °C, $T_{r,md}$ = 11.7 °C, $T_{r,bt}$ = 9.1 °C) and a receiving efficiency of 25% (η_{tp} = 10.8%, η_{md} = 8%, η_{bt} = 6.2%), indicating that the primary absorption of solar radiation occurred predominantly in the upper region. In contrast, the 0.1 wt% ATO nanofluid exhibited an average temperature increase of 11.7 °C ($T_{r,tp}$ = 15.1 °C, $T_{r,md}$ = 11.8 °C, $T_{r,bt}$ = 9.5 °C) and a receiving efficiency of 24.82% (η_{tp} = 10.2%, η_{md} = 8.05%, η_{bt} = 6.48%), demonstrating its capacity to absorb solar energy more evenly across the receiver, including the lower regions.

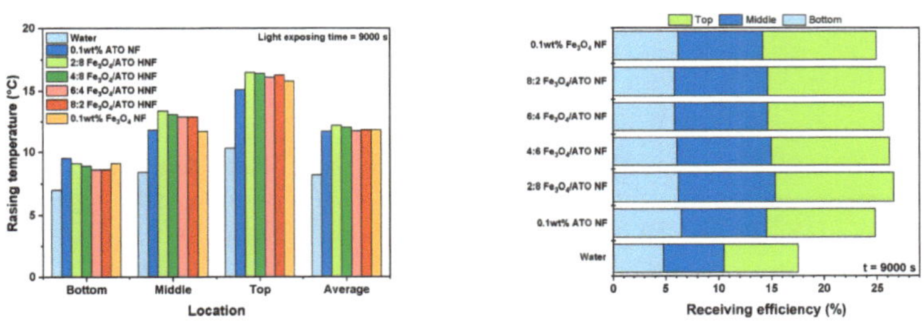

(a) (b)

Figure 8. (**a**) Rising temperature and (**b**) local receiving efficiency at t = 9000 s.

The 0.1 wt% Fe_3O_4/ATO hybrid nanofluid showed a further improvement in collection efficiency over the single-component nanofluids. Notably, the 0.1 wt% Fe_3O_4/ATO hybrid nanofluid with m_{Fe3O4}/m_{Total} = 0.2 recorded the highest performance, with an average temperature increase of 12.2 °C ($T_{r,tp}$ = 16.5 °C, $T_{r,md}$ = 16.5 °C, $T_{r,bt}$ = 13.4 °C) and a receiving efficiency of 26.6% (η_{tp} = 11.3%, η_{md} = 11.3%, η_{bt} = 9.14%). This enhancement is attributed to the combined presence of Fe_3O_4 and ATO nanoparticles, which effectively broadens the solar energy absorption spectrum, thereby facilitating more efficient solar energy capture compared to single-component nanofluids. The 0.1 wt% Fe_3O_4/ATO hybrid nanofluid demonstrated superior solar energy absorption performance, particularly in the upper and middle sections of the receiver.

Interestingly, the Fe_3O_4/ATO hybrid nanofluid with m_{Fe3O4}/m_{Total} = 0.2, despite having a relatively lower absorption rate for solar energy at wavelengths above 800 nm, exhibited the most efficient photothermal conversion. This outcome is likely due to its lower reflectance at 400 nm and the optical path length of the receiver, which was adequate for optimal solar absorption. These results indicate that for the efficient performance of direct absorption solar collectors (DASCs), it is crucial to optimize both the optical properties of the nanofluids and the design parameters of the receiver. However, a temperature gradient still develops between the top and bottom of the receiver, as light absorption primarily occurs at the upper section. This temperature gradient leads to localized heating of the nanofluid, which may negatively impact its dispersion stability. Therefore, mitigating stratification through flow disturbances within the receiver may be beneficial to improve the performance of conventional direct absorption solar collectors using hybrid nanofluids [40].

Figure 9 shows the photothermal conversion efficiency and the solar-weighted absorption coefficient of 0.1 wt% Fe_3O_4/ATO hybrid nanofluids as a function of m_{Fe3O4}/m_{Total} at an optical path length of 0.04 m. The photothermal conversion efficiency of the 0.1 wt% Fe_3O_4/ATO hybrid nanofluid reached its peak value of 0.932 at m_{Fe3O4}/m_{Total} = 0.2. Beyond this ratio, a decline in photothermal conversion efficiency was observed. This trend is consistent with the behavior of the solar-weighted absorption coefficient, which also attained its maximum value of 0.918 at m_{Fe3O4}/m_{Total} = 0.2 and subsequently decreased.

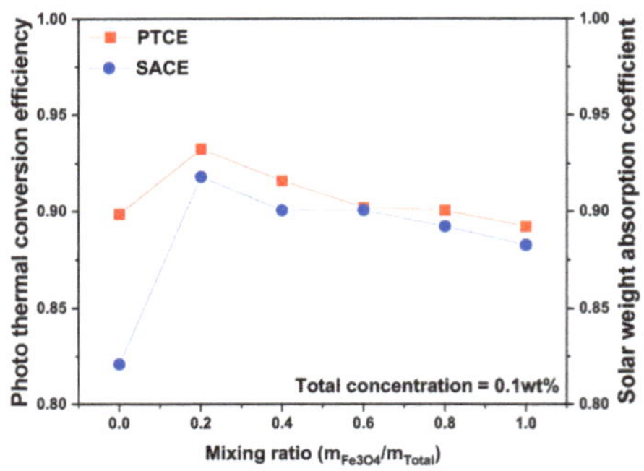

Figure 9. Comparison of photothermal conversion efficiency and solar-weighted absorption efficiency.

To enhance photothermal conversion performance using hybrid nanofluids, it is crucial to achieve a complementary interaction between the optical properties of the coexisting nanoparticles. While Fe_3O_4 nanofluids exhibit high scattering reflection in the ultraviolet and near-infrared regions, the coexistence of Fe_3O_4 and ATO nanoparticles in the base fluid reduces scattering reflection, thereby allowing for increased optical absorption and, consequently, higher photothermal conversion efficiency. Therefore, by carefully designing the m_{Fe3O4}/m_{Total} of the hybrid nanofluid in consideration of the optical path length of the direct absorption solar collector, the optimal performance of the system can be achieved.

Table 2 compares the solar-weighted absorption coefficient and photothermal conversion efficiency of single-component nanofluids, including MWCNT, Fe_3O_4, and ATO, with those of hybrid nanofluids. The Fe_3O_4/ATO hybrid nanofluid achieved a photothermal conversion efficiency of 93.2%, closely matching the 93.3% observed for MWCNT. Additionally, the solar-weighted absorption coefficient of the Fe_3O_4/ATO hybrid nanofluid was 0.938, which is comparable to the 0.935 recorded for MWCNT. While MWCNT exhibits excellent photothermal conversion efficiency, its practical use is limited due to agglomeration and sedimentation at elevated temperatures [41]. Therefore, the Fe_3O_4/ATO hybrid nanofluid presents itself as a potential alternative to MWCNT.

Table 2. Comparison of photothermal conversion efficiency and solar-weighted absorption coefficient for different nanofluids.

Nanofluid	Concentration	SWAC	PTEC
MWCNT [28]	0.002 wt%	0.935	0.933
Fe_3O_4 [28]	0.075 wt%	0.891	0.9
	0.1 wt%	0.888	0.898
ATO [28]	0.1 wt%	0.904	0.892
Fe_3O_4/ATO hybrid nanofluid	0.1 wt% (m_{Fe3O4}/m_{Total} = 0.2)	0.938	0.932

4. Conclusions

This study experimentally examined the optical properties and photothermal conversion performance of m_{Fe3O4}/m_{Total} nanofluids to improve the efficiency of direct absorption solar collectors. While Fe_3O_4 nanofluids demonstrated superior optical absorption compared to ATO nanofluids, they also exhibited significant scattering reflection at wavelengths below 600 nm, limiting their ability to absorb additional solar radiation. In contrast, ATO nanofluids, despite having slightly lower optical absorption, showed reduced scattering reflection. The solar-weighted absorption coefficient of the Fe_3O4/ATO nanofluid notably increased with the optical path length, reaching a peak of 0.938 at an optical path length of 0.06 m and a mixing ratio of m_{Fe3O4}/m_{Total} = 0.2

The Fe_3O_4/ATO hybrid nanofluid achieved more uniform optical absorption within the receiver, owing to the reduced scattering reflection and lower optical transmittance provided by the ATO component. This led to an enhancement in photothermal conversion efficiency. At a mixing ratio of m_{Fe3O4}/m_{Total} = 0.2, the Fe_3O_4/ATO hybrid nanofluid attained a photothermal conversion efficiency of 0.932, surpassing the efficiencies of Fe_3O_4 and ATO nanofluids, which were 0.883 and 0.821, respectively.

To effectively harness the benefits of hybrid nanofluids, it is vital to integrate complementary optical absorption characteristics that mitigate the limitations of single-component nanofluids. The contrasting optical properties of Fe_3O_4 and ATO nanofluids can enhance both the solar-weighted absorption coefficient and the photothermal conversion efficiency. However, sufficient optical path length is necessary for this effect to be significant. At shorter optical path lengths, the impact of scattering reflection is minimal, thereby reducing the improvement offered by the hybrid nanofluid. Thus, in designing direct absorption solar collectors with hybrid nanofluids, it is crucial to carefully consider both the optical path length and the nanofluid mixing ratio to achieve optimal performance.

Author Contributions: Conceptualization, H.C. and J.H.; methodology J.H.; validation, J.H. and H.C.; formal analysis, J.H. and H.K.; investigation, J.H. and H.K.; resources, J.H.; writing—original draft preparation, J.H. and H.K.; writing—review and editing, H.C. and J.H.; supervision, H.C. All authors have read and agreed to the published version of the manuscript.

Funding: This research was supported by "Regional Innovation Strategy (RIS)" through the National Research Foundation of Korea (NRF) funded by the Ministry of Education (MOE) (2021RIS-002).

Data Availability Statement: The data presented in this study are openly available at https://doi.org/10.1016/j.applthermaleng.2024.123005 [28].

Conflicts of Interest: The authors declare no conflicts of interest. The funders had no role in the design of the study; in the collection, analyses, or interpretation of data; in the writing of the manuscript; or in the decision to publish the results.

Abbreviations

Nomenclature		t	Time (s)
A	Area (m^2)		
$A(\lambda)$	Absorbance	**Greek symbols**	
ATO	Antimony-doped tin oxide	η	Efficiency
B	Heat dissipation rate (1/s)	λ	Wavelength (nm)
c_p	Specific heat (J/kg·°C)	ρ	Interface reflectance
EG	ethylene glycol		
I	Incident intensity (W/m^2)	**Subscript**	
k_{ex}	Extinction coefficient (1/m)	AM	Ambient
L_{op}	Optical path length (m)	AM1.5	Am 1.5 global
m	Mass (kg)	Btm	Bottom
NF	Nanofluid	eq	Equilibrium stage
NP	Nanoparticle	mid	Middle
$R(\lambda)$	Reflectance	PTC	Photothermal conversion
$S(L_{op})$	Solar weight absorption coefficient	top	Top
T	Temperature (°C)	rec	Receiving
$T(\lambda)$	Transmittance		

References

1. Aissa, A.; Qasem, N.A.A.; Mourad, A.; Laidoudi, H.; Younis, O.; Guedri, K.; Alazzam, A. A Review of the Enhancement of Solar Thermal Collectors Using Nanofluids and Turbulators. *Appl. Therm. Eng.* **2023**, *220*, 119663. [CrossRef]
2. Miglioli, A.; Aste, N.; Del Pero, C.; Leonforte, F. Photovoltaic-Thermal Solar-Assisted Heat Pump Systems for Building Applications: Integration and Design Methods. *Energy Built Environ.* **2023**, *4*, 39–56. [CrossRef]
3. Saini, P.; Ghasemi, M.; Arpagaus, C.; Bless, F.; Bertsch, S.; Zhang, X. Techno-Economic Comparative Analysis of Solar Thermal Collectors and High-Temperature Heat Pumps for Industrial Steam Generation. *Energy Convers. Manag.* **2023**, *277*, 116623. [CrossRef]
4. Vahidhosseini, S.M.; Rashidi, S.; Hsu, S.H.; Yan, W.M.; Rashidi, A. Integration of Solar Thermal Collectors and Heat Pumps with Thermal Energy Storage Systems for Building Energy Demand Reduction: A Comprehensive Review. *J. Energy Storage* **2024**, *95*, 112568. [CrossRef]
5. Qin, C.; Kim, J.B.; Lee, B.J. Performance Analysis of a Direct-Absorption Parabolic-Trough Solar Collector Using Plasmonic Nanofluids. *Renew. Energy* **2019**, *143*, 24–33. [CrossRef]
6. Howe, M.L.; Paul, T.C.; Khan, J.A. Radiative Properties of Al2O3 Nanoparticles Enhanced Ionic Liquids (NEILs) for Direct Absorption Solar Collectors. *Sol. Energy Mater. Sol. Cells* **2021**, *232*, 111327. [CrossRef]
7. Vital, C.V.P.; Farooq, S.; de Araujo, R.E.; Rativa, D.; Gómez-Malagón, L.A. Numerical Assessment of Transition Metal Nitrides Nanofluids for Improved Performance of Direct Absorption Solar Collectors. *Appl. Therm. Eng.* **2021**, *190*, 116799. [CrossRef]
8. Sreekumar, S.; Joseph, A.; Kumar, C.S.S.; Thomas, S.; Sujith Kumar, C.S.; Thomas, S. Investigation on Influence of Antimony Tin Oxide/Silver Nanofluid on Direct Absorption Parabolic Solar Collector. *J. Clean. Prod.* **2020**, *249*, 119378. [CrossRef]
9. Gupta, H.K.; Agrawal, G.D.; Mathur, J. An Experimental Investigation of a Low Temperature Al_2O_3-H_2O Nanofluid Based Direct Absorption Solar Collector. *Solar Energy* **2015**, *118*, 390–396. [CrossRef]
10. El-dabe, N.T.M.; Abou-zeid, M.Y.; Younis, Y.M. Magnetohydrodynamic Peristaltic Flow of Jeffry Nanofluid with Heat Transfer through a Porous Medium in a Vertical Tube. *Appl. Math. Inf. Sci.* **2017**, *11*, 1097–1103. [CrossRef]
11. Abuiyada, A.; Eldabe, N.T.; Abouzeid, M.; Elshaboury, S. Significance of Heat Source and Activation Energy on MHD Peristaltic Transport of Couple Stress Hyperbolic Tangent Nanofluid through an Inclined Tapered Asymmetric Channel. *Egypt. J. Chem.* **2023**, *66*, 417–436. [CrossRef]
12. Otanicar, T.P.; Phelan, P.E.; Prasher, R.S.; Rosengarten, G.; Taylor, R.A. Nanofluid-Based Direct Absorption Solar Collector. *J. Renew. Sustain. Energy* **2010**, *2*, 33102. [CrossRef]
13. Kalidoss, P.; Venkatachalapathy, S.G.; Suresh, S. Photothermal Performance of Hybrid Nanofluids with Different Base Fluids for Solar Energy Applications. *Energy Sources Part A Recovery Util. Environ. Eff.* **2021**, 1–16. [CrossRef]
14. Vallejo, J.P.; Sani, E.; Żyła, G.; Lugo, L. Tailored Silver/Graphene Nanoplatelet Hybrid Nanofluids for Solar Applications. *J. Mol. Liq.* **2019**, *296*, 112007. [CrossRef]
15. Huminic, G.; Vărdaru, A.; Huminic, A.; Fleacă, C.; Dumitrache, F. Broad-Band Absorption and Photo-Thermal Conversion Characteristics of RGO-Ag Hybrid Nanofluids. *J. Mol. Liq.* **2024**, *408*, 125347. [CrossRef]
16. Fang, J.; Xuan, Y. Investigation of Optical Absorption and Photothermal Conversion Characteristics of Binary CuO/ZnO Nanofluids. *RSC Adv.* **2017**, *7*, 56023–56033. [CrossRef]
17. Zhang, C.; Gao, L.; Zhou, X.; Wu, X. Stability and Photothermal Properties of Fe_3O_4-H_2O Magnetic Nanofluids. *Nanomaterials* **2023**, *13*, 1962. [CrossRef]
18. Ham, J.; Shin, Y.; Cho, H. Comparison of Thermal Performance between a Surface and a Volumetric Absorption Solar Collector Using Water and Fe_3O_4 Nanofluid. *Energy* **2022**, *239*, 122282. [CrossRef]
19. Han, X.; Lu, L.; Yan, S.; Yang, X.; Tian, R.; Zhao, X. Stability, Thermal Conductivity and Photothermal Conversion Performance of Water-Based ZnO Nanofluids. *J. Therm. Sci.* **2021**, *30*, 1581–1595. [CrossRef]
20. Shin, Y.; Ham, J.; Boldoo, T.; Cho, H. Magnetic Effect on the Enhancement of Photo-Thermal Energy Conversion Efficiency of MWCNT/Fe_3O_4 Hybrid Nanofluid. *Sol. Energy Mater. Sol. Cells* **2020**, *215*, 110635. [CrossRef]
21. Li, X.; Zeng, G.; Lei, X. The Stability, Optical Properties and Solar-Thermal Conversion Performance of SiC-MWCNTs Hybrid Nanofluids for the Direct Absorption Solar Collector (DASC) Application. *Sol. Energy Mater. Sol. Cells* **2020**, *206*, 110323. [CrossRef]
22. Qu, J.; Zhang, R.; Shang, L.; Wang, Z. Graphene Oxide/Multi-walled Carbon Nanotube—Therminol® 66 Hybrid Nanofluids for Low-to-medium Temperature Volumetric Solar Collectors. *Int. J. Energy Res.* **2020**, *44*, 7216–7228. [CrossRef]
23. Chen, L.; Xu, C.; Liu, J.; Fang, X.; Zhang, Z. Optical Absorption Property and Photo-Thermal Conversion Performance of Graphene Oxide/Water Nanofluids with Excellent Dispersion Stability. *Sol. Energy* **2017**, *148*, 17–24. [CrossRef]
24. Zuo, X.; Yang, W.; Shi, M.; Yan, H.; Guan, C.; Wu, S.; Zhang, Z.; Li, X.; Li, Z. Experimental Investigation on Photothermal Conversion Properties of Lampblack Ink Nanofluids. *Sol. Energy* **2021**, *218*, 1–10. [CrossRef]
25. Ni, Z.; Cao, X.; Wang, X.; Zhou, S.; Zhang, C.; Xu, B.; Ni, Y. Facile Synthesis of Copper (I) Oxide Nanochains and the Photo-Thermal Conversion Performance of Its Nanofluids. *Coatings* **2021**, *11*, 749. [CrossRef]
26. Zhang, H.; Wang, K.; Wang, L.; Xie, H.; Yu, W. Mesoporous CuO with Full Spectrum Absorption for Photothermal Conversion in Direct Absorption Solar Collectors. *Sol. Energy* **2020**, *201*, 628–637. [CrossRef]
27. Wang, H.; Li, X.; Luo, B.; Wei, K.; Zeng, G. The MXene/Water Nanofluids with High Stability and Photo-Thermal Conversion for Direct Absorption Solar Collectors: A Comparative Study. *Energy* **2021**, *227*, 120483. [CrossRef]

28. Tong, Y.; Ham, J.; Cho, H. Investigation of Thermo-Optical Properties and Photothermal Conversion Performance of MWCNT, Fe3O4, and ATO Nanofluid for Volumetric Absorption Solar Collector. *Appl. Therm. Eng.* **2024**, *246*, 123005. [CrossRef]
29. Chen, Z.; Chen, M.; Yan, H.; Zhou, P.; Chen, X.Y. Enhanced Solar Thermal Conversion Performance of Plasmonic Gold Dimer Nanofluids. *Appl. Therm. Eng.* **2020**, *178*, 115561. [CrossRef]
30. Shang, L.; Qu, J.; Wang, Z.; Zhang, M.; Li, C. Optical Absorption Property and Photo-Thermal Conversion Performance of Ag@Al2O3 Plasmonic Nanofluids with Al_2O_3 Nano-Shell Fabricated by Atomic Layer Deposition. *J. Mol. Liq.* **2021**, *326*, 115388. [CrossRef]
31. Joseph, A.; Sreekumar, S.; Kumar, C.S.S.; Thomas, S. Optimisation of Thermo-Optical Properties of SiO_2/Ag–CuO Nanofluid for Direct Absorption Solar Collectors. *J. Mol. Liq.* **2019**, *296*, 111986. [CrossRef]
32. Hazra, S.K.; Michael, M.; Nandi, T.K. Investigations on Optical and Photo-Thermal Conversion Characteristics of BN-EG and BN/CB-EG Hybrid Nanofluids for Applications in Direct Absorption Solar Collectors. *Sol. Energy Mater. Sol. Cells* **2021**, *230*, 111245. [CrossRef]
33. Tong, Y.; Boldoo, T.; Ham, J.; Cho, H. Improvement of Photo-Thermal Energy Conversion Performance of MWCNT/Fe_3O_4 Hybrid Nanofluid Compared to Fe_3O_4 Nanofluid. *Energy* **2020**, *196*, 117086. [CrossRef]
34. Wang, R.; Xing, L.; Ha, Y.; Zhong, P.; Wang, Z.; Cao, Y.; Li, Z. Photothermal Conversion and Thermal Management of Magnetic Plasmonic Fe_3O_4@ Au Nanofluids. *Solar RRL* **2023**, *13*, 2300269. [CrossRef]
35. Qu, D.; Cheng, L.; Bao, Y.; Gao, Y.; Zheng, X.; Qin, G. Enhanced Optical Absorption and Solar Steam Generation of CB-ATO Hybrid Nanofluids. *Renew. Energy* **2022**, *199*, 509–516. [CrossRef]
36. Xiao, Y.; Tian, W.; Yu, L.; Chen, M.; Zheng, X.; Qin, G. Tunable Optical Properties of ATO-CuO Hybrid Nanofluids and the Application as Spectral Beam Splitters. *Energy* **2023**, *289*, 129964. [CrossRef]
37. Xu, Y.-Y.; Zhou, M.; Geng, H.-J.; Hao, J.-J.; Ou, Q.-Q.; Qi, S.-D.; Chen, H.-L.; Chen, X.-G. A Simplified Method for Synthesis of Fe3O4@ PAA Nanoparticles and Its Application for the Removal of Basic Dyes. *Appl. Surf. Sci.* **2012**, *258*, 3897–3902. [CrossRef]
38. *ASTM G173-03*; Standard Tables for Reference Solar Spectral Irradiances: Direct Normal and Hemispherical on 37° Tilted Surface. American Society for Testing and Materials: West Conshohocken, PA, USA, 2012.
39. Qu, J.; Zhang, R.; Wang, Z.; Wang, Q. Photo-Thermal Conversion Properties of Hybrid CuO-MWCNT/H_2O Nanofluids for Direct Solar Thermal Energy Harvest. *Appl. Therm. Eng.* **2019**, *147*, 390–398. [CrossRef]
40. Struchalin, P.G.; Yunin, V.S.; Kutsenko, K.V.; Nikolaev, O.V.; Vologzhannikova, A.A.; Shevelyova, M.P.; Gorbacheva, O.S.; Balakin, B.V. Performance of a Tubular Direct Absorption Solar Collector with a Carbon-Based Nanofluid. *Int. J. Heat. Mass. Transf.* **2021**, *179*, 121717. [CrossRef]
41. Bortolato, M.; Dugaria, S.; Agresti, F.; Barison, S.; Fedele, L.; Sani, E.; Del Col, D. Investigation of a Single Wall Carbon Nanohorn-Based Nanofluid in a Full-Scale Direct Absorption Parabolic Trough Solar Collector. *Energy Convers. Manag.* **2017**, *150*, 693–703. [CrossRef]

energies

MDPI

Article

A Pore Scale Study on Heat Transfer Characteristics of Integrated Thermal Protection Structures with Phase Change Material

Ziyuan Huang [1,2], Hongming Zhang [2,3], Chao Zhang [2,3], Wei Tang [1,*], Guangming Xiao [2,*] and Yanxia Du [2]

1 State Key Laboratory of Environment-Friendly Energy Materials, Southwest University of Science and Technology, Mianyang 621010, China; huangziyuan233@163.com
2 State Key Laboratory of Aerodynamics, China Aerodynamics Research and Development Center, Mianyang 621000, China; tri-ng@stu.xjtu.edu.cn (H.Z.); zhangchao007@stu.xjtu.edu.cn (C.Z.); yanxiadu@163.com (Y.D.)
3 School of Aerospace, Xi'an Jiaotong University, Xi'an 710049, China
* Correspondence: tangwei@swust.edu.cn (W.T.); xiaogm.2004@tsinghua.org.cn (G.X.)

Abstract: Phase change material (PCM) are characterized by their high latent heat and low density. Combining PCM with building walls, aircraft fuselages, and other structures can significantly enhance the thermal sink capability of these structures. In order to address the issue of low heat storage efficiency resulting from the low thermal conductivity of PCM, a novel integrated thermal protection structure (ITPS) architecture with a supportive structure based on a porous lattice has been designed. Experimental and numerical methods were employed to investigate the thermal response characteristics of the ITPS with and without PCM, the melting behavior of PCM within the porous lattice, and the effects of lattice configuration and pore size on the PCM melting rate. The current ITPS study includes evaluation of two types of lattice configurations and three different pore sizes. The results indicate that the inclusion of PCM reduces the internal panel temperature of the ITPS by approximately 15%. The melting of PCM occurs primarily at the central region of the porous lattice and gradually spreads towards the periphery until complete melting is achieved. Specifically, the Gibson–Ashby lattice configuration enhances the PCM melting rate by 43.5%, while the tetradecahedron lattice configuration yields a 53.1% improvement. Furthermore, for PCM with different pore sizes, smaller pores exhibit faster melting rates during the early and intermediate stages, whereas larger pores exhibit faster melting rates in the later stages as the proportion of liquid PCM increases. The conclusions of this study provide valuable insights for the application of PCM in the field of thermal management.

Keywords: phase change material; porous lattice; numerical simulation; heat conduction; natural convection

1. Introduction

PCM possess numerous outstanding characteristics, such as high latent heat, low density, and sustained stability during phase transition [1–5]. Integrating PCM with structures like building walls and aircraft shells can significantly enhance the thermal sink capabilities of these structures [6–8]. This integrated approach enables the structures to regulate their internal temperature by storing and releasing solar energy when exposed to sunlight, thereby achieving temperature control and energy conservation.

However, conventional PCM often exhibit low thermal conductivity, which limits their heat storage efficiency when used directly as heat-absorbing materials. To enhance the heat storage efficiency of PCM, researchers have employed various optimization methods. Among them, the preparation of phase change microcapsules [9,10], the addition of high thermal conductivity materials [11,12], and the insertion of metallic fins [13,14] are commonly used methods. However, compared to these methods, embedding high thermal conductivity porous foam as a thermal conductive framework [15–19] has shown more significant and stable performance improvements [20]. To investigate the impact of porous metal foam on heat storage performance, Yang et al. [21] conducted numerical simulations and found that the application of porous metal foam can significantly improve the heat transfer efficiency and temperature uniformity of PCMs. In addition, Xiao et al. [22] prepared foam copper/paraffin composite phase change materials (PCMs) using a vacuum impregnation method and measured their effective thermal conductivity, showing that the thermal conductivity of PCMs is 15 times higher than that of pure paraffin. Lafdi et al. [23] experimentally studied the influence of pore size and porosity of porous metal foam on the PCM melting process, revealing that an increase in aluminum foam porosity leads to faster attainment of steady-state temperature for the PCMs. In conclusion, embedding high thermal conductivity porous foam as a thermal conductive framework can effectively improve the heat storage rate of PCMs, reduce the melting time, and decrease temperature gradients. However, porous foam does possess some limitations, such as a closed-cell structure, irregular pore shapes, and variable pore sizes, which impede further enhancement of the heat storage capacity of PCMs and optimization possibilities.

The porous lattice manufactured using 3D printing technology overcomes the drawbacks of traditional porous foam structures such as non-connectivity and irregularity, while offering strong design flexibility, making it an ideal thermal enhancement material [24]. By optimizing geometric parameters such as lattice configuration, porosity, and pore size, the heat storage capacity of PCMs can be further improved. Hu et al. [25] conducted numerical simulations to investigate the impact of a cubic lattice configuration on the heat transfer characteristics of paraffin. The numerical results demonstrate that paraffin impregnated in an aluminum lattice exhibits a significantly shortened melting time and more uniform temperature distribution compared to pure paraffin. Zhang et al. [26] prepared PCMs with different types and pore sizes, studying their heat storage performance on a visualized phase change heat storage experimental platform. Experimental results indicate that the introduction of lattice structures can enhance the thermal conductivity and heat storage density of PCMs. Qureshi et al. [27] fabricated PCMs with four different structures and measured their effective thermal conductivity. The research findings reveal that the thermal conductivity of the diamond structure increased by 11.9 times compared to pure paraffin, the gyro structure increased by 10.4 times, and the original structure increased by 7.3 times. Sriharsha et al. [28] investigated the influence of pore size and porosity of porous lattices on the heat transfer characteristics of PCMs. The results demonstrate that smaller pore sizes and higher porosity can make higher heat storage efficiency of PCMs.

In summary, the lattice configurations of interest in the current study include Kelvin, body-centered cubic, face-centered cubic, primitive, gyroid, and Gibson–Ashby structures et al. Pore sizes of porous lattice structures in practical applications typically fall within the millimeter range. As a result, millimeter-sized pore sizes are commonly employed in related studies [29–31]. Furthermore, the addition of a porous lattice has been found to effectively enhance the heat storage efficiency of PCMs. Therefore, this study combines the metallic shell structure of a certain aircraft to design an ITPS containing a periodic porous lattice. The lattice units serve as both the supporting structure and the thermal conductivity enhancement material for PCMs. The thermal response characteristics of the ITPS samples with and without PCMs were obtained through experiments. A numerical model of the ITPS containing PCMs was established, and the melting process of the PCMs inside the ITPS was numerically simulated using numerical methods. Additionally, an analysis of the factors influencing the melting rate of PCMs was conducted. The conclusions of this research provide important references for the application of PCMs in energy storage and thermal management fields.

The paper is organized as follows: Section 2 describes the experimental models, numerical models, and materials employed for the analysis. Section 3 presents the experimental methods, numerical methods, and validation results used for the analysis. Section 4 presents and discusses the obtained results, while Section 5 provides the conclusions.

2. Models and Materials

2.1. Experimental Models

The experiments are divided into two cases: with and without PCMs. Additionally, in order to compare the influence of different lattice structure on the heat transfer characteristics of ITPS, an experimental model as shown in Figure 1 was created. The left cavity of the metallic base was filled with an alumina ceramic component containing a Gibson–Ashby lattice (GA) to obtain GA-ITPS. The right cavity of the metallic base was filled with an alumina ceramic component containing a tetradecahedron lattice [14] to obtain 14-ITPS. The porous lattice units in the alumina ceramic components had a porosity (ε) of 0.75, and the cell size was 8 mm. The flat plate in the alumina ceramic component served as the inner plate of ITPS, while the bottom panel of the metallic base served as the outer plate of ITPS. The thickness of both plates was 2 mm. The dimensions of the two cavities in the metallic base were 76 mm × 36 mm × 10 mm. PCM was filled through the holes on the inner plate, resulting in GA-PCMs when combined with the GA lattice, and 14-PCMs when combined with the tetradecahedron lattice.

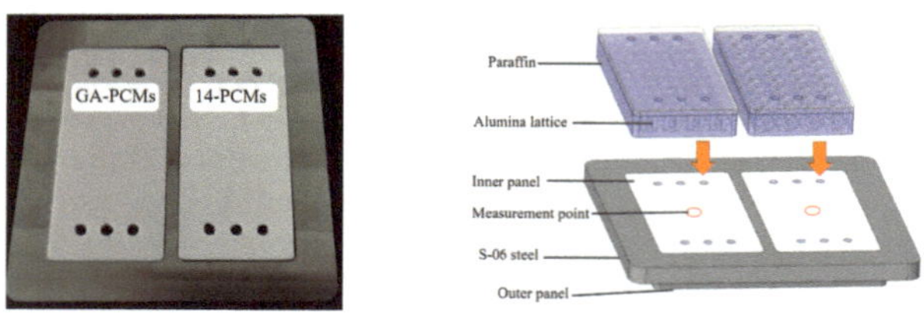

(a) (b)

Figure 1. Experimental model. (**a**) Photos of ITPS; (**b**) ITPS schematic.

2.2. Simulation Models

To further investigate the PCM melting process inside the ITPS and factors affecting the melting rate of PCMs, a simulation model was established as shown in Figure 2a. In the left cavity of the metallic base, the PCM is integrated with porous lattices of varying structures (Gibson–Ashby and tetradecahedron) and diverse pore sizes (8 mm, 6 mm, and 4 mm), while the right cavity was filled with pure PCM. The primary role of PCM is to provide latent heat, while the lattice structure serves to enhance thermal conduction. Therefore, the model was designed with one side featuring the lattice structure and the other side without it. This configuration aims to demonstrate the ability of the lattice structure to enhance thermal conduction and thereby accelerate the melting of PCM. The simulation model was subjected to boundary conditions as shown in Figure 2b: a constant temperature boundary (Th) of 353.15 K was applied to the lower surface of the model, while the walls surrounding the model were treated as adiabatic boundaries. The ambient temperature was set to 298.15 K.

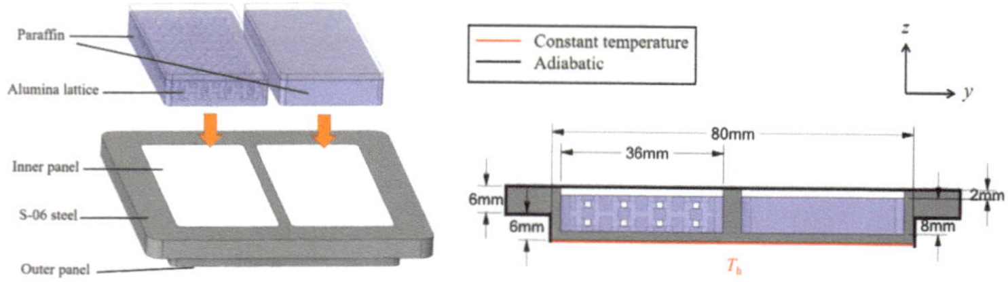

(a) (b)

Figure 2. Simulation model and boundary conditions. (**a**) Photos of ITPS; (**b**) ITPS schematic.

2.3. Material Properties Parameters

In this study, the material of the metal base is S-06 stainless steel, and the material of the lattice is alumina ceramic. Considering requirements such as phase transition temperature, latent heat of fusion, density, and chemical stability, slice paraffin (China National Pharmaceutical Group Chemical Reagent Co., Ltd., Shanghai, China) with a phase transition temperature ranging from 324.45 K to 334.25 K was chosen as the PCM. First, the paraffin was subjected to DSC testing using a DSC instrument (NETZSCH, DSC 200 F3). Then, the thermal conductivity of paraffin, aluminum oxide ceramic, and S-06 stainless steel was measured using a thermal conductivity analyze, and the material properties are listed in Table 1 [32].

Table 1. Thermophysical properties of the materials.

Material	ρ (kg/m^3)	c_p (J/(kg·K))	λ (W/(m·K))	L (kJ/kg)	T_s (K)	μ (kg/(m·s))	α (K^{-1})
Paraffin	824	2200	0.2977	213.05	324.45	0.00689	0.00583
Steel (S-06)	8030	502.48	16.27	-	-	-	-
Alumina ceramic	3500	880	27.5	-	-	-	-

3. Methods and Verifications

3.1. Experimental Methods

Two alumina ceramic components containing different lattice units (Gibson–Ashby lattice and tetradecahedron lattice) were manufactured through 3D printing. These components were then filled into the respective grooves of the metallic base to create ITPS experimental specimens with different lattice structures. For the thermal response experiments of ITPS with PCM, the following steps were taken to prepare the ITPS specimens containing PCMs:

(1) The 3D-printed alumina ceramic components were combined with the metallic base to form the ITPS experimental specimens.
(2) The paraffin and the ITPS specimens were heated to 353.15 K to completely liquefy paraffin.
(3) The liquefied paraffin was poured into the porous lattice units through pre-designed holes, and the system was allowed to thermally equilibrate for a certain period to ensure complete filling of the porous lattice.
(4) The paraffin was cooled and solidified at room temperature (298.15 K), resulting in the formation of ITPS specimens containing PCMs.

A phase change thermal storage experimental platform was set up to conduct heating experiments on the prepared specimens. As shown in Figure 3, the platform consists of a graphite heating plate, insulation cotton, a thermocouple, a temperature tester, and a computer. The process of this heating experiment is as follows:

(1) First, heat the temperature of the graphite heating plate to 353.15 K.
(2) Then, wrap the experimental sample with thermal insulation cotton around its surroundings. And place a thermocouple at the center point on the surface of the inner plate.
(3) Next, place the experimental sample on a graphite heating platform at a temperature of 353.15 K. The temperature data are transmitted to the temperature tester through the thermocouple.
(4) Finally, the collected temperature data was processed and analyzed using a computer.

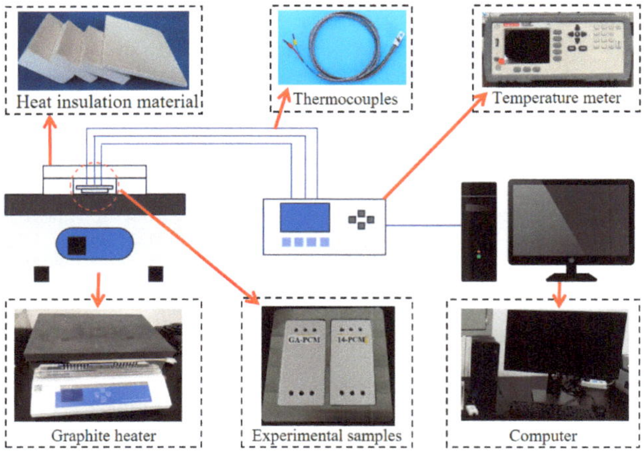

Figure 3. Phase change thermal storage experimental platform.

3.2. Numerical Methods

The heat transfer in PCMs during the phase transition process is highly complex and involves heat conduction between the lattice framework and the solid PCM, convective heat transfer between the lattice framework and the liquid PCM and natural convection within the liquid PCM, among others. Therefore, to simplify the analysis and solution, the following assumptions are introduced in the computational model [33]:

(1) The molten liquid paraffin is considered as an incompressible Newtonian fluid, and its flow within the enclosed space is assumed to be laminar.
(2) Except for density differences caused by the Boussinesq assumption, all other thermophysical properties of the material are assumed to be constant.
(3) Thermal radiation within the PCMs is neglected.
(4) The distribution of paraffin and aluminum oxide is assumed to be uniform and isotropic.

This study employs the enthalpy-porosity model in Fluent 2022 R1 software package to simulate the melting process of PCM [34]. In this model, the solid–liquid mixture region of PCM is treated as a porous medium, and the liquid fraction β of PCM is equivalent to the porosity of the porous medium. $\beta = 0$ indicates that PCM is in the solid state, $0 < \beta < 1$ indicates the solid–liquid mixture state of PCM, and $\beta = 1$ indicates that PCM is in the liquid state. The liquid fraction β is defined as follows [29]:

$$\beta = \begin{cases} 0 & T < T_s \\ \frac{T-T_s}{T_l-T_s} & T_s < T < T_l \\ 1 & T > T_l \end{cases} \tag{1}$$

Based on the assumptions mentioned above, the governing equations can be expressed as follows:

Continuity equation:

$$\frac{\partial \rho}{\partial t} + \nabla \cdot (\rho v) = 0 \tag{2}$$

Momentum equation:

$$\frac{\partial (\rho v)}{\partial t} + \nabla \cdot (\rho v_i v) = \nabla \cdot (\mu \nabla v_i) - \frac{\partial p}{\partial x_i} + S \tag{3}$$

Energy equation:

$$\frac{\partial(\rho H)}{\partial t} + \nabla \cdot (\rho v H) = \frac{\lambda}{c_p} \nabla^2 \cdot H + F \tag{4}$$

In Equations (2)–(4), ρ represents the density of the PCM, kg/m^3; v represents the fluid velocity, m/s; μ represents the liquid phase viscosity of the PCM; p represents the pressure, Pa; S represents the source term in the momentum equation; λ represents the thermal conductivity, W/(m·K); H represents the specific enthalpy; c_p represents the specific heat; F represents the source term in the energy equation.

$$H = h + \Delta L \tag{5}$$

$$h = h_{ref} + \int_{T_{ref}}^{T} c_p dT \tag{6}$$

$$\Delta L = \beta L \tag{7}$$

In Equations (5)–(7), h represents the sensible enthalpy, J/(kg·K); ΔL represents the released latent heat, kJ/kg; L represents the latent heat, kJ/kg.

3.3. Method Verification

3.3.1. Numerical Method Verification

To verify the accuracy of the numerical simulation method employed in this study, we conducted numerical simulations of the PCM melting experiment performed by Babak et al. [34]. The numerically simulated PCM liquid fraction variation curve was compared with Babak's experimental results, as shown in Figure 4. The two curves exhibited a good agreement, with a relative error of less than 2% for the liquid fraction. Next, the numerically simulated PCM phase interface evolution was compared with the experimental results, as illustrated in Figure 5. The transverse melting rate obtained from the numerical simulation was faster than the experimental rate, while the longitudinal melting rate was slower than the experimental rate. The main source of error was attributed to slight heat losses during the experimental process. However, the deviations between the numerical simulation and experimental results were relatively small. These findings demonstrate that the proposed numerical simulation method can accurately solve PCM melting-related problems.

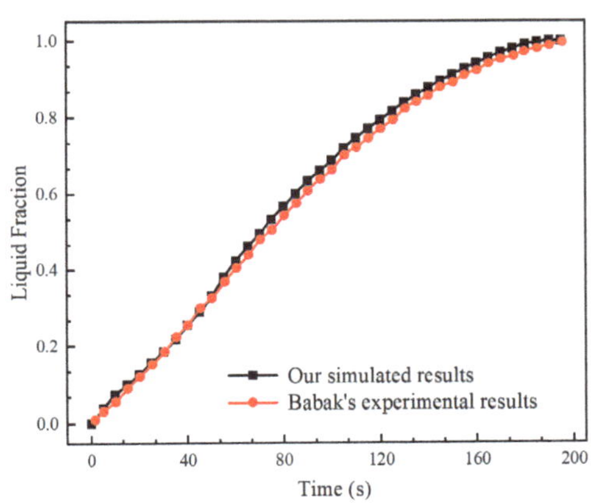

Figure 4. Comparison of liquid fractions: Babak's experimental results [35] and our simulated results.

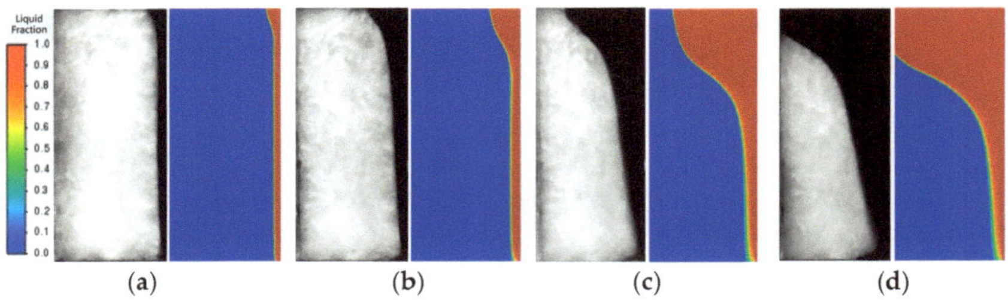

Figure 5. Comparison of solid–liquid phase distribution: Babak's experimental results [35] (left) and our simulated results (right). (**a**) t = 10 min; (**b**) t = 20 min; (**c**) t = 40 min; (**d**) t = 60 min.

3.3.2. Mesh Independence Analysis

In this study, we employed a GA-ITPS model with a lattice pore of 8 mm. Three simulation models were established with grid quantities of 4.2×10^5, 1.2×10^6, and 2.0×10^6, respectively. The purpose was to validate the mesh independence by comparing the time-dependent liquid fraction curves obtained from these models. The results are presented in Figure 6, where the liquid fraction curves of the three grid quantities exhibit similar trends. However, upon closer examination in the zoomed-in section, a more pronounced change in the liquid fraction can be observed when the grid quantity increases from 418,286 to 1,192,057. Nevertheless, further increasing the grid quantity beyond 1,192,057 has minimal effect on the numerical simulation results. Therefore, we selected the model with a grid quantity of 1,192,057 for conducting the numerical simulations.

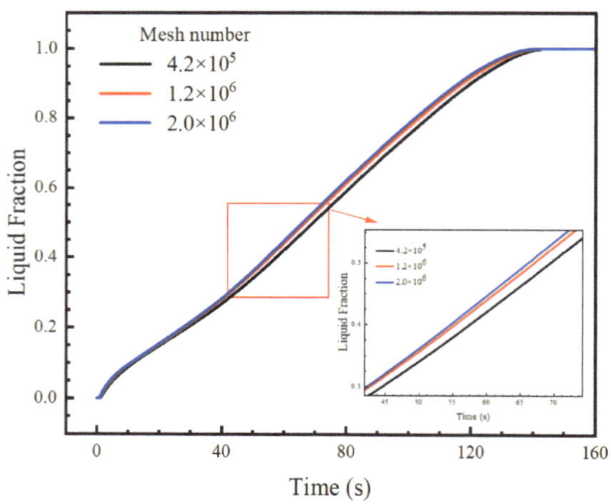

Figure 6. Mesh independence analysis.

4. Results and Discussion

4.1. Heat Transfer Characteristics of ITPS

Heating experiments were conducted on ITPS specimens with and without paraffin using a heating platform. The ITPS specimens comprised both a G-A lattice and tetradecahedron lattice. The temperature response characteristics of GA-ITPS with and without paraffin are shown in Figure 7a. During the heat storage process, the maximum temperature decrease on the inner plate surface was 16 K, with an average decrease of 8.3 K, resulting in an overall temperature reduction of approximately 15.1%. The temperature response characteristics of 14-ITPS with and without paraffin are shown in Figure 7b. During the heat storage process, the maximum temperature decrease on the inner plate surface was 17.4 K, with an average decrease of 8.8 K, resulting in an overall temperature reduction of approximately 15.6%. The experimental results indicate that the addition of paraffin can effectively reduce the inner plate surface temperature of the ITPS specimens. Therefore, paraffin can be considered as a heat-absorbing layer material in ITPS.

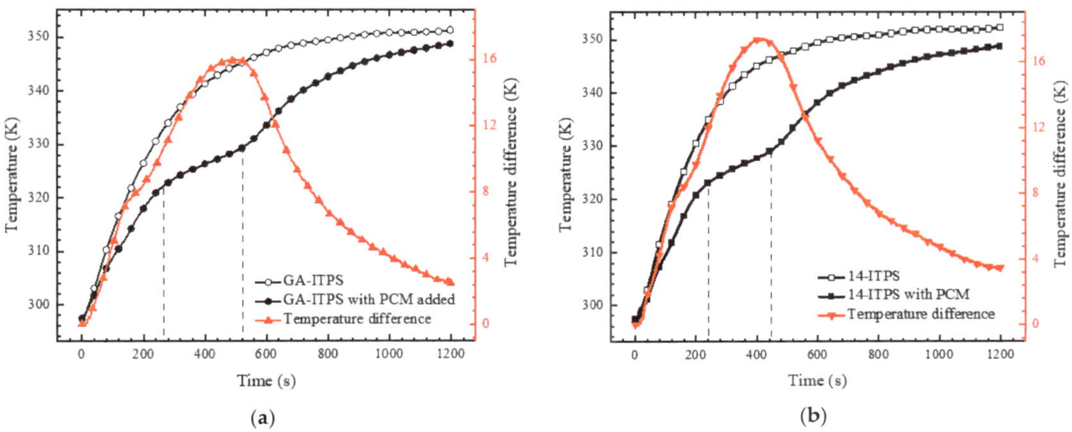

Figure 7. Temperature response of ITPS. (**a**) GA-ITPS; (**b**) 14-ITPS.

The temperature rise process of the ITPS specimens containing paraffin can be divided into three stages. The first stage is the sensible heat storage stage before paraffin melts, during which heat energy is stored through the material's sensible heat, leading to a rapid temperature increase. The second stage is the latent heat storage stage during paraffin melting, where a significant amount of heat is absorbed as paraffin undergoes a phase change, resulting in a gradual temperature increase. The third stage is the sensible heat storage stage after the complete melting of paraffin, during which heat energy is stored again through the sensible heat of paraffin, leading to an accelerated temperature increase that eventually approaches the heating temperature of the outer plate.

The temperature response characteristics of GA-ITPS and 14-ITPS with paraffin were compared, as shown in Figure 8. The inner plate surface temperature of GA-ITPS was consistently lower than that of 14-ITPS, with an average temperature difference of 3.1 K and a maximum difference of 6.7 K. Additionally, the latent heat storage stage of 14-ITPS with paraffin was shorter than that of GA-ITPS with the PCM. These results indicate that the lattice configuration has a significant impact on the temperature response of ITPS and the melting rate of the internal PCM.

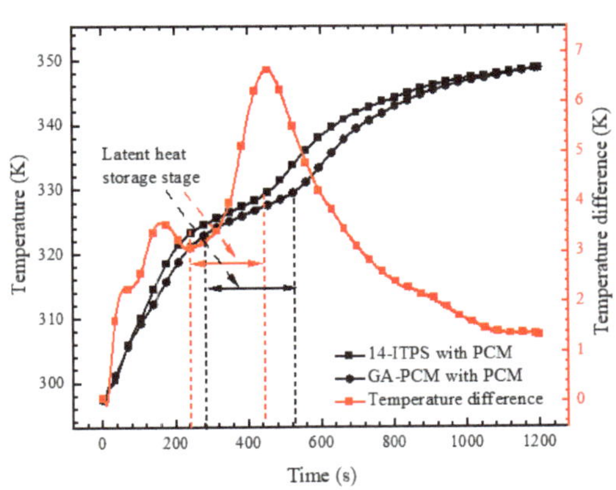

Figure 8. Comparison of temperature response between GA-ITPS with PCM and 14-ITPS with PCM.

4.2. Evolution of Solid–Liquid Interface

To investigate the melting process and flow behavior of paraffin, a selected row of lattices in the middle of the ITPS configuration was chosen. Three interfaces at X = 50, 52, and 54 mm were extracted, representing the 1/2, 3/4, and 1 positions of the selected lattice, as shown in Figure 9.

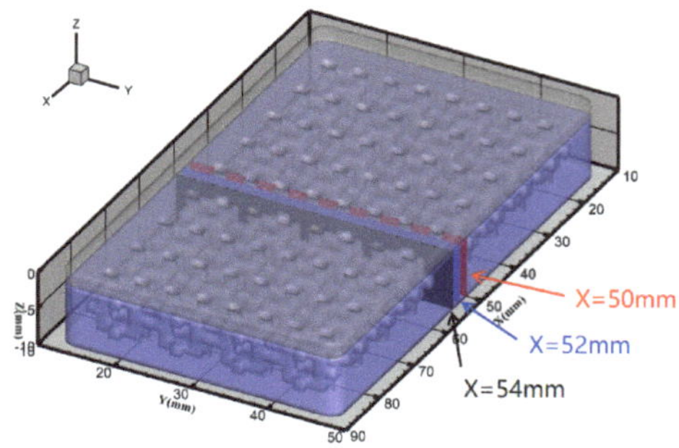

Figure 9. Position as displayed in the contour map.

The melting process of pure paraffin is depicted in Figure 10a, where the solid–liquid interface is approximately planar at t = 20 s. Due to uneven heating from the base, the left and right sides of the paraffin melt first after t = 20 s. At this stage, heat conduction from the metal base to paraffin dominates, resulting in a trapezoidal approximation of the solid–liquid interface. As the proportion of liquid paraffin increases within the chamber, natural convection starts to strengthen. Combining the velocity contour map of pure paraffin in Figure 11a, multiple upward and downward convection loops are observed at the bottom, with stronger flow closer to the metal base around the periphery. This natural convection path leads to a transition of the solid–liquid interface from trapezoidal to an irregular shape.

During the melting process of PCMs shown in Figure 10b,c, the solid–liquid interface spreads gradually from the porous lattice at the center towards the surrounding area until the paraffin is completely transformed into a liquid state. The reason behind this phenomenon is that the heat from the metal base is rapidly transferred to the porous lattice, which has a high thermal conductivity. Subsequently, the heat is conducted to the paraffin encapsulated near the porous lattice, leading to rapid melting of paraffin. Additionally, as evident from Figure 11b,c, in PCMs, natural convection only occurs within the pore channels of the porous lattice, and its intensity is significantly lower than that observed in pure paraffin. This indicates that natural convection is restrained by the porous lattice, thereby weakening its impact on the solid–liquid interface.

4.3. Effects of Structure

The effects of lattice structure on the melting rate of PCMs was investigated through numerical simulations. The liquid fraction variation of paraffin in different lattice structures is shown in Figure 12. Pure paraffin completely melts at t = 254 s. GA-PCMs reach complete melting at t = 144 s, reducing the melting time by 43.5% compared to pure paraffin. Similarly, 14-PCMs achieve complete melting at t = 119 s, shortening the melting time by 53.1% compared to pure paraffin. Despite the paraffin content in PCMs being only 25% lower than pure paraffin, the melting time differs by nearly twice as much. Therefore, incorporating a porous lattice in paraffin significantly reduces the required melting time.

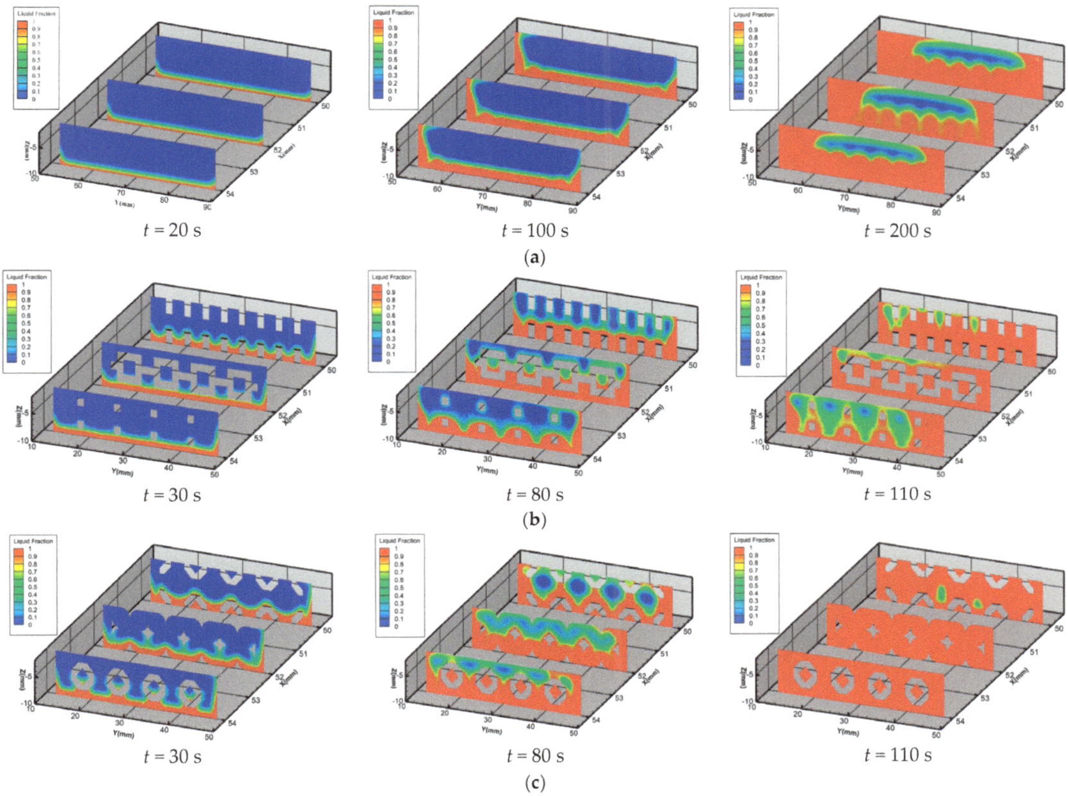

Figure 10. Evolution of solid–liquid interface during melting of paraffin. (**a**) Pure PCM; (**b**) GA-PCMs; (**c**) 14-PCMs.

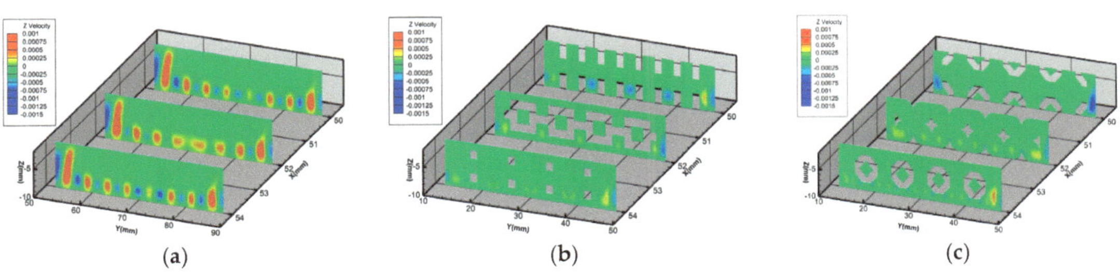

Figure 11. Flow velocity of paraffin at $\beta = 0.6$. (**a**) Pure PCM; (**b**) GA-PCMs; (**c**) 14-PCMs.

The melting time of 14-PCMs is 17.36% shorter than that of GA-PCMs. From Figure 11b,c, it can be observed that the strength of natural convection inside 14-PCMs and GA-PCMs is similar. Therefore, the primary factor that affects the rate of paraffin melting is the thermal conductivity between porous lattice and paraffin. The thermal conductivity between porous lattice and paraffin depends on the contact area between porous lattice and paraffin. As shown in Table 2, since the specific surface area (A) of GA-PCMs and 14-PCMs is 2.61 mm^{-1} and 3.29 mm^{-1}, respectively, the melting time of 14-PCMs is shorter than that of GA-PCMs.

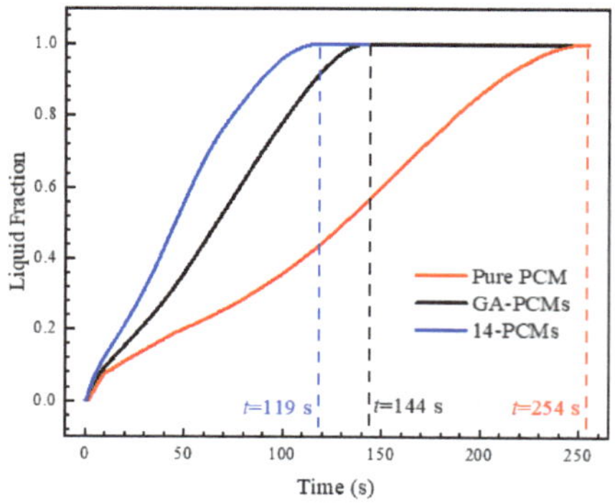

Figure 12. Liquid fraction variation in PCMs with different pore structure.

Table 2. Geometric parameters of a porous lattice.

Lattice Structure	d_p (mm)	ε	A (mm^{-1})	l (mm)
Tetradecahedron	8	0.75	3.29	8
Gibson–Ashby	8	0.75	2.61	8
Gibson–Ashby	6	0.75	3.48	6
Gibson–Ashby	4	0.75	5.12	4

4.4. Effects of Pore Size

The effects of lattice pore size (d_p) on the melting rate of PCMs was investigated through numerical simulations. The GA lattice was selected as the research object, and three different pore sizes of 8, 6, and 4 mm were designed for the GA lattice, with their geometric parameters shown in Table 2. The interfaces at 1/2, 3/4, and 1 positions of the single cell in the center row of the porous lattice were chosen as the display interfaces for demonstration.

The variation of liquid fraction with different pore sizes of GA-PCMs is depicted in Figure 13. During the early to mid-stage of melting, smaller pore sizes result in faster melting rates and higher liquid fractions. For instance, at $t = 50$ s, the PCMs with an 8 mm pore size had a liquid fraction of 0.36, while the PCMs with a 6 mm pore size had a liquid fraction of 0.39, and the PCMs with a 4 mm pore size had a liquid fraction of 0.42. This can be attributed to the scarcity of liquid paraffin during the early to mid-stage of melting, which weakens the natural convection effect within the PCMs, as illustrated in Figure 14a. Consequently, heat transfer primarily occurs through conduction, and thus, smaller pore sizes with a larger surface area exhibit stronger heat transfer capability. In the later stage of melting, as the proportion of liquid paraffin increases, natural convection becomes more prominent. As shown in Figure 14b, lattices with larger pore sizes exhibit stronger natural convection. Therefore, in the later stage of melting, PCMs with larger pore sizes melt at a faster rate.

The Rayleigh number (*Ra*) represents the ratio between buoyancy and viscous forces in a fluid and is primarily used to characterize the intensity of natural convection. A higher *Ra* indicates a more pronounced convective behavior and more vigorous fluid flow. It is mathematically defined as shown in Equation (8).

$$Ra = \frac{g\alpha\Delta T l^3}{\mu\kappa} \tag{8}$$

In Equation (8), g represents the acceleration due to gravity, m/s^2; α represents the thermal expansion coefficient, K^{-1}; l represents the characteristic length, mm; μ represents the kinematic viscosity, m^2/s; and κ represents the thermal diffusivity, m^2/s.

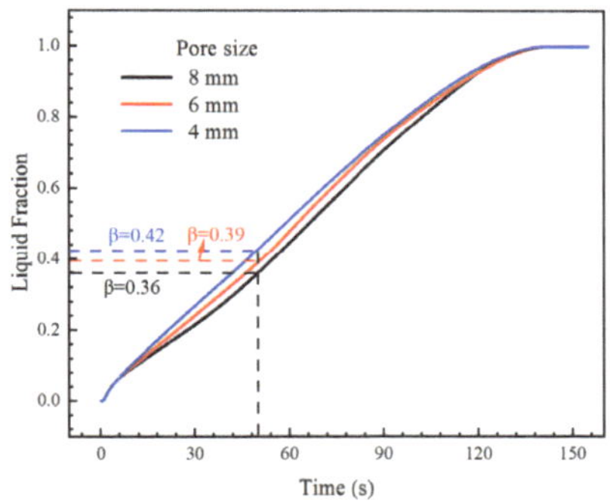

Figure 13. Liquid fraction variation in PCMs with different pore sizes.

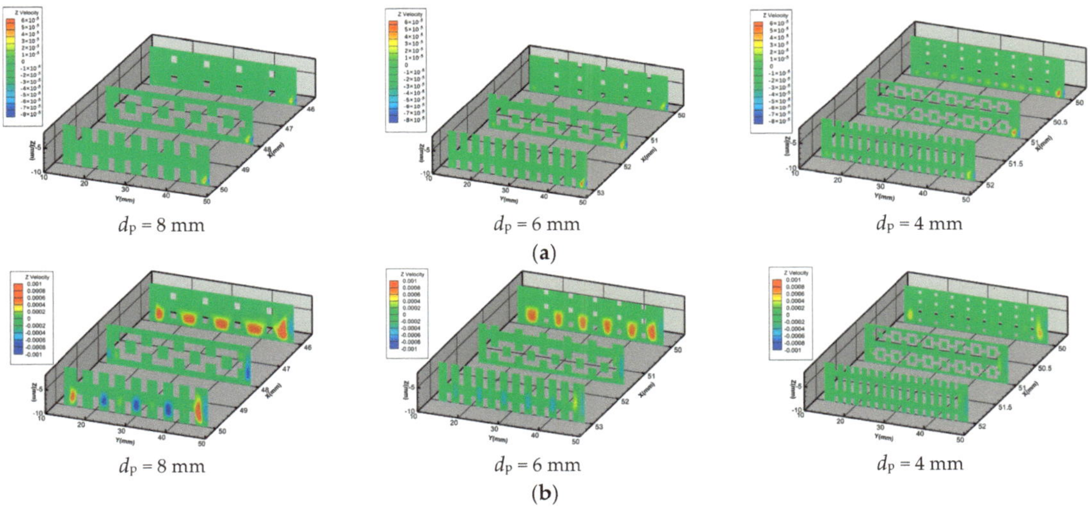

Figure 14. Velocity distribution of GA-PCMs with different pore. (**a**) t = 30 s; (**b**) t = 100 s.

For porous lattices, the characteristic length is consistent with the pore size. Therefore, as the characteristic length of the porous lattice increases, the *Ra* also increases, indicating a more pronounced natural convection phenomenon. Additionally, Table 2 reveals that smaller pore sizes in the porous lattice correspond to larger surface areas, resulting in increased contact area with the liquid paraffin. This increased contact area leads to an increase in viscous resistance against the liquid paraffin within the porous lattice. Consequently, as the pore size decreases, the natural convection effect becomes weaker.

5. Conclusions

This paper based on experimental and numerical methods, investigated the influence of PCMs on the thermal insulation performance of ITPS, as well as the simulation and analysis of the PCM melting process and factors affecting its melting rate. The concluding remarks can be drawn as follows:

(1) PCMs can effectively enhance the thermal insulation capability of ITPS. The addition of paraffin can decrease the average temperature of the inner plate of ITPS by approximately 15%.

(2) Porous lattices can significantly improve the heat storage efficiency of PCMs. The incorporation of porous lattices can increase the melting rate of paraffin wax by around 50%.
(3) The geometrical characteristics of the porous lattices have a significant effect on the melting rate of PCMs, and the heat storage efficiency of PCMs can be accurately adjusted by modifying these geometrical characteristics. Among different lattice structures, a larger specific surface area results in a faster melting rate of PCMs. Regarding different pore sizes, a smaller pore size leads to a faster melting rate when thermal conduction dominates heat transfer. However, a larger pore size results in a faster melting rate when natural convection dominates the heat transfer.

Author Contributions: Conceptualization, Z.H., W.T. and G.X.; methodology, Z.H.; software, Z.H.; validation, Z.H., H.Z. and C.Z.; formal analysis, Z.H.; investigation, Z.H.; resources, Z.H.; data curation, Z.H.; writing—original draft preparation, Z.H.; writing—review and editing, Z.H., H.Z. and C.Z.; visualization, Z.H.; supervision, W.T. and G.X.; project administration, G.X. and Y.D.; funding acquisition, Y.D. All authors have read and agreed to the published version of the manuscript.

Funding: This research was supported by the National Key Research and Development Program of China (No. 2019YFA0405202).

Data Availability Statement: The data that support the findings of this study are available from the first author upon reasonable request.

Conflicts of Interest: The authors declare no conflicts of interest.

Nomenclature

A	Specific surface area (mm^{-1}).
d_p	Pore size (mm).
c_p	Specific heat (J/(kg·K)).
F	Source term used in the energy equation (W/m^3).
g	Acceleration due to gravity (m/s^2).
H	Specific enthalpy (kJ/kg).
ΔL	Released latent heat (kJ/kg).
h	Sensible enthalpy (J/(kg·K)).
h_{ref}	Reference sensible enthalpy (J/(kg·K)).
L	Latent heat (kJ/kg).
l	Characteristic length (mm).
p	Pressure (Pa).
Ra	Rayleigh number.
S	Source term used in the momentum equation (N/m^3).
T	Temperature (K).
T_h	Constant temperature (K).
T_l	Liquidus temperature (K).
T_s	Solidus temperature (K).
T_{ref}	Reference temperature (K).
ΔT	Temperature difference (K).
t	Time (s).
v	Velocity (m/s).
α	Thermal expansion coefficient (K^{-1}).
β	Liquid fraction.
ε	Porosity.
κ	Thermal diffusivity (m^2/s).
λ	Thermal conductivity (W/(m·K)).
μ	Kinematic viscosity (m^2/s).
ρ	Density (kg/m^3).

References

1. Yuan, Y.P.; Xiang, B.; Cao, X.L.; Zhang, N.; Sun, L.L. Research status and development on latent energy storage technology of building. *J. Southwest Jiaotong Univ.* **2016**, *51*, 585–598.
2. Jiang, Z.; Zou, B.Y.; Cong, L.; Xie, C.; Li, C.; Qiao, G.; Zhao, Y.; Nie, B.; Zhang, T.; Ge, Z.; et al. Research progress and outlook of thermal energy storage technologies. *Energy Storage Sci. Technol.* **2022**, *11*, 2746–2771.
3. Li, Z.; Li, B.R.; Chen, H.Z.; Wen, B.; Du, X.Z. State of the art review on phase change thermal energy storage technology. *Chem. Ind. Eng. Prog.* **2020**, *39*, 5066–5085.
4. Sun, W.C.; Feng, J.X.; Zhang, Z.G.; Fang, X.M. Research progress of phase change heat storage technology for passive energy conservation in buildings. *Chem. Ind. Eng. Prog.* **2020**, *39*, 1824–1834.
5. Shang, B.; Hu, J.; Hu, R.; Cheng, J.; Luo, X. Modularized thermal storage unit of metal foam/paraffin composite. *Int. J. Heat Mass Transf.* **2018**, *125*, 596–603. [CrossRef]
6. Verdier, D.; Falcoz, Q.; Ferriere, A. Design of a protection thermal energy storage using phase change material coupled to a solar receiver. *High Temp. Mater. Process.* **2014**, *33*, 509–523. [CrossRef]
7. Cao, C.Y.; Wang, R.X.; Xing, X.D.; Song, H.W.; Huang, C.G. Design and experimental investigation of integrated thermal protection structure containing phase change material layer. *J. Astronaut.* **2019**, *40*, 352–361.
8. Du, Y.X.; Xiao, G.M.; Gui, Y.W.; He, L.X.; Liu, L. Heat transfer characteristics inside phase change thermal control device. *J. Chem. Ind. Eng.* **2012**, *63*, 107–113.
9. Zhao, C.Y.; Zhang, G.H. Review on microencapsulated phase change materials (MEPCMs): Fabrication, characterization and applications. *Renew. Sustain. Energy Rev.* **2011**, *15*, 3813–3832. [CrossRef]
10. Zhang, N.; Yuan, Y. Synthesis and thermal properties of nanoencapsulation of paraffin as phase change material for latent heat thermal energy storage. *Energy Built Environ.* **2020**, *1*, 410–416. [CrossRef]
11. Zhang, Y.; Ding, J.; Wang, X.; Yang, R.; Lin, K. Influence of additives on thermal conductivity of shape-stabilized phase change material. *Sol. Energy Mater. Sol. Cells* **2006**, *90*, 1692–1702. [CrossRef]
12. Motahar, S.; Alemrajabi, A.A.; Khodabandeh, R. Experimental study on solidification process of a phase change material containing TiO2 nanoparticles for thermal energy storage. *Energy Convers. Manag.* **2017**, *138*, 162–170. [CrossRef]
13. Agyenim, F.; Eames, P.; Smyth, M. A comparison of heat transfer enhancement in a medium temperature thermal energy storage heat exchanger using fins. *Sol. Energy* **2009**, *83*, 1509–1520. [CrossRef]
14. Yang, X.; Lu, Z.; Bai, Q.; Zhang, Q.; Jin, L.; Yan, J. Thermal performance of a shell-and-tube latent heat thermal energy storage unit: Role of annular fins. *Appl. Energy* **2017**, *202*, 558–570. [CrossRef]
15. Meng, Z.N.; Zhang, P. Experimental and numerical investigation of a tube-in-tank latent thermal energy storage unit using composite PCM. *Appl. Energy* **2017**, *190*, 524–539. [CrossRef]
16. Zhu, F.; Zhang, C.; Gong, X. Numerical analysis and comparison of the thermal performance enhancement methods for metal foam/phase change material composite. *Appl. Therm. Eng.* **2016**, *109*, 373–383. [CrossRef]
17. Xiao, G.M.; Du, Y.X.; Liu, L.; Wei, D.; Yang, X.F.; Gui, Y.W. Numerical simulation and experimental study on the thermal characteristics of composite phase change materials for heat storage. *J. Exp. Fluid Mech.* **2018**, *32*, 6.
18. Zhang, N.; Liu, H.; Xia, Q.; Zhang, Z.; Tong, Z.; Du, Y.; Yuan, Y. Experimental investigation of the effect of copper foam pore structure on the thermal performance of phase change material in different centrifugal force fields. *Int. J. Heat Mass Transf.* **2023**, *206*, 123945. [CrossRef]
19. Chen, H.; Liu, X.L.; Yang, Y.X.; Zhong, L.Q.; Wang, L.; Gao, N. Numerical simulation of foam metal copper/paraffin phase change thermal storage process. *CIESC J.* **2019**, *70*, 86–92.
20. Chintakrinda, K.; Weinstein, R.D.; Fleischer, A.S. A direct comparison of three different material enhancement methods on the transient thermal response of paraffin phase change material exposed to high heat fluxes. *Int. J. Therm. Sci.* **2011**, *50*, 1639–1647. [CrossRef]
21. Yang, X.; Yu, J.; Guo, Z.; Jin, L.; He, Y.-L. Role of porous metal foam on the heat transfer enhancement for a thermal energy storage tube. *Appl. Energy* **2019**, *239*, 142–156. [CrossRef]
22. Xiao, X.; Zhang, P.; Li, M. Preparation and thermal characterization of paraffin/metal foam composite phase change material. *Appl. Energy* **2013**, *112*, 1357–1366. [CrossRef]
23. Lafdi, K.; Mesalhy, O.; Shaikh, S. Experimental study on the influence of foam porosity and pore size on the melting of phase change materials. *J. Appl. Phys.* **2007**, *102*, 759. [CrossRef]
24. Al-Ketan, O.; Rowshan, R.; Abu Al-Rub, R.K. Topology-mechanical property relationship of 3D printed strut, skeletal, and sheet based periodic metallic cellular materials. *Addit. Manuf.* **2018**, *19*, 167–183. [CrossRef]
25. Hu, X.; Gong, X. Pore-scale numerical simulation of the thermal performance for phase change material embedded in metal foam with cubic periodic cell structure. *Appl. Therm. Eng.* **2019**, *151*, 231–239. [CrossRef]
26. Zhang, T.; Liu, F.; Deng, X.; Zhao, M.; Zhou, H.; Zhang, D.Z. Experimental study on the thermal storage performance of phase change materials embedded with additively manufactured triply periodic minimal surface architected lattices. *Int. J. Heat Mass Transf.* **2022**, *199*, 123452. [CrossRef]
27. Qureshi, Z.A.; Omari, S.A.; Elnajjar, E.; Mahmoud, F.; Al-Ketan, O.; Abu Al-Rub, R. Thermal characterization of 3D-Printed lattices based on triply periodic minimal surfaces embedded with organic phase change material. *Case Stud. Therm. Eng.* **2021**, *27*, 101315. [CrossRef]

28. Sundarram, S.S.; Li, W. The effect of pore size and porosity on thermal management performance of phase change material infiltrated microcellular metal foams. *Appl. Therm. Eng.* **2014**, *64*, 147–154. [CrossRef]
29. Yang, Z.; Liu, F.; Zhang, T.; Deng, X.; Zhang, Z.W. Numerical simulation and experiment of heat storage process of TPMS porous aluminum-paraffin composite phase change material. *Chem. Ind. Eng. Prog.* **2022**, *41*, 4918–4927.
30. Caket, A.G.; Wang, C.; Nugroho, M.A.; Celik, H.; Mobedi, M. Recent studies on 3D lattice metal frame technique for enhancement of heat transfer: Discovering trends and reasons. *Renew. Sustain. Energy Rev.* **2022**, *167*, 112697. [CrossRef]
31. Molteni, M.; Candidori, S.; Graziosi, S.; Gariboldi, E. Improving the thermal response flexibility of 2- and 3-phase composite phase change materials by metallic triply periodic minimal surface structures. *J. Energy Storage* **2023**, *72*, 108185. [CrossRef]
32. Cai, W.; Yang, W.; Zhang, K.; Fan, J.H.; Wu, J.Y.; Xing, T.; He, R. Experimental and numerical simulation of the melting process of paraffin phase change material. *J. Southwest Univ. Sci. Technol.* **2018**, *33*, 6–9+81.
33. Qureshi, Z.A.; Elnajjar, E.; Al-Ketan, O.; Al-Rub, R.A.; Al-Omari, S.B. Heat transfer performance of a finned metal foam-phase change material (FMF-PCM) system incorporating triply periodic minimal surfaces (TPMS)—ScienceDirect. *Int. J. Heat Mass Transf.* **2021**, *170*, 121001. [CrossRef]
34. Voller, V.R.; Prakash, C. A fixed grid numerical modelling methodology for convection-diffusion mushy region phase-change problems. *Int. J. Heat Mass Transf.* **1987**, *30*, 1709–1719. [CrossRef]
35. Kamkari, B.; Shokouhmand, H.; Bruno, F. Experimental investigation of the effect of inclination angle on convection-driven melting of phase change material in a rectangular enclosure—ScienceDirect. *Int. J. Heat Mass Transf.* **2014**, *72*, 186–200. [CrossRef]

energies

Article

Solar-Thermal-Chemical Integrated Design of a Cavity-Type Solar-Driven Methane Dry Reforming Reactor

Zhou-Qiao Dai, Xu Ma, Xin-Yuan Tang, Ren-Zhong Zhang and Wei-Wei Yang *

Key Laboratory of Thermo-Fluid Science and Engineering of MOE, School of Energy and Power Engineering, Xi'an Jiaotong University, Xi'an 710000, China
* Correspondence: yangww@mail.xjtu.edu.cn

Abstract: In this work, the solar-thermal-chemical integrated design for a methane dry reforming reactor with cavity-type solar absorption was numerically performed. Combined with a multiphysical reactor model, the gradient optimization algorithm was used to find optimal radiation flux distribution with fixed total incident solar energy for maximizing overall hydrogen yield, defined as the ratio of molar flow of exported hydrogen to imported methane, which can be applied for guiding the optical property design of solar adsorption surface. The comprehensive performances of the reactor under the conditions of original solar flux and optimal solar flux were analyzed and compared. The results show that for the inlet volume flow rate of 8–14 $L\cdot min^{-1}$, the hydrogen production rate was increased by up to 5.10%, the energy storage efficiency was increased by up to 5.55%, and the methane conversion rate was increased by up to 6.01%. Finally, the local absorptivities of the solar-absorptive coating on the cavity walls were optimized and determined using a genetic algorithm, which could realize the predicted optimal radiation flux distribution.

Keywords: solar-thermal-chemical integrated design; dry reforming of methane; gradient optimization algorithm; genetic algorithm; optimal solar radiation heat flux distribution

1. Introduction

To combat global warming and fossil energy shortages, it is important to increase the share of renewable energy consumption to achieve sustainable development [1–3]. It is noteworthy that the solar-driven dry reforming of methane (DRM) is a promising clean energy technology that integrates hydrogen production, carbon sequestration, and solar energy utilization [4,5], which uses concentrating mirrors to collect light and heat to create a high-temperature environment and achieve an increase in the calorific value of fuel [5–7].

For solar thermochemical reactors and absorbers, the construction plays a major role in the energy utilization's efficiency and safety. According to the form of contact between fluid/particles and solar radiation, solar absorbers/reactors can be divided into the direct and indirect system [8–10]. For direct solar absorbers, much of the research has been devoted to maximizing the 'volumetric effect' and thus achieving higher utilization efficiency. This effect is defined as solar radiation penetrating the absorber and volumetric absorption, the lower frontal temperature of the absorber which is lower than the outlet fluid temperature due to convection cooling the inlet fluid, and finally the lower frontal face heat radiation loss [11,12]. The thermal performance of the silicon carbide foam solar absorber with four different geometric parameters was investigated by a combination of experiments and simulations by Wang et al. [13]. The results showed that the maximum thermal efficiency of this direct solar absorber is 72.48% at a porosity of 0.95, which proves that the increase in porosity has a positive effect on the 'volumetric effect'. In addition, for absorbers with the same porosity (a porosity of 0.85), the pore size reduction also helps to increase thermal efficiency. Unlike ref. [13], Du et al. [14] used super-alloy Inconel 718 as a material, combined with pore-scale reconstruction and 3D printing techniques, and fabricated several porous samples with different geometric parameters as porous volume solar receivers. The experimental results show that for a uniform solar radiation receiver, the small aperture sample is beneficial for enhancing convective heat transfer and reducing the front surface temperature and heat radiation loss, but weakens the radiation penetration depth. In addition, the researchers also proposed a radial gradient variation design, which improves the thermal efficiency of the bulk absorber and reduces the flow resistance. Kasaeian et al. [15] studied the performance of a multi-channel volumetric solar absorber with different aperture shapes of the same internal surface area by numerical simulation. The results showed that the triangular channel achieved the highest outlet gas temperature of 894.2 °C and an efficiency of 0.697, while the circular channel showed the lowest outlet temperature of 869.8 °C and an efficiency of 0.676. Li et al. [16] built a numerical model of solar-thermal integration for volumetric solar absorber and optimized the two parameters of porosity and pore size in radial and axial sub-regions by combining it with a genetic algorithm. The optimization results show that, compared with a uniform porous absorber, the thermal efficiency of an optimal sub-regional gradient porous absorber can be improved by 1.23~9.76%, and the pressure drop can be reduced by 7.88~55.73%. Raffaele et al. [17] creatively designed and fabricated a needle-type volumetric absorber and tested its performance, which showed that the absorber has an outlet fluid temperature higher than the front surface solid temperature and a thermal efficiency reaching 90%, much higher than other types of absorbers, achieving the so-called volumetric effect. Based on these studies, some scholars have also carried out a lot of work for the volumetric solar-thermal-chemical reactor. For instance, Shi et al. combined the biomimetic leaf-type [18] and biomimetic venous [19] concepts to propose two hierarchical porous structure reactors and showed that the methane conversion of the two new reactors increased by 4.5% and 5.9%, respectively, compared to the conventional homogeneous pore structure reactor. Correspondingly, Liu et al. [20] also proposed a foam reactor with concave geometry at the front end, and optimized the porosity, pore size, and reactor length parameters by numerical simulation combined with a multi-island genetic algorithm, and the results showed that the optimized solar fuel efficiency reached 53.3%.

Indirect solar heat absorbers/reactors have also been extensively studied by many scholars because of their simple structure, safety, and stability. Most indirect solar absorbers/reactors have a cavity structure because of the cavity effect (multiple reflections and the absorption of visible infrared radiation at the cavity wall), which enhances solar radiation utilization. Therefore, Bellos et al. [21] studied five different coil-type cavity receivers for solar dish concentrators, and the results showed that the cylindrical-conical structure has the best optical efficiency of 85.42% and the highest thermal efficiency of 67.95%. Beyond that, a novel construction for a gas-phase solar indirect irradiation receiver was also crafted by Sedighi et al. [22], combined with impact jet compound technology, and the findings revealed that the cylindrical cavity with an inverted conical base has the maximum optical efficiency of roughly 92%. Huang et al. [23] exploited a numerical model of radiation transfer within a coupled cavity to evaluate the absorption performance of solar radiation by both structures (direct/indirect). The results revealed that the indirect radiation receiver shows a higher solar thermal efficiency compared to the direct absorption radiation receiver.

Carbon deposition is an unavoidable problem during the thermochemical reaction of methane reforming. In commercial applications, carbon deposition in the chemical reaction imposes regular cleaning requirements for the glass window of direct solar thermal chemical reactors. The safety of the glass window also seems to be difficult to guarantee under a high temperature. Indirect solar thermochemical reactors can be designed without glass windows, and combined with the concept of cavity design, can also achieve a higher optical efficiency and energy storage efficiency. The indirect thermochemical reactor is also simpler in structure, which is convenient for early installation and later maintenance. In summary, indirect solar thermal chemical reactors seem to be more suitable for commercial application and promotion.

For cavity receivers, the concentrated solar radiation is absorbed by the surface of the cavity, and the heat flux distribution on the cavity wall is normally extremely uneven. Thus, when extreme operating conditions occur, the cavity walls can develop local hot spots due to large temperature gradients, resulting in large thermal stresses, which are a great challenge for the reliability of the receiver [24–26]. On the other hand, the radiant heat flux distribution on the cavity surface also affects the overall temperature field distribution, which in turn affects the overall energy storage efficiency of the device [27–30]. In order to solve the above problems, Yan et al. [25] proposed a structural optimization method for a disc concentrator in order to improve the heat flux uniformity at the wall of a solar cavity-type absorber. The results showed that the non-uniformity factor was reduced from 0.55~0.63 to 0.10~0.22 while ensuring excellent optical efficiency. Different from the method in ref. [25], Wang et al. [31] addressed the problem of the uneven distribution of solar flux on the solar tower receiver wall by optimizing the distribution of cavity wall coats in an attempt to achieve a uniformity of flux distribution while maintaining the lowest possible optical loss. This work shows that enhancing flux uniformity and reducing reflection losses are two conflicting objectives, and a trade-off choice needs to be made between them. Similarly, Tang et al. [32,33] achieved nearly uniform circumferential solar heat flow and optimal axial solar heat flow by arranging secondary mirrors and an adapted trough concentrator, improving the fuel conversion and energy storage efficiency for the straight-tube indirect solar methanol reforming reactor.

The distribution of the solar radiation heat flux on the cavity wall greatly influences the optical efficiency and the safety of a solar thermochemical reactor. Therefore, in this study, a numerical solar-thermal-chemical model is constructed for the cavity-type solar thermochemical reactor with a dish concentrator based on the previous study [34]. The gradient optimization algorithm is used to find the optimal solar radiation heat flux distribution with a constant total incident solar energy, using the optimization of the hydrogen yield as the optimization objective. In addition, the comprehensive performances of the reactor's original heat flux and optimal heat flux are compared and analyzed. Finally, this study optimizes the absorptivity of the solar absorption coating on the cavity wall using a genetic algorithm, which achieves the optimal solar radiation flux distribution.

2. Mathematical Model and Optimization Method

2.1. Model Description

As shown in Figure 1, the concentrator reactor involved in this study was mainly divided into two parts: one part was the dish concentrator system for the energy conversion of solar-to-thermal energy, and the other part was the thermochemical reactor for the energy conversion of thermal-to-chemical energy. As shown in Figure 2a, the dish concentrator system was simplified to a model which included a parabolic concentrating mirror and a light-receiving conical cavity. It was simulated by the non-sequential ray-tracing method, which has no predefined path for any rays. The ray is emitted and projected onto any object in the ray path, and then may be reflected, refracted, scattered, etc. Additionally, in this study, only the reflection and absorption of light were considered. During the optical simulations, the solar direct normal irradiance (DNI) was assumed to be stable at 1000 $W{\cdot}m^{-2}$, the radius of the dish concentrator was 1.2 m, and the reflectivity of the mirror was 0.9. Other relevant parameters are shown in Table 1.

The thermochemical reactor, due to the structural symmetry and computational consideration, was simplified to a two-dimensional (2D) model in this study, which is shown in Figure 2b. All solar energy collected by the dish concentrator was reflected into the cavity, absorbed by the surface, and converted into thermal energy. The porous media were arranged in the preheating area. Then, the gas mixture (CH_4 and CO_2) at the feed ratio of 1:1 entered the outer channel's preheating area. The inlet flow rate of the gas mixture was set to 10 $L{\cdot}min^{-1}$. Considering that too-high inlet volume flow rates would lead to the under-utilization of the inlet reactants and that too low inlet volume flow rates would lead to a sufficient supply of energy such that the change in flux distribution at the cavity wall would have almost no effect, an incorrect optimization solution would be obtained that would not be applicable for the whole operating field. Therefore, an appropriate inlet volume flow rate of 10 L·min^{-1} was chosen as the operating parameter in the subsequent optimization process. The gas mixture entered from the outer channel and was heated and then turned around to flow into the inner channel. It is noteworthy that the reaction area was a SiC porous ceramic coated with the catalyst Ni/Al_2O_3. The relevant structural parameters were determined according to those in ref. [34]. Moreover, the rear section of the inner channel was also filled with a porous medium to enhance heat transfer for the full utilization of solar sensible heat. The ambient temperature was set to 300 K, which was considered as the temperature of the reactants at the inlet.

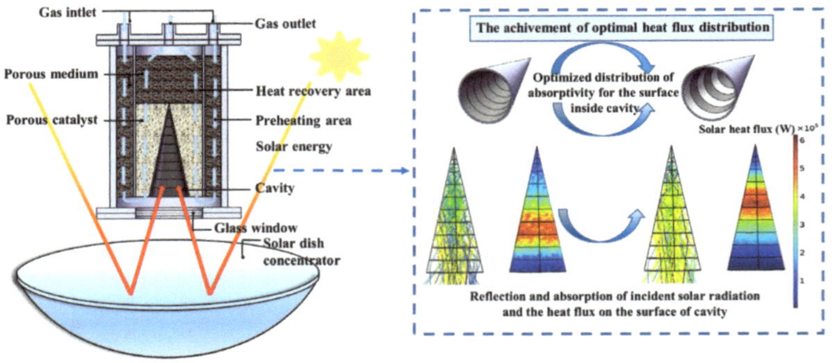

Figure 1. Schematic diagram of cavity-type solar thermochemical reactor with dish concentrator and the achievement of optimal heat flux distribution.

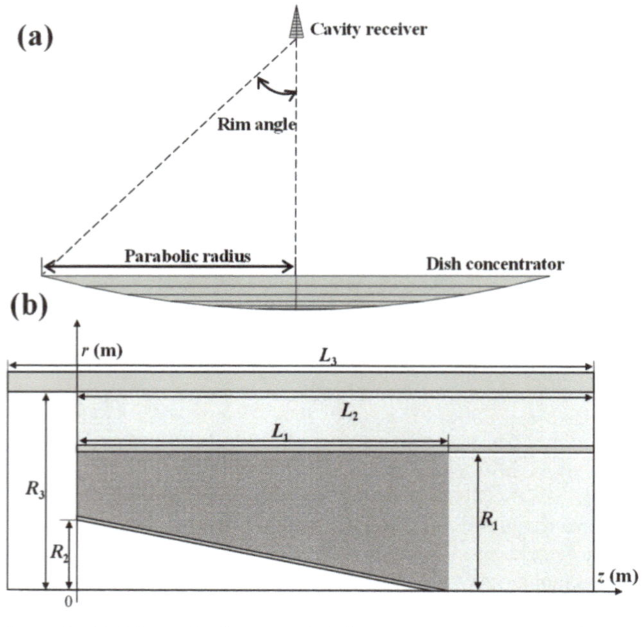

Figure 2. (**a**) Schematic diagram of dish concentrator system, and (**b**) 2D schematic diagram of cavity-type solar thermochemical reactor.

Table 1. Relevant parameters of dish concentrator and DRM reactor.

Parameter	Values	Parameter	Value
Dish diameter	1.20 (m)	R_3	0.047 (m)
Surface reflectivity	0.9	d_p	0.003 (m)
Slope error	2 (mrad)	L_1	0.086 (m)
Solar half angle	4.65 (mrad)	L_2	0.13 (m)
R_1	0.03 (m)	L_3	0.14 (m)
R_2	0.017 (m)	φ	0.8

2.2. Mathematical Model

2.2.1. Mass Conservation Equation

The equation for the mass conservation of the gas mixture can be expressed as

$$\nabla \cdot \left(\rho_f \vec{u}\right) = 0 \tag{1}$$

where $\vec{u}$ is the superficial velocity.

2.2.2. Momentum Conservation Equation

The momentum conservation equation for fluid flow in the porous reaction bed can be indicated as

$$\frac{1}{\varphi}\nabla\left(\rho_f \frac{\vec{u}\cdot\vec{u}}{\varphi}\right) = -\nabla p + \nabla \cdot \left(\frac{\mu_f}{\varphi}\nabla\vec{u}\right) + \vec{F} \tag{2}$$

where $\vec{F}$ is the resistance term, and μ_f is the dynamic viscosity of gas mixture which is described in the Wilke equation, which is

$$\mu_f = \sum_{i=1}^{n} \frac{m_i \mu_i}{\sum\limits_{j=1}^{n} m_j \phi_{ij}} \tag{3}$$

$$\phi_{ij} = \frac{1}{\sqrt{8}}\left[1 + \left(\frac{\mu_i}{\mu_j}\right)^{1/2}\left(\frac{M_i}{M_j}\right)^{1/4}\right]^2 \left(1 + \frac{M_i}{M_j}\right)^{-1/2} \tag{4}$$

where m_i, μ_i, and M_i refer to the mole fraction, viscosity, and molecular weight of species i, respectively.

The resistance term $\vec{F}$ in Equation (2) can be indicated as [35]

$$\vec{F} = -\frac{1039 - 1002\varphi}{d_p^2}\mu_f \vec{u} - \frac{0.5138\rho_f \varphi^{-5.739}}{d_p}\left|\vec{u}\right|\vec{u} \tag{5}$$

2.2.3. Energy Conservation Equation

Considering the temperature difference between the solid and the fluid inside the porous medium and the difference in physical properties, the local thermal non-equilibrium (LTNE) assumption [36] is used in this study, which can be indicated as the fluid phase:

$$\nabla\left(\rho_f C_{p,f} u T_f\right) = \nabla \cdot \left(\lambda_{e,f} \nabla T_f\right) + S_f \tag{6}$$

and the solid phase:

$$\nabla\left(\rho_s C_{p,s} T_s\right) = \nabla \cdot \left(\lambda_{e,s} \nabla T_s\right) + S_s \tag{7}$$

where subscripts "f" and "s" denote fluid and solid, respectively; $C_{p,f}$ and $C_{p,s}$ are the specific heats at a constant pressure of the fluid phase and solid phase; $\lambda_{e,f}$ and $\lambda_{e,s}$ are the effective thermal conductivities; S_f and S_s are the heat source terms.

Equations (8) and (9) show the effective thermal conductivities of solid and fluid from the Schuetz–Glicksman empirical formula [37].

$$\lambda_{e,f} = \varphi \lambda_f \tag{8}$$

$$\lambda_{e,s} = \frac{\lambda_s}{3}(1-\varphi) \tag{9}$$

The heat source terms S_f and S_s of the energy equations are calculated by

$$S_f = S_{con,f} + S_{chem} \tag{10}$$

$$S_s = S_{con,s} + S_r + S_{loss} \tag{11}$$

where S_{chem} is the heat generation and consumption source term caused by chemical reactions; $S_{con,f}$ and $S_{con,s}$ are the convective heat transfer terms between the fluid phase and solid phase, respectively; S_{loss} is the radiative heat dissipation.

The energy source, S_r, due to thermal radiation, is estimated by the Rosseland approximation [38], which can be expressed as

$$S_r = -\frac{16n^2\sigma T^3}{3\kappa_e}\frac{dT}{dz} \tag{12}$$

In addition, the radiation heat loss of the cavity radiation is assessed on the basis of surface-to-surface radiation. The cavity inner walls are divided into N discrete surface elements and then the radiative heat transfer among them is computed. For each surface k, the radiosity, J_k, is evaluated by the following equation:

$$J_k = \varepsilon_k \sigma T_k + \tau_k \sum_{j=1}^{N} J_j F_{kj} \tag{13}$$

where ε_k is the emissivity of the surface k, and τ_k is the reflectivity of the surface k. For opaque gray body surfaces, the reflectivity can be simply expressed as $\tau_k = 1 - \varepsilon_k$. The term F_{kj} represents the view factor between the surface k and the generic surface j.

The radiative heat loss from the front surface can be calculated by the following equation:

$$S_{loss} = -\varepsilon\sigma(T^4 - T_0{}^4) \tag{14}$$

where T_0 is the ambient temperature (300 K).

Furthermore, the relationship between the source terms $S_{con,f}$ and $S_{con,s}$ can be described as

$$S_{con,f} = -S_{con,s} = h_v(T_s - T_f) \tag{15}$$

Additionally, the two-phase heat exchange coefficient, h_v, obtained by Wu et al. [35] can be described as

$$h_v = \frac{\lambda_f(32.504\varphi^{0.38} - 109.94\varphi^{1.38} + 166.65\varphi^{2.38} - 86.98\varphi^{3.38})Re^{0.438}}{d_p{}^2} \tag{16}$$

where Re is the Reynolds number, which is defined by

$$Re = \frac{\rho_f u d_p}{\mu_f} \tag{17}$$

2.2.4. Species Conservation Equation

The mass transfer equation can be expressed as [39]

$$\rho_f(\vec{u}\cdot\nabla)\omega_i+\nabla\cdot(-\rho\omega_i\sum_{j=1}^{n}D_{ij}(\nabla m_j+(m_j-\omega_j)(\frac{\nabla p}{p}))+D_i^T(\frac{\nabla T}{T}))=S_i \tag{18}$$

where ω_i, D_{ij}, D_i^T and S_i are, respectively, the mass fraction for the gas of species *i*, the multicomponent Fick's diffusivity of components *i/j*, the thermal diffusion coefficient, and the species generation rate.

The specific heat and thermal conductivity of gas can be calculated by using the mixing law [40]:

$$C_{p,f}=\sum_{i}^{n}\omega_i C_{p,i} \tag{19}$$

$$\lambda_f=\sum_{i}^{n}m_i\lambda_i \tag{20}$$

where λ_i and $C_{p,i}$ represent the thermal conductivity coefficient and the heat capacity and enthalpy coefficients of species *i*. The relevant parameters from the COMSOL material library are used here.

2.2.5. Reaction Kinetics

The reaction kinetics of DRM are a very complex reaction system and the kinetic expression can be shown below.

(1) The dry reforming of methane [41]

$$CH_4+CO_2\rightarrow 2CO+2H_2\cdot\Delta H^o_{298}=+247.0\text{kJ/mol} \tag{21}$$

$$r_1=\rho_{cat}\frac{k_1K_{CO_2,1}K_{CH_4,1}P_{CH_4}P_{CO_2}}{(1+K_{CO_2,1}P_{CO_2}+K_{CH_4,1}P_{CH_4})^2}\left[1-\frac{(P_{CO}P_{H_2})^2}{K_{P_1}P_{CH_4}P_{CO_2}}\right] \tag{22}$$

(2) The reverse water-gas shift reaction (RWGS) [41]

$$CO_2+H_2\rightarrow CO+H_2O\cdot\Delta H^o_{298}=+41.7\text{kJ/mol} \tag{23}$$

$$r_2=\rho_{cat}\frac{k_2K_{CO_2,2}K_{H_2,2}P_{CO_2}P_{H_2}}{(1+K_{CO_2,2}P_{CO_2}+K_{H_2,2}P_{H_2})^2}\left[1-\frac{P_{CO}P_{H_2O}}{K_{P_2}P_{CO_2}P_{H_2}}\right] \tag{24}$$

(3) The methane cracking reaction [42]

$$CH_4\rightarrow C+2H_2\cdot\Delta H^o_{298}=+74.0\text{kJ/mol} \tag{25}$$

$$r_3=\rho_{cat}\frac{k_3K_{CH_4,3}(P_{CH_4}-\frac{P_{H_2}^2}{K_{P_3}})}{(1+K_{CH_4,3}P_{CH_4}+\frac{P_{H_2}^{1.5}}{K_{H_2,3}})^2} \tag{26}$$

where the relevant kinetic parameters are listed in Table 2.

The reaction source term can be expressed as

$$S_{chem}=\sum_{i=1}^{3}r_i\Delta H \tag{27}$$

where ΔH is the reaction enthalpy.

The hydrogen production rate, r_{H_2}, can also be expressed as

$$r_{H_2}=2r_1+2r_3-r_2 \tag{28}$$

Furthermore, the methane conversion, η_{CH_4}, and the carbon dioxide, η_{CO_2}, can be expressed as

$$\eta_{CH_4} = \frac{F_{CH_4,in} - F_{CH_4,out}}{F_{CH_4,in}} \tag{29}$$

$$\eta_{CO_2} = \frac{F_{CO_2,in} - F_{CO_2,out}}{F_{CO_2,in}} \tag{30}$$

The hydrogen yield characterizes the performance in terms of hydrogen production, which can be defined as

$$Y_{H_2} = \frac{F_{H_2,out}/M_{H_2}}{F_{CH_4,in}/M_{CH_4}} \tag{31}$$

Typically, the product is stored under environmental conditions, so the chemical storage energy is defined by reaction enthalpy. The chemical energy storage efficiency is defined as

$$\eta_{ch} = \frac{Q_{ch}}{Q_{so}} = \frac{\oiiint \sum_{i=1}^{3} r_i \Delta H_{i,298} dV}{\oiiint I(z) dV} \tag{32}$$

When the outlet product is cooled to ambient temperature, the released sensible heat, Q_{se}, can be indicated as

$$Q_{se} = \sum_{i} \int_{T_s}^{300K} F_{i,out} C_{p,i} dT \tag{33}$$

According to the above definition, we can observe that the system energy loss Q_{lo} can be expressed as

$$Q_{lo} = Q_{so} - Q_{ch} - Q_{se} \tag{34}$$

Table 2. Chemical reaction kinetic parameters.

Parameter	Values	Parameter	Value
k_1	$1.29 \times 10^6 e^{-102,065/(RT)}$	$K_{CO_2,1}$	$2.61 \times 10^{-2} e^{+37,641/(RT)}$
k_2	$0.35 \times 10^6 e^{-81,030/(RT)}$	$K_{CH_4,1}$	$2.60 \times 10^{-2} e^{+40,684/(RT)}$
k_3	$6.95 \times 10^3 e^{-58,893/(RT)}$	$K_{CO_2,2}$	$0.5771 e^{+9262/(RT)}$
K_{P_1}	$6.78 \times 10^{14} e^{-259,660/(RT)}$	$K_{H_2,2}$	$1.494 e^{+6025/(RT)}$
K_{P_2}	$56.4971 e^{-36,580/(RT)}$	$K_{CH_4,3}$	$0.21 e^{-567/(RT)}$
K_{P_3}	$2.98 \times 10^5 e^{-84,400/(RT)}$	$K_{H_2,3}$	$5.18 \times 10^7 e^{-133,210/(RT)}$

2.3. Optimization Method for Solar Flux and Absorptivity

The hydrogen yield, Y_{H_2}, was considered to be an important performance parameter indicator. In the optimization process, the highest hydrogen yield production was obtained by optimizing the solar flux distribution in the cavity, and it was taken as the objective function. For solar flux distribution, it was described by the Bezier curve, which is widely used in scenarios such as shape-optimized design [43,44] and robot motion control [45,46]. In addition, the curve was very suitable for the gradient optimization algorithm in this study due to its high order derivability. All optimization-related processes are shown in Figure 3. In the first stage, the gradient optimization algorithm searches for the optimal radiation flux distribution mathematically described by the 5th-order Bernstein basis function around the objective of maximizing the hydrogen yield. In the second stage, the objective is to minimize the difference between the true flux distribution and the optimal radiation flux distribution are which mathematically described by least squares. The cavity wall absorbance was varied to achieve the optimal flux distribution in the joint simulation of MATLAB and COMSOL.

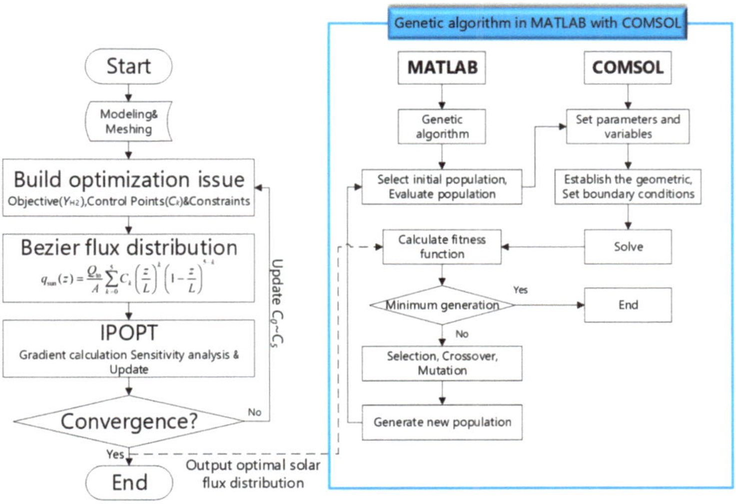

Figure 3. Overall simulation and optimization flow chart.

The interior point optimizer (IPOPT) in the COMSOL software was used in this study to find the optimal solar flux distribution. Based on the full-space in-point method, IPOPT can efficiently solve large-scale nonlinear optimization problems while maintaining excellent convergence characteristics, the suitability of which will be demonstrated later [47]. In this study, considering the amount of computation and the accuracy of the mathematical description, the 5th-order Bernstein basis function was used to describe the flux distribution, which is expressed as

$$q_{\text{sun}}(z) = \frac{Q_{\text{to}}}{A}\sum_{k=0}^{5} C_k\left(\frac{z}{L}\right)^k\left(1-\frac{z}{L}\right)^{5-k} \tag{35}$$

where Q_{to} is the total incident solar energy, A is the cavity surface area, and C_k is the control point of the curve. Two constraints are considered in this study: a solar heat flux greater than or equal to 0, and a fixed total incident solar energy, as shown in Equation (36).

$$\begin{cases} \text{Objective function}: \ \max(Y_{\text{H}_2}) \\ \text{s.t}: \ q_{\text{sun}}(z) \geq 0 \\ \quad\quad \oiint q_{\text{sun}}(z)dA = Q_{\text{to}} \end{cases} \tag{36}$$

After solving the optimal radiation flux distribution, the cavity wall was divided into 10 zones at a radial spacing of 0.0017 m. Additionally, the optimal solar radiation flux distribution was obtained from the optical model by optimizing the wall absorptivity. This part of the optimization was implemented in a joint MTLAB and COMSOL simulation by means of the genetic algorithm. The optimal solar radiation flux distribution obtained from the IPOPT solution was imported into MATLAB. Then, the least square objective (l_s) of the optimal solar radiation flux distribution (q_{op}) and the real radiation distribution (q_{re}) generated by changing the wall absorptivity was set in MATLAB, which can be expressed as below:

$$l_s = \sum_{k=1}^{n}\frac{[q_{\text{re}}(r_k) - q_{\text{op}}(r_k)]^2}{n} \tag{37}$$

The solar absorptivity values, α_1~α_{10}, of each zone are optimized by the genetic algorithm. Therefore, the problem of optimizing the absorptivities of the cavity walls by means of the genetic algorithm to achieve the optimal solar radiation flux can be described as follows:

$$\begin{cases} \text{Find } X = [\alpha_1, \alpha_2 \cdots\cdots \alpha_{10}] \\ \text{Minimize } ls = \sum_{k=1}^{n} \frac{[q_{\text{re}}(r_k) - q_{\text{op}}(r_k)]^2}{n} \\ \text{s.t. } 0 \le \alpha_1, \alpha_2 \cdots\cdots \alpha_{10} \le 1 \end{cases} \quad (38)$$

In the genetic algorithm optimization constraint, the absorptivity of the cavity walls for solar radiation is considered to be always greater than 0 and less than 1.

3. Model Validation

In this study, the optical model of a dish concentrator system (3D) was implemented in the ray optics module of commercial software COMSOL, which was solved by a non-sequential ray tracing method, and the GMRES solver is used to solve this module [48]. Additionally, the finite element method was used to solve this multiphysics computational fluid dynamics model of the thermochemical reactor (2D) in a fully coupled manner. The PARDISO solver in the COMSOL software was chosen to solve this part [49]. Then, the grid independence test showed that 108,000 cells met the requirement of a 10^{-5} relative error. Additionally, in this study, the optical model of the dish concentrator was validated, and the relevant results were compared with the date in ref. [50]. As shown in Figure 4a, the solar radiation distributions on the receiving surface were well-matched for all three deflection angles (30°, 45°, and 60°). The results of the methane dry reforming kinetic model calculations were compared with those of the experimental study [51]. As shown in Figure 4b, the absolute error did not exceed 5.5% in the conversions of CH_4 and CO_2, so the kinetic model of this study can be considered feasible and accurate.

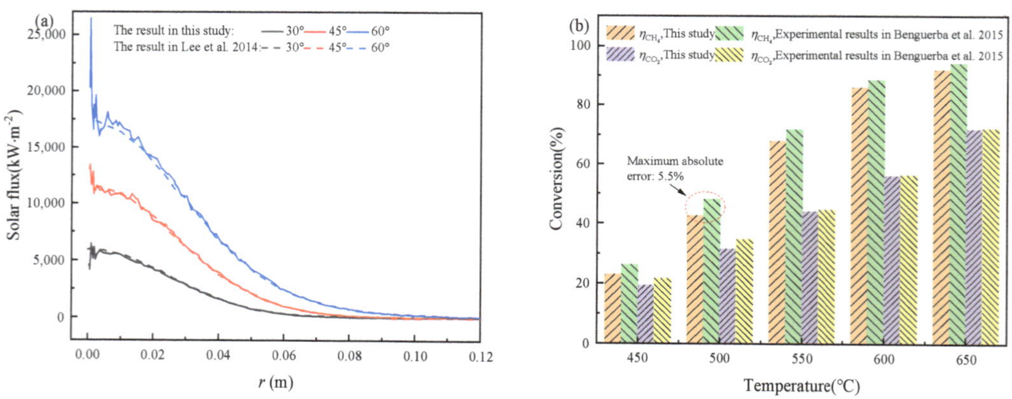

Figure 4. Verifications of (**a**) optical model of dish concentrator, (**b**) and of chemical reaction [50,51].

4. Results and Discussion

4.1. Analysis of Optimization Results and Comprehensive Performance of Cases

4.1.1. Optimal Flux Distribution Optimization Results

Before solving the optimal flux distribution, it is necessary to know the value of total incident solar energy (Q_{to}) that is concentrated and reflected into the cavity by the dish concentrator. With a constant parabolic radius of the dish concentrator (1.2 m), it can be assumed that the energy received by the concentrator is unchanged. The effect of different cavity absorptivities of solar radiation and the effect of the dish concentrator rim angle on Q_{to} were studied and analyzed, the results of which are shown in Figure 5a. Firstly, the total incident solar energy increased as the absorptivities of the cavity walls increased, and a high absorptivity always helps to reduce the reflective losses of solar radiation. When the absorptivity was less than or equal to 0.6, the rim angle showed an opposite increasing trend to the total incident solar energy. Therefore, when the absorptivity is small, the small rim angle is more conducive to obtain higher total incident solar energy for the cavity. In addition, when the absorptivity was larger than or equal to 0.8, the maximum total incident solar energy was achieved at a rim angle of 45°, followed by a deflection angle of 30°, and the minimum energy was obtained at a rim angle of 65°. The maximum total incident solar energy was 1031.8 W at an absorptivity of 1 and a rim angle of 45°. Considering the actual case where the cavity wall material was silicon carbide, a deflection angle of 45° and an absorptivity of 0.92 was used as the original case (Q_{to} = 1031.7 W) for analysis and comparison. Q_{to} = 1031.7 W was used as the total incident solar energy in the gradient optimization algorithm IPOPT solution.

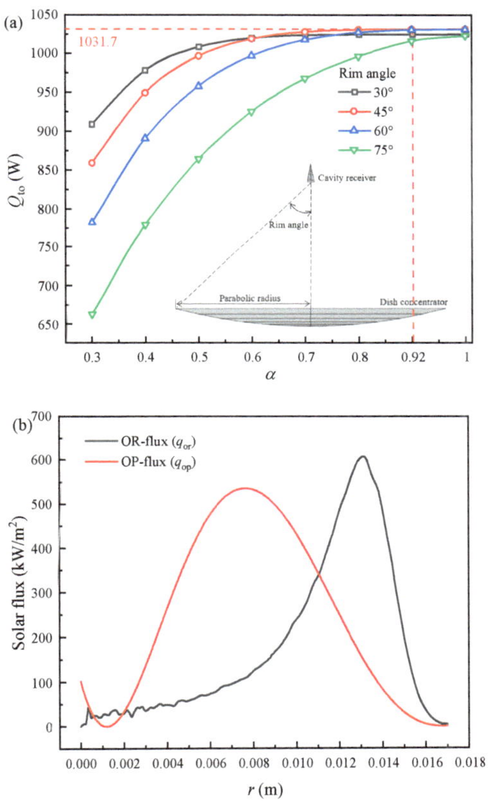

Figure 5. (**a**) Effect of rim angle and absorptivity on total incident solar energy, and (**b**) comparison of two flux distributions.

Combined with the gradient optimization algorithm, the optimal solar flux distribution was obtained at the reactant inlet volume flow rate of 10 L·min^{-1}. The optimization tolerance in the optimization solver was set to 0.0001, and the results converged after 158 iterations, which took almost 34 h. The optimization calculation results are shown in Figure 5b. The Bernstein polynomial control variable search results are $C_0 = -0.562$, $C_1 = -16.691$, $C_2 = 67.655$, $C_3 = -4.534$, $C_4 = -5.258$, and $C_5 = -0.997$. In the original flux distribution (OR-flux) case, the solar radiation was concentrated at the radius of 0.013 m near the front face, with the maximum radiation heat flux of approximately 600 kW·m^{-2}. Additionally, at the end of the reaction area, almost no solar energy reached the cavity wall. The peak optimal flux distribution (OP-flux) was about 550 kW·m^{-2}, which is slightly lower compared to the original distribution. The solar radiation was mainly concentrated at

r = 0.008 m. The results of the gradient optimization solution indicate the inward migration of the concentrated solar radiation, which is favorable for hydrogen production.

The result of the comprehensive comparative performance analysis of the two flux distributions is shown in Figure 6a. As shown in the graph, the optimal flux distribution as an optimization target had a 4.00% increase in hydrogen yield compared to the original flux distribution. Additionally, an unexpected increase in reactant conversion and chemical energy storage efficiency was also observed. The optimal flux distribution increased CH_4 conversion and CO_2 conversion by 5.36% and 5.77%, respectively, compared to the original flux distribution. One thing that can be noticed is that the increase in reactant conversion was somewhat higher than the hydrogen yield. The increase of methane conversion means that more of reaction (1) and reaction (3) took place. However, the hydrogen yield did not increase to the corresponding percentage, which means that reaction (2), which consumes hydrogen in under an optimal flux distribution, proceeded more strongly. The increase in the mass fraction of water at the outlet confirms this idea, as shown in Figure 6a, with an increase of 1.09% under the optimal flux distribution. It is also known that the carbon monoxide mass fraction increases by 3.22% at the optimal flux distribution, which means that more carbon dioxide is converted by the reduction reaction.

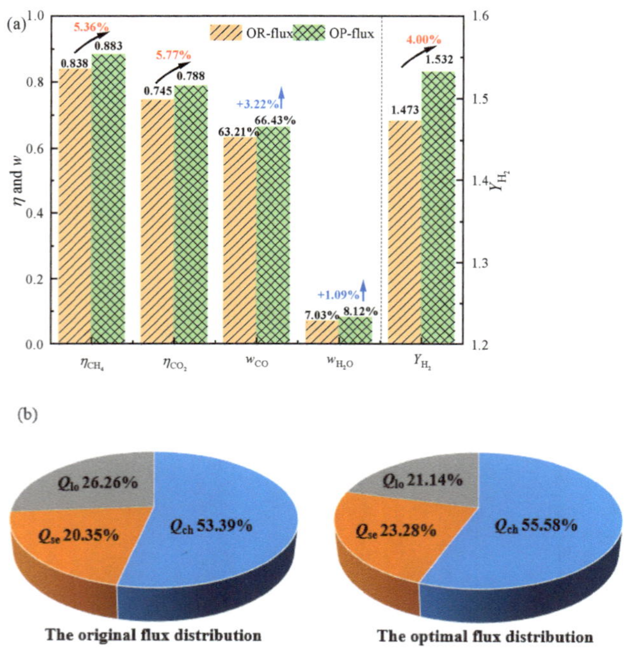

Figure 6. (**a**) Comprehensive performance and (**b**) energy utilization comparison at two solar fluxes.

Energy utilization under the two flux distributions was also analyzed and compared in this study, as shown in Figure 6b. Chemical energy storage had the largest share of total energy at 53.39% in the original flux distribution and further increased to 55.58% in the optimal flux distribution, an increase of 2.19%. In addition, the share of exported sensible heat increased from 20.34% to 23.28%, which means that more of the sensible heat energy could be utilized. The increase in chemical energy storage and sensible heat energy can be attributed to the decrease in energy loss, which decreased by 5.12% from 26.25% to 21.13%.

4.1.2. Analysis of Solid–Liquid Phase Temperature Field Distribution

The distribution of solar flux is directly related to the temperature field of the thermochemical reactor, which affects the thermochemical conversion process in the reactor, and therefore it is necessary to analyze and study. As shown in Figure 7a, under the original flux distribution, the high temperature region was found at the front end close to the cavity wall due to the solar radiation concentrated at the front of cavity. Additionally, the maximum solid phase temperature was 1377 K. Additionally, the high temperatures in the front section area mean large radiation heat loss. As shown in Figure 7b, affected by the temperature of the solid phase, although the temperature of the fluid phase in the first half of the reaction area was high, the temperature was low in the second half, even less than 1060 K in under the original flux distribution. Therefore, it is known that the reactor's inability to obtain enough energy for the second half resulted in a decrease in fluid temperature.

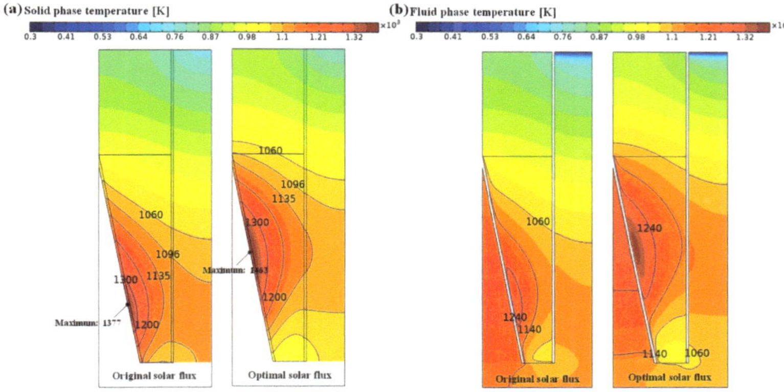

Figure 7. Temperature distributions in two cases: (a) solid phase, and (b) fluid phase.

For the optimal flux distribution, as most of the solar radiation was concentrated at $r = 0.008$ m, the high temperature region was displaced backwards compared to the original flux distribution. The maximum solid phase temperature was 1463 K, an increase of 86 K compared to the original flux distribution, and the entire solid phase area was almost always higher than 1060 K. Due to the heat exchange with the solid phase, the fluid temperature distribution was very similar to the solid phase temperature distribution. As shown in Figure 7b, for the original distribution, although the gas phase temperature in the reaction area was still acceptable in the front section where the solar radiation accumulates, the temperature remained mostly below 1060 K in the rear reaction area. Under the optimal flux distribution, the gas temperature was almost always greater than 1060 K throughout the reaction region. However, the gas phase temperature was less than 1060 K in a small part of the inlet area, which may have been caused by the chemical reaction's energy consumption and insufficient energy supply. This may mean that the chemical reaction's performance under the optimal flux distribution is not as good as it is under the original flux distribution at this area.

4.1.3. Comparison and Analysis of Hydrogen Production

In order to explore the changes in the hydrogen production rate and fluid temperature, the average parameter values on the axial cross-section were used for analysis in this study. In the solution of gradient optimization algorithms, the hydrogen production is used as the only optimization objective. Therefore, it was necessary to analyze the hydrogen production rate and the hydrogen mass flow rate, as shown in Figure 8.

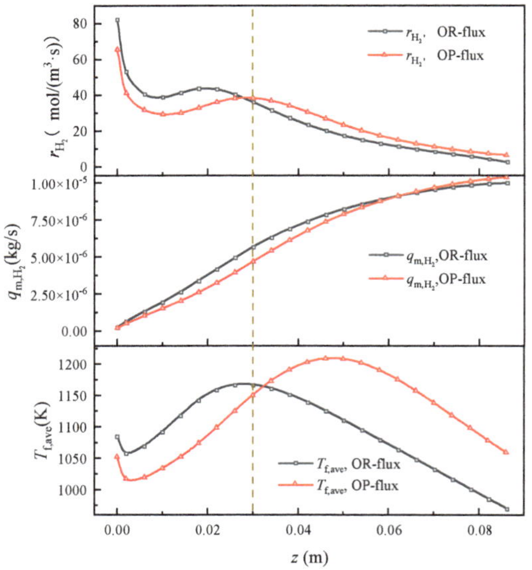

Figure 8. Comparison and analysis of the hydrogen production rate, hydrogen mass flow, and fluid temperature.

The trend in fluid temperature was the same for both flux distributions. The temperature of the fluid rose from the inlet due to a continuous supply of solar energy, and then gradually decreased because of the energy consumed by the chemical reaction. Additionally, when z was greater than 0.03 m, the optimal flux distribution corresponded to a higher fluid temperature than in the original case. Conversely, the fluid temperature under the optimal flux distribution was lower than the original flux distribution when z was less than 0.03 m, as shown in Figure 8.

Temperature distribution has a very close relationship with the rate of hydrogen production. The lower the reaction temperature, the smaller the hydrogen production rate in this study, as shown in Figure 8. Therefore, when z was larger than 0.03 m, the hydrogen production rate corresponding to the original flux distribution was the smallest, while the hydrogen production rate corresponding to the optimal solar flux distribution was the largest. When z was less than 0.03 m, the hydrogen production rate under the original flux distribution was larger than under the optimal flux distribution.

However, high hydrogen production rates do not necessarily mean high hydrogen yields, and the impact of the reaction volume on hydrogen production can be significant. Therefore, the hydrogen mass flow rate at the cross section was also analyzed and compared in this study. It is worth noting that the difference between the hydrogen mass flow rate of the original flux distribution and that of the optimal flux distribution increased gradually until z = 0.03 m, reaching the maximum around z = 0.03 m, as shown in Figure 8. After that, the gap of the hydrogen mass flow rate between the two gradually decreased, and the hydrogen mass flow rate under the optimal flux distribution became gradually higher than it did in the original case ($z \geq 0.06$ m), which benefited from the large volume of the reaction zone in the rear region.

4.1.4. Performance Comparison under Different Inlet Volume Flow Rates

It is also necessary to analyze the performance of the original flux distribution compared to the optimal flux distribution for different inlet volume flow conditions. As shown in Figure 9, under the optimized working condition of the inlet volume flow of 10 L·min^{-1}, the hydrogen yield increased by 4.00%, while when v_{in} = 12 L·min^{-1} and 14 L·min^{-1}, the hydrogen yield increased even more, by 5.11% and 4.92%, respectively. In addition, at

v_{in} = 8 L·min^{-1}, the hydrogen production also increased by 0.13%. This means that the optimal flux distribution could feed more energy into the chemical reaction when the energy supply was not sufficient. Additionally, the optimal flux distribution of the methane conversion and carbon dioxide conversion increased by 2.44~5.79% and 4.32~6.61% at v_{in} = 8~10 L·min^{-1}, respectively. The optimal flux distribution obtained by the gradient optimization for a single operating condition showed a significant performance improvement from that under the original flux distribution for all four operating conditions.

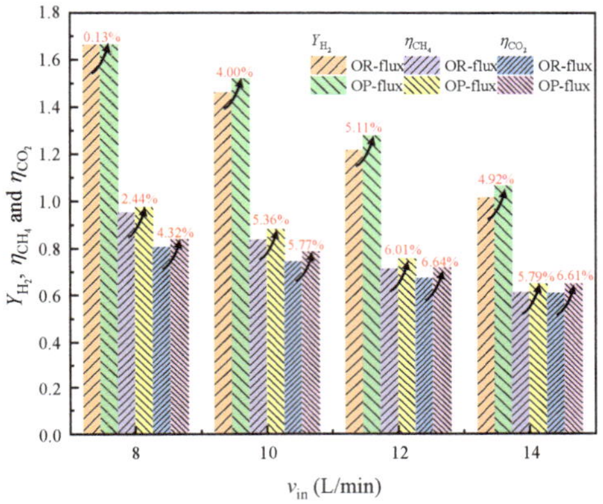

Figure 9. Performance comparison at different inlet volume flows for the optimal solar flux.

4.2. Achievement of Optimal Solar Flux Distribution

In order to achieve an optimal solar flux distribution with the aim of maximizing hydrogen production, the method of optimizing the absorptivity of solar radiation of the cavity walls was used. Firstly, it was important to ensure that the total incident solar energy is not lost too much before achieving the goal of optimal flux distribution. Then, from Figure 5a, it can be seen that for the rim angle of 30°, the change in wall absorptivity had the least effect on the total incident solar energy. Therefore, the dish concentrator of the 30° rim angle was used in this part of the optimization study to minimize the solar energy loss due to the change in absorptivity.

Genetic algorithms are algorithms that simulate natural biological evolutionary mechanisms and are used to find the optimal solution to real world problems. They consist of three basic processes: selection, crossover, and mutation. In the selection step, the winning dominant individual is selected and passed on to the next generation, which is based on assessing the individual's fitness function. In the crossover step, selected individuals exchange portions of genes in random patterns based on crossover scores. In the mutation step, whether each individual changes is based on the mutation probability. Then, one or several genes of the mutant individual are selected to change in a random pattern. The optimization process includes the above three genetic operators. The calculation of genetic optimization converges to the optimal solution when the fitness of the optimal individual is equal to the fitness of the population and remains constant. In this study, the optimization process was terminated by the maximum number of generations. The crossover fraction, mutation rate, and population size were set to 0.8, 0.05, and 100. The maximum generation was set to 20.

As shown in Figure 10a, after the number of generations was greater than 13, the difference between the mean fitness value of the population and the best individual fitness value were not significant, and the results could be considered to have basically reached convergence. At the end of the 20th generation, the average fitness value of the population was 7.68 × 109, while the optimal individual fitness value was 6.58 × 109, with the difference of only 1.1 × 109. The results which took almost 219 h to obtain can be considered reliable. The results of the optimization of the absorptivity of solar radiation of the cavity surface are $\alpha_1 = 0.97$, $\alpha_2 = 0.99$, $\alpha_3 = 0.99$, $\alpha_4 = 0.85$, $\alpha_5 = 0.75$, $\alpha_6 = 0.65$, $\alpha_7 = 0.56$, $\alpha_8 = 0.48$, $\alpha_9 = 0.38$, and $\alpha_{10} = 0.36$. The real flux distribution (RE-flux) achieved by varying the absorptivity is shown in Figure 10b, where a good fit to the optimal flux distribution can be seen over the entire cavity wall. When the rim angle was 30° and the absorptivity was 0.92, the solar flux was mainly concentrated around $r = 0.012$ m. By setting a lower absorptivity at r larger than 0.012 m, more light was reflected into the cavity and absorbed, thus achieving the optimal flux distribution. Another point to note is that, in the genetic algorithm optimization of absorptivity to achieve the optimal flux distribution, the total incident solar energy was 1023.8 W, which is only 0.69% lower than the 1031 W set in the optimal flux finding process, and can be considered as an achievement of the optimal flux distribution without energy loss.

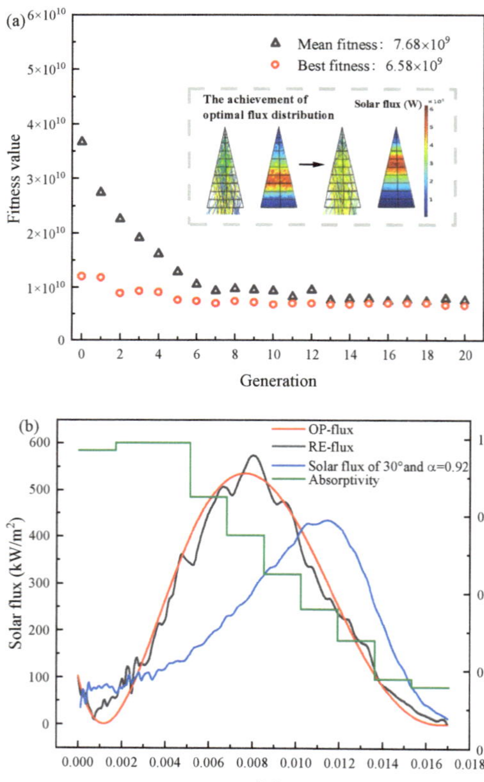

Figure 10. (**a**) Genetic algorithm optimization iteration and (**b**) achievement of optimal solar flux and optimization results of cavity wall absorptivity.

In this study, the comprehensive performance of the real solar flux distribution achieved by changing the wall absorption rate was compared with other cases under different inlet volume flow rates, as shown in Figure 11. The results show that the hydrogen yield under the real flux distribution decreased slightly compared to the optimal flux distribution, but was still higher than the original solar flux, with a 0.10~3.90% improvement under the four inlet flow conditions. In addition, the methane conversion at the real flux distribution also decreased slightly compared to the optimal flux distribution, but still increased by 2.13~4.85% compared to the original flux distribution. This was due to the loss of total incident solar energy and the difference between the real flux distribution and the optimal flux distribution. For the real flux distribution achieved, the increase in carbon dioxide conversion was the largest, with an increase of 3.97~5.78% at v_{in} = 8~10 L·min^{-1} compared to the original flux distribution, even if it was somewhat lower than the optimal flux distribution.

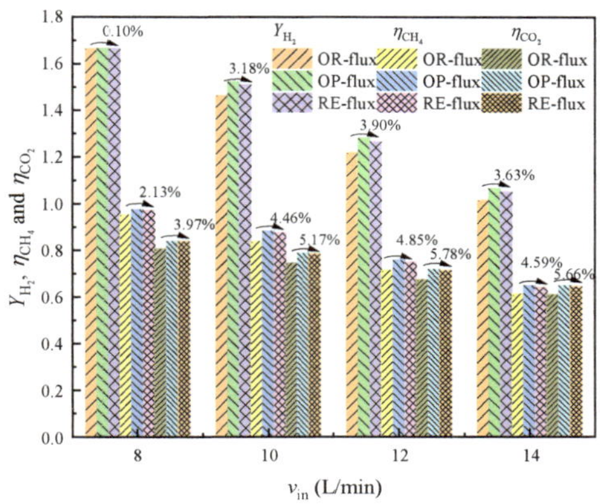

Figure 11. Performance comparison at different inlet volume flows for the real flux distribution achieved by varying the absorptivity.

5. Conclusions

In this study, a numerical model of a preheated cavity solar thermochemical reactor-dish methane dry reformer was constructed based on previous research. The optimal solar radiation heat flux distribution described by the 5th-order Bernstein polynomial was solved with a constant value of total incident solar energy and a hydrogen production rate as the optimization objective by combining it with the gradient optimization algorithm.

The comprehensive performances of the two solar flux distributions were compared and analyzed. In addition, the absorptivity of the solar-absorbing coating on the cavity wall was optimized by using the genetic algorithm in an attempt to achieve the optimal solar radiation flux distribution. The salient findings are as follows:

(1) Taking 1031.7 W as the total incident solar energy, the gradient optimization algorithm performed 158 iterations to obtain the optimal solar flux distribution. The Bernstein polynomial control variables solved by IPOPT are $C_0 = -0.562$, $C_1 = -16.691$, $C_2 = 67.655$, $C_3 = -4.534$, $C_4 = -5.258$, and $C_5 = -0.997$. The main energy of the optimal solar flux distribution was concentrated deep in the cavity, which greatly reduced the loss of heat radiation and resulted in more efficient energy use.

(2) Under the optimal solar flux, the hydrogen production rate increased by 0.13~5.10%, the energy storage efficiency increased by 1.09~5.55%, and the methane conversion rate increased by 2.43~6.01% for the inlet volume flow rate of 8~14 L·min^{-1} compared to the original flux distribution.

(3) The absorptivity of the ten-segment cavity wall for solar radiation was optimized by the genetic algorithm, and the optimal solar flux distribution was roughly achieved, while the energy loss was only 0.6%. Although the performance of the real flux distribution was not as good as that of the optimal flux distribution, it was also much better than the original flux distribution.

There are also some shortcomings of this study. The optimal flux distribution obtained does not seem to be the final ideal flux distribution, as it is only for one hydrogen production target. In subsequent studies, different performance evaluation indicators, such as conversion rate, carbon deposition, etc., can be taken into account to find the most suitable flux distribution. The aim of this study is to provide a research idea for an integrated solar-thermal-chemical design. Furthermore, the optimal flux distribution found in this study was achieved by the adjustment of wall absorption rate, which also places high demands on the coating materials.

Author Contributions: Conceptualization, Z.-Q.D. and W.-W.Y.; methodology, Z.-Q.D.; software, R.-Z.Z.; validation, X.M. and W.-W.Y.; data curation, X.-Y.T. and R.-Z.Z.; writing—original draft preparation, Z.-Q.D. and X.M.; writing—review and editing, X.M. and W.-W.Y.; supervision, W.-W.Y.; All authors have read and agreed to the published version of the manuscript.

Funding: This work was financially sponsored by the National Natural Science Foundation of China (no. 52090063).

Data Availability Statement: The data presented in this study are available on request from the corresponding author.

Conflicts of Interest: The authors declare no conflict of interest.

Nomenclature

C_p	specific heat ($J \cdot kg^{-1} \cdot K^{-1}$)
dp	aperture size (m)
Dij	binary diffusivity coefficient ($m^2 \cdot s^{-1}$)
Fi	mass flow of the species i ($kg \cdot s^{-1}$)
hv	local volumetric heat transfer coefficient ($W \cdot m^{-3} \cdot K^{-1}$)
ΔH	enthalpy change of the reaction ($kJ \cdot mol^{-1}$)
I	radiation intensity ($MW \cdot m^{-2} \cdot sr^{-1}$)
k	reaction rate constant
K	adsorption constant
Kp	equilibrium constant for reaction
L	length of reactor (m)
mi	mole fraction of species i
M	mole weight ($kg \cdot mol^{-1}$)
p	pressure (Pa)
Pi	partial pressure of species i (bar)
q	energy flux ($MW \cdot m^{-2}$)
Q	energy (W)
R	universal gas constant ($J \cdot mol^{-1} \cdot K^{-1}$)
ri	chemical reaction rate of the i-th reaction ($mol \cdot m^{-3} \cdot s^{-1}$)
r_{H_2}	hydrogen production rate ($mol \cdot m^{-3} \cdot s^{-1}$)
T	temperature (K)
$\vec{u}$	velocity vector ($m \cdot s^{-1}$)
Y_{H_2}	hydrogen yield
z	axial coordinate (m)
Greeks	
α	absorptivity
ε	emissivity
η	energy efficiency or reactant conversion
κ_e	extinction coefficient of porous media (m^{-1})
λ	thermal conductivity ($W \cdot m^{-1} \cdot K^{-1}$)
μ	dynamic viscosity ($Pa \cdot s$)
ρ	density ($kg \cdot m^{-3}$)
σ	Stefan-Boltzmann constant ($W \cdot m^{-2} \cdot K^{-4}$)
τ	reflectivity
φ	porosity
ω	mass fraction
α	absorptivity
Superscripts and subscripts	
a	actual value
ch	chemical storage energy

cat	catalyst
eff	effective value
f	fluid
i, *j*	species (CH_4, H_2O, H_2, CO, CO_2)
in	inlet
max	maximum value
out	outlet
so	solar energy
s	solid
se	sensible thermal energy
Abbreviations	
DRM	dry reforming of methane

References

1. Wen, H.; Liang, W.; Lee, C.-C. China's progress toward sustainable development in pursuit of carbon neutrality: Regional differences and dynamic evolution. *Environ. Impact Assess. Rev.* **2023**, *98*, 106959. [CrossRef]
2. Gunnarsdottir, I.; Davidsdottir, B.; Worrell, E.; Sigurgeirsdottir, S. Sustainable energy development: History of the concept and emerging themes. *Renew. Sustain. Energy Rev.* **2021**, *141*, 110770. [CrossRef]
3. Anam, M.Z.; Bari, A.B.M.M.; Paul, S.K.; Ali, S.M.; Kabir, G. Modelling the drivers of solar energy development in an emerging economy: Implications for sustainable development goals. *Resour. Conserv. Recycl. Adv.* **2022**, *13*, 200068. [CrossRef]
4. He, C.; Wu, S.; Wang, L.; Zhang, J. Recent advances in photo-enhanced dry reforming of methane: A review. *J. Photochem. Photobiol. C Photochem. Rev.* **2022**, *51*, 100468. [CrossRef]
5. Su, B.; Wang, Y.; Xu, Z.; Han, W.; Jin, H.; Wang, H. Novel ways for hydrogen production based on methane steam and dry reforming integrated with carbon capture. *Energy Convers. Manag.* **2022**, *270*, 116199. [CrossRef]
6. Wang, Z.; Han, W.; Zhang, N.; Gan, Z.; Sun, J.; Jin, H. Energy level difference graphic analysis method of combined cooling, heating and power systems. *Energy* **2018**, *160*, 1069–1077. [CrossRef]
7. Ling, Y.; Li, W.; Jin, J.; Yu, Y.; Hao, Y.; Jin, H. A spectral-splitting photovoltaic-thermochemical system for energy storage and solar power generation. *Appl. Energy* **2020**, *260*, 113631. [CrossRef]
8. Agrafiotis, C.; von Storch, H.; Roeb, M.; Sattler, C. Solar thermal reforming of methane feedstocks for hydrogen and syngas production—A review. *Renew. Sustain. Energy Rev.* **2014**, *29*, 656–682. [CrossRef]
9. Villafán-Vidales, H.I.; Arancibia-Bulnes, C.A.; Riveros-Rosas, D.; Romero-Paredes, H.; Estrada, C.A. An overview of the solar thermochemical processes for hydrogen and syngas production: Reactors, and facilities. *Renew. Sustain. Energy Rev.* **2017**, *75*, 894–908. [CrossRef]
10. Sheu, E.J.; Mokheimer, E.M.A.; Ghoniem, A.F. A review of solar methane reforming systems. *Int. J. Hydrogen Energy* **2015**, *40*, 12929–12955. [CrossRef]
11. Kribus, A.; Gray, Y.; Grijnevich, M.; Mittelman, G.; Mey-Cloutier, S.; Caliot, C. The promise and challenge of solar volumetric absorbers. *Sol. Energy* **2014**, *110*, 463–481. [CrossRef]
12. Gomez-Garcia, F.; González-Aguilar, J.; Olalde, G.; Romero, M. Thermal and hydrodynamic behavior of ceramic volumetric absorbers for central receiver solar power plants: A review. *Renew. Sustain. Energy Rev.* **2016**, *57*, 648–658. [CrossRef]
13. Wang, P.; Li, J.B.; Xu, R.N.; Jiang, P.X. Non-uniform and volumetric effect on the hydrodynamic and thermal characteristic in a unit solar absorber. *Energy* **2021**, *225*, 120130. [CrossRef]
14. Du, S.; Xia, T.; He, Y.-L.; Li, Z.-Y.; Li, D.; Xie, X.-Q. Experiment and optimization study on the radial graded porous volumetric solar receiver matching non-uniform solar flux distribution. *Appl. Energy* **2020**, *275*, 115343. [CrossRef]
15. Kasaeian, A.; Barghamadi, H.; Pourfayaz, F. Performance comparison between the geometry models of multi-channel absorbers in solar volumetric receivers. *Renew. Energy* **2017**, *105*, 1–12. [CrossRef]
16. Li, J.B.; Wang, P.; Liu, D.Y. Optimization on the gradually varied pore structure distribution for the irradiated absorber. *Energy* **2022**, *240*, 122787. [CrossRef]
17. Capuano, R.; Fend, T.; Stadler, H.; Hoffschmidt, B.; Pitz-Paal, R. Optimized volumetric solar receiver: Thermal performance prediction and experimental validation. *Renew. Energy* **2017**, *114*, 556–566. [CrossRef]
18. Shi, X.; Wang, F.; Cheng, Z.; Liang, H.; Dong, Y.; Chen, X. Numerical analysis of the biomimetic leaf-type hierarchical porous structure to improve the energy storage efficiency of solar driven steam methane reforming. *Int. J. Hydrogen Energy* **2021**, *46*, 17653–17665. [CrossRef]
19. Shi, X.; Zhang, X.; Wang, F.; Yang, L.; Dong, Y.; Shuai, Y. Thermochemical analysis of dry methane reforming hydrogen production in biomimetic venous hierarchical porous structure solar reactor for improving energy storage. *Int. J. Hydrogen Energy* **2021**, *46*, 7733–7744. [CrossRef]
20. Liu, X.; Cheng, B.; Zhu, Q.; Gao, K.; Sun, N.; Tian, C.; Wang, J.; Zheng, H.; Wang, X.; Dang, C.; et al. Highly efficient solar-driven CO_2 reforming of methane via concave foam reactors. *Energy* **2022**, *261*, 125141. [CrossRef]
21. Bellos, E.; Bousi, E.; Tzivanidis, C.; Pavlovic, S. Optical and thermal analysis of different cavity receiver designs for solar dish concentrators. *Energy Convers. Manag. X* **2019**, *2*, 100013. [CrossRef]

22. Sedighi, M.; Taylor, R.A.; Lake, M.; Rose, A.; Izadgoshasb, I.; Vasquez Padilla, R. Development of a novel high-temperature, pressurised, indirectly-irradiated cavity receiver. *Energy Convers. Manag.* **2020**, *204*, 112175. [CrossRef]
23. Huang, H.; Lin, M. Optimization of solar receivers for high-temperature solar conversion processes: Direct vs. Indirect illumination designs. *Appl. Energy* **2021**, *304*, 117675. [CrossRef]
24. Sánchez-González, A.; Santana, D. Solar flux distribution on central receivers: A projection method from analytic function. *Renew. Energy* **2015**, *74*, 576–587. [CrossRef]
25. Yan, J.; Peng, Y.; Cheng, Z. Optimization of a discrete dish concentrator for uniform flux distribution on the cavity receiver of solar concentrator system. *Renew. Energy* **2018**, *129*, 431–445. [CrossRef]
26. Fang, J.; Tu, N.; Wei, J.; Du, X. Effects of surface optical and radiative properties on the thermal performance of a solar cavity receiver. *Sol. Energy* **2018**, *171*, 157–170. [CrossRef]
27. Chen, F.; Li, M.; Hassanien Emam Hassanien, R.; Luo, X.; Hong, Y.; Feng, Z.; Ji, M.; Zhang, P. Study on the Optical Properties of Triangular Cavity Absorber for Parabolic Trough Solar Concentrator. *Int. J. Photoenergy* **2015**, *2015*, 1–9. [CrossRef]
28. Xiao, G.; Yang, T.; Ni, D.; Cen, K.; Ni, M. A model-based approach for optical performance assessment and optimization of a solar dish. *Renew. Energy* **2017**, *100*, 103–113. [CrossRef]
29. Tu, N.; Wei, J.; Fang, J. Numerical investigation on uniformity of heat flux for semi-gray surfaces inside a solar cavity receiver. *Sol. Energy* **2015**, *112*, 128–143. [CrossRef]
30. Wang, F.; Shuai, Y.; Yuan, Y.; Yang, G.; Tan, H. Thermal stress analysis of eccentric tube receiver using concentrated solar radiation. *Sol. Energy* **2010**, *84*, 1809–1815. [CrossRef]
31. Wang, K.; He, Y.-L.; Li, P.; Li, M.-J.; Tao, W.-Q. Multi-objective optimization of the solar absorptivity distribution inside a cavity solar receiver for solar power towers. *Sol. Energy* **2017**, *158*, 247–258. [CrossRef]
32. Tang, X.; Yang, W.; Dai, Z.; Yang, Y. Inverse Design of Local Solar Flux Distribution for a Solar Methanol Reforming Reactor Based on Shape Optimization. *Front. Energy Res.* **2022**, *10*, 881822. [CrossRef]
33. Tang, X.Y.; Yang, W.W.; Yang, Y.; Jiao, Y.H.; Zhang, T. A design method for optimizing the secondary reflector of a parabolic trough solar concentrator to achieve uniform heat flux distribution. *Energy* **2021**, *229*, 120749. [CrossRef]
34. Dai, Z.-Q.; Yang, Y.-J.; Yang, W.-W.; Rong, L.; Tang, X.-Y. A thermochemical reactor design with better thermal management and improved performance for methane/carbon dioxide dry reforming. *Int. J. Hydrogen Energy* **2022**, *47*, 34794–34809. [CrossRef]
35. Wu, Z.; Caliot, C.; Flamant, G.; Wang, Z. Coupled radiation and flow modeling in ceramic foam volumetric solar air receivers. *Sol. Energy* **2011**, *85*, 2374–2385. [CrossRef]
36. Zheng, L.; Zhang, C.; Zhang, X.; Zhang, J. Flow and radiation heat transfer of a nanofluid over a stretching sheet with velocity slip and temperature jump in porous medium. *J. Frankl. Inst.* **2013**, *350*, 990–1007. [CrossRef]
37. Wu, Z.; Wang, Z. Fully coupled transient modeling of ceramic foam volumetric solar air receiver. *Sol. Energy* **2013**, *89*, 122–133. [CrossRef]
38. Wang, F.; Tan, J.; Yong, S.; Tan, H.; Chu, S. Thermal performance analyses of porous media solar receiver with different irradiative transfer models. *Int. J. Heat Mass Transf.* **2014**, *78*, 7–16. [CrossRef]
39. Lee, B.; Yun, S.-W.; Kim, S.; Heo, J.; Kim, Y.-T.; Lee, S.; Lim, H. CO_2 reforming of methane for H2 production in a membrane reactor as CO_2 utilization: Computational fluid dynamics studies with a reactor geometry. *Int. J. Hydrogen Energy* **2019**, *44*, 2298–2311. [CrossRef]
40. Zhang, H.; Shuai, Y.; Lougou, B.G.; Jiang, B.; Wang, F.; Cheng, Z.; Tan, H. Effects of multilayer porous ceramics on thermochemical energy conversion and storage efficiency in solar dry reforming of methane reactor. *Appl. Energy* **2020**, *265*, 114799. [CrossRef]
41. Richardson, J.T.; Paripatyadar, S.A. Carbon dioxide reforming of methane with supported rhodium. *Appl. Catal.* **1990**, *61*, 293–309. [CrossRef]
42. Snoeck, J.-W.; Froment, G.F.; Fowles, M. Kinetic Study of the Carbon Filament Formation by Methane Cracking on a Nickel Catalyst. *J. Catal.* **1997**, *169*, 250–262. [CrossRef]
43. Tsai, C.-Y. Design of free-form trough reflector for solar thermal concentrator system based on quadratic Bézier curves. *Opt. Commun.* **2022**, *511*, 128024. [CrossRef]
44. Tang, X.-Y.; Dou, P.-Y.; Dai, Z.-Q.; Yang, W.-W. Structural design and analysis of a solar thermochemical reactor partially filled with phase change material based on shape optimization. *Sol. Energy* **2022**, *236*, 613–625. [CrossRef]
45. Duraklı, Z.; Nabiyev, V. A new approach based on Bezier curves to solve path planning problems for mobile robots. *J. Comput. Sci.* **2022**, *58*, 101540. [CrossRef]
46. Xu, L.; Cao, M.; Song, B. A new approach to smooth path planning of mobile robot based on quartic Bezier transition curve and improved PSO algorithm. *Neurocomputing* **2022**, *473*, 98–106. [CrossRef]
47. Biegler, L.T.; Zavala, V.M. Large-scale nonlinear programming using IPOPT: An integrating framework for enterprise-wide dynamic optimization. *Comput. Chem. Eng.* **2009**, *33*, 575–582. [CrossRef]
48. Saad, Y.; Schultz, M.H. GMRES: A Generalized Minimal Residual Algorithm for Solving Nonsymmetric Linear Systems. *SIAM J. Sci. Stat. Comput.* **1986**, *7*, 856–869. [CrossRef]
49. Schenk, O.; Gartner, K. Solving unsymmetric sparse systems of linear equations with PARDISO. *Futur. Gener. Comp. Syst.* **2004**, *20*, 475–487. [CrossRef]

50. Lee, H. The geometric-optics relation between surface slope error and reflected ray error in solar concentrators. *Sol. Energy* **2014**, *101*, 299–307. [CrossRef]
51. Benguerba, Y.; Dehimi, L.; Virginie, M.; Dumas, C.; Ernst, B. Modelling of methane dry reforming over Ni/Al_2O_3 catalyst in a fixed-bed catalytic reactor. *Reac. Kinet. Mech. Cat.* **2015**, *114*, 109–119. [CrossRef]

energies

MDPI

Article

A Novel Pressure-Controlled Molecular Dynamics Simulation Method for Nanoscale Boiling Heat Transfer

Cong Wang [1], Yalong Kong [1], Zhigang Liu [1], Lin Guo [1,*] and Yawei Yang [2,*]

[1] Energy Research Institute, Qilu University of Technology, Jinan 250014, China
[2] Electronic Materials Research Laboratory, Key Laboratory of the Ministry of Education, International Center for Dielectric Research, Shaanxi Engineering Research Center of Advanced Energy Materials and Devices, School of Electronic Science and Engineering, Xi'an Jiaotong University, Xi'an 710049, China
* Correspondence: linguo@sderi.cn (L.G.); ywyang@xjtu.edu.cn (Y.Y.)

Abstract: Pool boiling, enabling remarkable phase-change heat transfer, has elicited increasing attention due to its ubiquitous applications in solar thermal power stations. An explicit understanding of the effect of system pressure on pool boiling is required to enhance the phase-change heat transfer. Despite its wide application when exploring the potential mechanism of boiling, the molecular dynamics method still needs to be improved when discussing the working mechanism of system pressure. Therefore, in the present study, a novel molecular dynamics simulation method of nanoscale pool boiling was proposed. This method provides a way to change and control pressure during the phase-change process. Furthermore, the bubble nucleation and growth in nanoscale pool boiling are quantitatively investigated through pressure-control molecular dynamics simulations. We expect that this study will improve the present simulation method of pool boiling and provide useful insights to the physics of the process.

Keywords: molecular dynamics; nucleate boiling; pressure control

1. Introduction

Solar energy is available in abundance in many parts of the Earth and has zero global warming potential, unlike thermal power plants where the main heat source can contribute to climate change [1,2]. Sonawane et al., analyzing a number of published studies, found that solar desalination plants powered by solar energy have increased to 76% [3]. The highest solar energy and exergy production still using black toner is 26.9% and 27.0%, respectively, higher than that of a conventional solar still [4]. Considering the unavailability of the solar source at nighttime and intermittency during the day [5], latent heat storage that offers great storage density per unit volume is a promising solution [6]. Therefore, an explicit understanding of boiling is required to enhance the phase-change heat transfer and the efficient operation of a whole power plant [7].

Boiling is a kind of thermal phenomenon in which liquid steam latent heat transfers heat at a low surface temperature [8], which is common in the field of industrial production and thermal management [9]. The enhancement of boiling heat transfer is of great significance because it can improve, for example, the efficiency and economy of heat engines and reduce energy consumption. Therefore, it has received extensive attention in energy and environmental protection-related fields [10]. When the surface comes in contact with liquid, boiling will occur if the temperature is higher than a certain threshold, while pool boiling will occur when the heating surface is lower than the free surface of the liquid [11]. With the increased cooling capacity of facilities, the cooling fluid parameters can be better optimized, which can improve pool boiling heat transfer performance [12].

Pool boiling is a typical two-phase flow phenomenon that includes the following stages [13–15]. First, single-phase convective heat transfer, followed by bubble nucleation. In the process of increasing superheat, the heat transfer efficiency continues to improve, and the heat flux reaches the critical value under certain conditions. Nucleate boiling changes to film boiling, which has a negative impact on heat transfer, and then leads to minimum heat flux in a short time [16]. After a stable film is generated near the surface, the heat flux increases again [17]. However, due to the influence of the gas film, the surface temperature is very high, which may be harmful to the equipment [18]. The characteristics of the nucleate boiling process are controllable and efficient, so it can meet the requirements of industrial production well. In order to improve the performance of the cooling system, it is necessary to increase the initial superheat of nucleate boiling, which is another research hotspot in this field [19].

At present, research on nanoscale boiling is increasing, and a series of achievements have been reached [20]. Cleanroom-processed micro- and nanostructures have been exploited for systematically studying structural effects on boiling heat transfer enhancement [21]. Some scholars have found that boiling heat transfer can be enhanced in different ways, mainly by modifying the heating surface through micro/nano structures (such as microporous layers [22,23], micro-fins [15,24,25], micro-channels [23,26,27], nanowires [28], and nanoparticles [29,30]). The micro/nanostructures discussed before can achieve one or more goals, such as increasing the heat transfer coefficient or critical heat flux. However, research results show that it is generally difficult for existing technologies to achieve all the above goals at the same time [31].

The adjustment of liquid supplement and bubble coalescence can significantly affect the heat transfer performance of such structures. In the traditional experimental mode, only the macroscopic analysis of the bubble dynamic behavior can be carried out, so the microscopic mechanism of the heat transfer process cannot be clarified. The boiling process is very complex, so it is difficult to accurately observe bubble behavior and changes [32]. During boiling, heat transfer performance is easily affected by roughness and wettability, which is unfavorable to the accuracy of experimental results. Due to the complexity of the mechanism and the difficulty of observation, the bubble nucleation process cannot be accurately studied by conventional experimental techniques.

Molecular dynamics (MD) simulation method is currently attracting attention in the heat transfer mechanism research [33,34]. Its advantage is that it can accurately analyze the microscopic behavior of molecular motion, thus providing support for the study of heat transfer enhancement [35,36]. Therefore, it is widely concerned at present. Some scholars have studied the nanoscale boiling and bubble nucleation process based on MD technology, and found that it can meet the application requirements well [37]. The MD technology can accurately control the size, wettability, and roughness of liquid surface structure, and improve the reference value of the results obtained. It can analyze the interface evolution in the nucleation process from a microscopic perspective, which provides support for the in-depth exploration of the nucleation mechanism [38]. Some scholars have studied the nucleate boiling phenomenon caused by the solid wall under heated liquid, and discussed the correlation between the surface wettability and explosive boiling when the liquid is heated. The results show that boiling intensity increases with increased wettability. The MD simulation results also found that the energy transfer rate efficiency would be greatly improved under the conditions of strong wettable surface.

Nevertheless, many problems still exist that need to be solved to clarify MD simulations. The boiling pressure affects the phase-change heat transfer performance. Bubbles are expected to exhibit various dynamic behaviors during different environmental pressures [39]. However, on reviewing the literature, the pressure effects has not been discussed in previous studies. Specifically, all the cited works that were found employed MD simulations that apply the fixed volume method, where the change of the fluid temperature induced by the heat transfer results in a change of the system pressure. Therefore, a novel pressure-controlled MD simulation method for nanoscale boiling heat transfer was proposed in the present paper. The main aim of the proposed method is to provide a way to change and control the pressure during the nanoscale phase-change process. Moreover, the bubble nucleation and growth in nanoscale pool boiling can also be quantitatively studied based on the proposed numerical method.

2. Computational Methods

2.1. MD Simulation System

All simulations were performed using the software package for the large-scale atomic/molecular massively parallel simulator (LAMMPS). Figure 1a below shows the initial configuration of this system, which includes a 16 nm (x) × 5 nm (y) × 60 nm (z) cuboid. The working gas is set as argon (Ar), which is used as the simulation medium in this experiment. The solid wall is made of copper, which has good thermal conductivity and is easy to process, and has been widely used in many engineering fields and experimental research. The bottom solid wall is composed of three layers of Cu atoms, which are arranged in a face-centered cubic (FCC) lattice structure with a lattice constant of 0.35 nm. Liquid Ar atoms are initially arranged on the surface of ordinary copper in the form of FCC, and the lattice constant is 0.58 nm. The thickness of liquid Ar film is 10 nm, and the top wall consists of two layers of copper atoms. In order to facilitate simulation analysis, periodic boundary conditions are set in the vertical and horizontal directions of the simulation box.

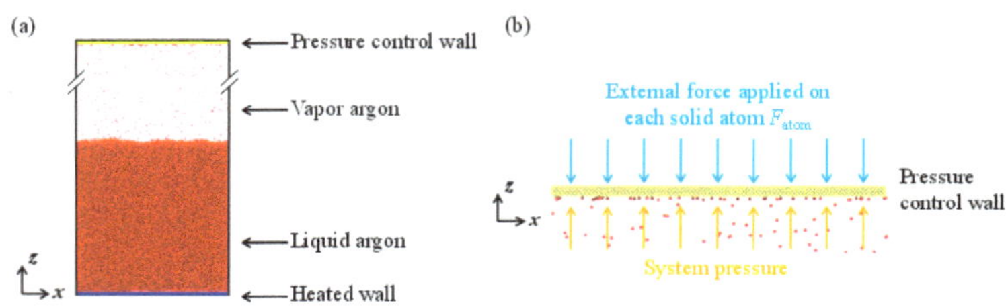

Figure 1. (**a**) Initial configuration of the simulation system; (**b**) illustration of the pressure control method.

2.2. Simulation Method and Procedure

In this paper, the interactions between argon–argon (Ar–Ar) atoms, copper–copper (Cu–Cu) atoms and argon–copper (Ar–Cu) atoms are expressed as follows [40]:

$$E_{\text{Ar-Ar,Cu-Cu,Ar-Cu}} = 4\varepsilon\left[\left(\frac{\sigma}{r}\right)^{12} - \left(\frac{\sigma}{r}\right)^{6}\right], r < r_c \quad (1)$$

where ε represents energy, σ represents distance; r is inter-atomic distance, and r_c is the cut-off radius. The interactions were determined using Equations (2) and (3) in terms of the combined Lorentz-Berthelot rule:

$$\varepsilon_{\text{Ar-Cu}} = \alpha\sqrt{\varepsilon_{\text{Ar-Ar}}\varepsilon_{\text{Cu-Cu}}} \quad (2)$$

$$\sigma_{\text{Ar-Cu}} = \frac{\sigma_{\text{Cu-Cu}} + \sigma_{\text{Ar-Ar}}}{2} \quad (3)$$

The copper–argon interaction strength can be adjusted based on the energy parameter of α in the present research. Energy parameters used in the present study are detailed in Table 1.

Table 1. Potential energy parameters.

Inter–Particle Interaction	**Potential Energy Parameter**	
	ε**/eV**	σ**/Å**
Ar–Ar	0.0104	3.405
Ar–Cu_{bottom}	Variable	2.87
Ar–Cu_{up}	0.02	2.87
Cu_{bottom}–Cu_{bottom}, Cu_{up}–Cu_{up}	0.406	2.338
Cu_{bottom}–Cu_{up}, Cu_{bottom}–Cu_{up}	0	0

In the simulation research, the large-scale atom/molecule parallel simulator is applied, and the atomic parameters are updated based on the velocity Verlet algorithm. In order to meet the temperature control requirements, the Langevin thermostat is applied. The phase-change heat transfer process can be divided into equilibrium stage and wall temperature-change stage, which are simulated separately. First, the temperature of the whole system is controlled at 80 K and kept in balance when it lasts for 0.5 ns. Then, the temperature of the bottom wall is heated appropriately to reach 145 K. The microcanonical ensemble (NVE) is adopted and the simulation lasts for 5 ns. The data is acquired simultaneously in these two stages. The data are output every 1000 time-steps, and the atom trajectories are visualized by the open-source visualization software OVITO [41].

The absorbed heat flux q can be determined as follows [42]:

$$q = \frac{1}{V}\left[\sum_{i=1}^{N}\left(v_i \cdot e_i + \frac{1}{2}\sum_{j=1, j\neq i}^{N} r_{ij}\left(F_{ij} \cdot v_i\right)\right)\right] \tag{4}$$

where V is the volume of liquid computation domain, N denotes the number of liquid argon atoms, v_i and e_i represent the thermal velocity and the total energy (including kinetic energy and potential energy) of atom i, respectively, and F_{ij} refers to the force acting on atom i from atom j. To constrict the noise influence associated with oscillations in the original data, the exhibited value of q in the present paper is the average of the data every 10,000 timesteps.

2.3. Pressure Control Method

Usually, in the MD simulation of pool boiling, the size of the simulation box remains unchanged; however, during the boiling process, the liquid will undergo a gas–liquid phase change process under the heating of the wall, and the system pressure will change if the simulation area remains unchanged, which will influence the whole boiling process. Therefore, a systemic pressure-control method is proposed herein to keep the pressure in the simulated area constant in the boiling process. Specifically, a control wall is set above the simulation area, as shown in Figure 1. In the simulation process, firstly, the force on the solid wall in z-direction is set to 0, and then an external force is applied to each solid particle on the control wall. The external force value is calculated by a given system pressure, as shown in Figure 1b. The applied external force in -z-direction is balanced with the pressure applied by the system to the pressure control wall in z-direction, to keep the whole system pressure constant and avoid the influence of pressure change on the bubble dynamics.

Considering the wall-fix method can only simulate the atmospheric pressure, 1 atm was set in the pressure-control method. The force applied on every single atom of the pressure-control wall F_{atom} is calculated by

$$F_{\text{atom}} = \frac{F_{\text{total}}}{N_{\text{up}}} = \frac{p \times S_{xy}}{N_{\text{up}}} \tag{5}$$

where F_{total} is the total force applied on the pressure-control wall, N_{up} is the number of atoms on the pressure-control wall, p denotes the pressure set in the simulation, and S_{xy} is the cross-sectional area in the x-y plane. Moreover, the wall-fixing method was also calculated and compared setting and freezing an upper wall to the effect of phase-change on the system pressure. Different lengths of the simulation box in the z-direction were set and computed.

2.4. Quantitative Description of Bubble Dynamic Behavior

To quantify the changes in bubble shape and size with time in the boiling process, a bubble-related statistical method was used. Statistical calculations are divided into three steps. First, the calculation file is exported, as shown in Figure 2a; the area is calculated and divided into cubic bins along the three principal directions using a custom MathWorks MATLAB script. Second, the number of argon molecules in each bin is counted, and the bin is marked with the number of molecules less than a certain value (Figure 2b). Third, the processed file is exported, the marked bin is read in OVITO, and the bubble shape and volume are calculated (Figure 2c).

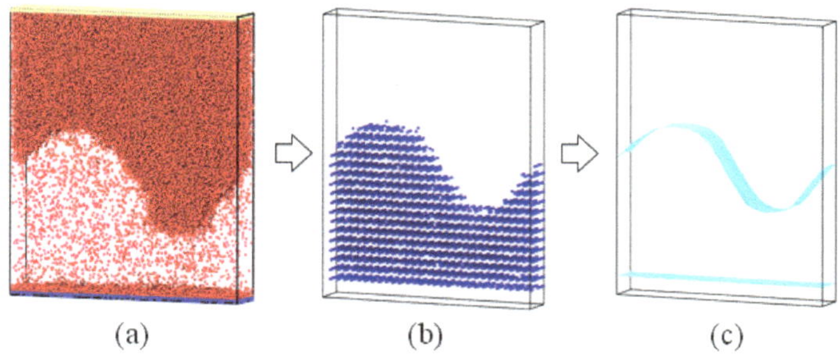

Figure 2. Three steps of the bubble calculation method. (**a**) Bins partition; (**b**) liquid/vapor identification; (**c**) interface recognition.

3. Results and Discussion

3.1. Visualization of the System Equilibrium Stage

Four different kinds of boiling systems were constructed and simulated in the present study, which can be seen in Figure 3. The *x*-*z* plane view is exhibited to observe the interface evolution process. The pressure-control method was applied in the pressure-control case, shown in Figure 3a. The initial height of the simulation box *H* in case (a) equals 433 Å. The other three cases apply the wall-fix method, which means the simulation box remains unchanged during the simulation process. The heights of the simulation box *H* in cases (b), (c), and (d) are 216 Å, 433 Å, and 650 Å, respectively. There are two equilibrium processes in the pressure-control case. The liquid is controlled at 80 K to reach an equilibrium state. Under the given system pressure, atoms of the upper control wall are subject to external forces F_{atom}, which can be calculated by Equation (5). Therefore, to maintain the set system pressure, the upper control wall is observed to move in the -*z*-direction in Figure 3a. As the equilibrium process continues, the upper wall reaches a certain position and remains stationary. Finally, both the liquid and the system box reach the equilibrium state at 2500 ps. In contrast, there is only one equilibrium process in the wall-fixing cases. The box heights parameter is kept constant and only the liquid shifts to the equilibrium state.

3.2. Bubble Dynamic Behaviors of the Phase-Change Stage

Under the condition of entering equilibrium state, the temperature of the bottom wall increases from 80 K to 160 K to simulate the heating process. Thus, the liquid is heated and the phase-change phenomena can be observed. Representative snapshots of the four cases during the phase-change process are chosen and illustrated in Figure 4. This stage starts at 2500 ps, which is the time marking completion of the first equilibrium stage.

The simulation box height keeps increasing with time during the phase-change process, as shown in Figure 4a. The heights of the other three cases remain constant, as shown in Figure 4b–d. These results are consistent with the analysis in Section 2.3. The system pressure increases with the continuous occurrence of the vapor–liquid phase change process. The upper control wall moves upwards as a result. The model can also be verified by combining the results of Figures 3a and 4a. Specifically, the initial height of the simulation box is 433 Å, and the system pressure is set as one bar. As the system moves through the equilibrium stage, the upper wall moves downwards because the argon atoms are in the liquid state under 80 K and there is no argon vapor at this stage. In contrast, when the argon atoms absorb heat from the contact wall, the phase-change process occurs and the volume of the vapor increases. As a result, the upper control wall moves upwards. The upper wall moves along with the system pressure changes, which verifies the pressure control method.

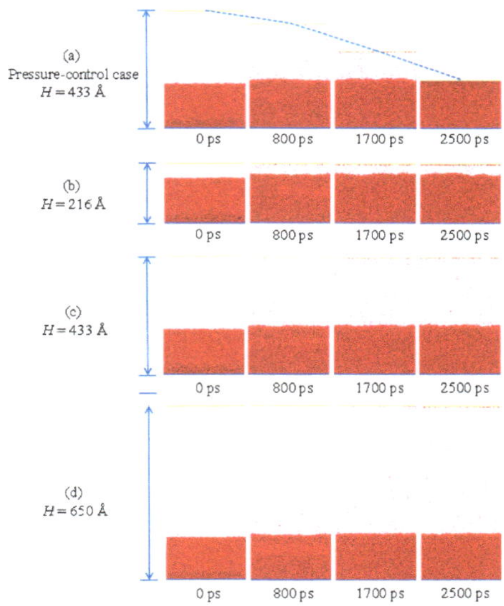

Figure 3. Snapshots of the system equilibrium stage. (**a**) The pressure-control case; (**b**) the simulation box height H equals 216 Å; (**c**) the simulation box height H equals 433 Å; (**d**) the simulation box height H equals 650 Å.

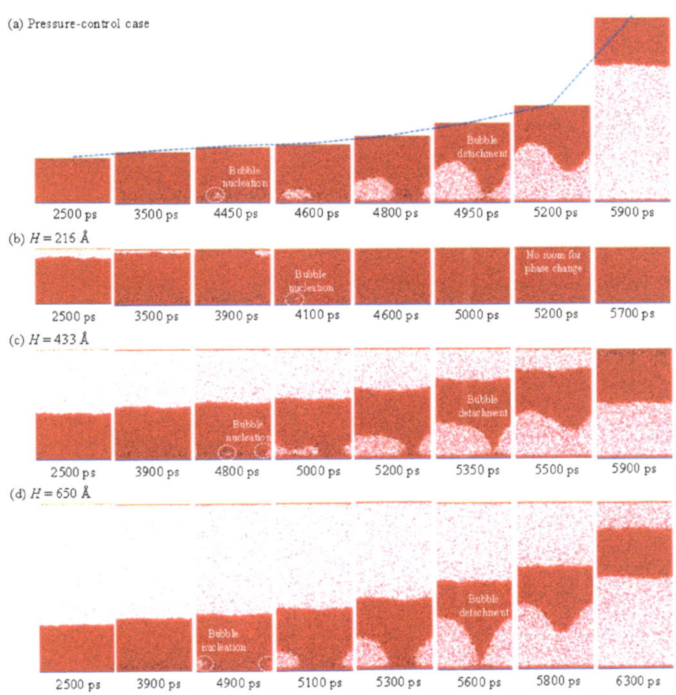

Figure 4. Snapshots of nucleate boiling of liquid argon on smooth surfaces with (**a**) pressure control method, (**b**) simulation box height H = 216 Å, (**c**) H = 433 Å, and (**d**) H = 650 Å. Bubble nucleation and detachment during the boiling process are also marked correspondingly.

The bubble dynamic behavior was also investigated. As illustrated in Figure 4a, a suitable bubble nucleus is produced on the wall surface at about 4450 ps. This indicates that the tin is 1950 ps. According to the experimental results, small vesicular embryos also appeared at the base center. However, their size is small and does not reach the critical bubble core size, so its stability is poor and it will disappear after a period of time. In this process, the stable bubble nucleus continues to grow upward and laterally, finally merging to form a large bubble and generate a stable vapor film at 4950 ps.

The effects of the simulation box height on bubble dynamic behavior were also examined. Figure 4b shows bubble nucleation at 4100 ps; however, the box height is relatively small for the vapor-liquid phase change. No room remained for subsequent bubble growth. Both cases c and d were observed to have undergone complete bubble nucleation, bubble growth, and bubble detachment on the heated surface, as marked in Figure 4. Nevertheless, different droplet dynamic behaviors were observed. Bubble nucleation was observed at 4800 ps and 4900 ps in cases c and d, respectively, which were later than the time when they were observed in the pressure-control case. The bubble detachment time also varied in different cases (at 4950 ps, 5350 ps, and 5600 ps). The height of the simulation box of the wall-fix method affected the bubble dynamic behavior. Furthermore, it is very necessary to research the mechanisms of the interface evolution process, which will be discussed later in this paper.

3.3. *Quantitative Study of Bubble Dynamics Behavior*

In Section 3.2, the boiling heat transfer over the copper surface is exhibited intuitively in terms of representative snapshots. The snapshots in Figure 4 show that the simulation box height of the wall-fix method affected bubble dynamic behavior, which was also different from the results of the pressure-control case. To reveal the potential mechanisms of this process, more quantitative studies of bubble dynamic behavior are examined herein.

The bubble volume was determined to quantify the bubble nucleation process. The growth curves of bubbles on surfaces with and without pressure control are presented in Figure 5a,b. The bubble growth process includes two stages based on its volumetric variations: the nucleation and growth stages. In the first stage, the size and volume of bubbles are increasing. When heat is introduced, the stable bubble nucleus begins to enter the rapid growth stage; under this condition, the inflection point of the bubble growth curve corresponds to t_{in}. When the solid–liquid interaction intensity is low, the nucleation process is slow. After that, the bubbles begin to grow rapidly and maintain a certain stability.

The liquid near the bottom wall is heated by coming in contact with the substrate, which is closely related to energy transfer. Therefore, the liquid here is of great significance in the study of the boiling heat transfer mechanism. Figure 5c shows the change trend of liquid film temperature near the substrate during the time increasing process (z = 20–40 Å). The curves include three stages, marked in Figure 5a: Stage I refers to the equilibrium stage of the simulation. The temperature near the heated wall is balanced at 80 K in this stage. The bottom wall is heated at 2500 ps and the phase-change process begins. The liquid temperature increased quickly in the initial stage. Then in Stage II, horizontal fluctuation appeared. After the bubble grew and detached from the heated wall, there was no heat transfer between the near wall region and the bottom wall. The temperature fluctuated horizontally, and this period of activity is referred to as Stage III.

To compare the heat transfer of the four cases, the absorbed heat flux q in the near wall region of liquid argon was determined. The heat flux curves with and without pressure control are illustrated in Figure 5d. In the initial period, the heat flux is basically kept constant during this process. When the liquid is continuously heated, a group of liquid atoms absorb more heat, and bubble nucleation will occur when a certain critical value is reached. After that, the bubble volume increases continuously, and its interface with the liquid needs to absorb energy to evaporate, which will greatly increase the heat flux. It also shows that nucleate pool boiling can promote heat transfer on the solid surface, and it has the advantages of high efficiency and safety.

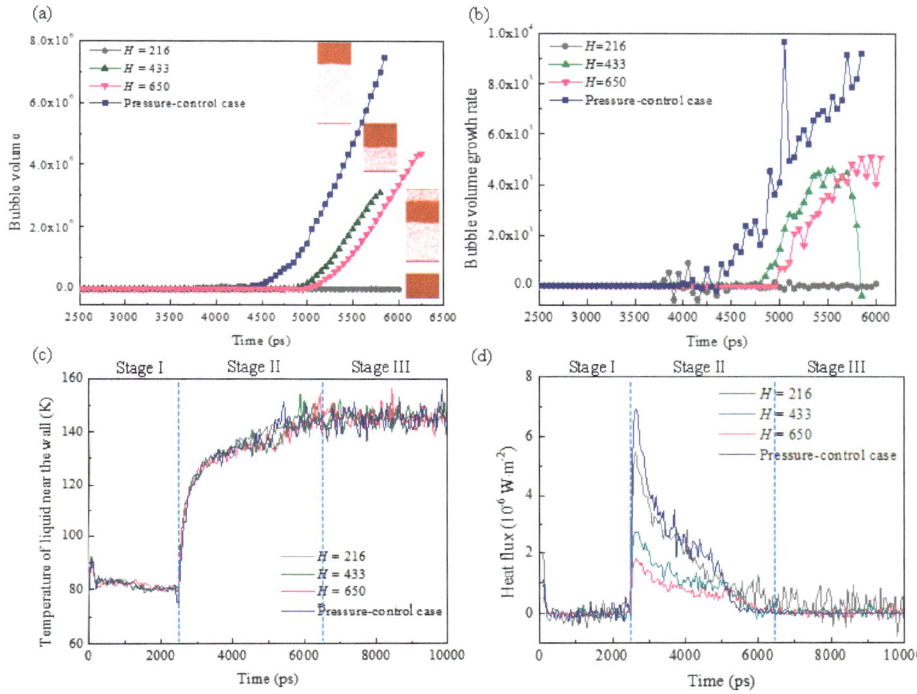

Figure 5. Quantitative investigation of the boiling process. (**a**) The time evolution of the bubble growth curve; (**b**) the bubble growth rate; (**c**) the time evolution of temperature in the liquid film near the bottom wall; (**d**) the average heat flux.

3.4. Mechanisms of Nucleate Boiling with Pressure-Control

3.4.1. Illustrations of Nucleate Boiling from the Perspective of Energy

To quantify the bubble nucleation of the pressure-control case, the energy was computed. The simulation domain was divided into bins with a size of 0.5 Å (x) × 0.5 Å (y) × 0.5 Å (z). The number density, E_k and E_p of argon atoms in each cell were determined. The mean total energy E of atoms can be calculated by adding E_k and E_p. According to the above discussion results, on the macro scale, the energy barrier of liquid gas phase transition belongs to latent heat, which can be overcome to produce phase transition after continuous heating and absorption of sufficient heat. According to the same idea, the macro energy theory and the micro kinetic energy related theoretical tools are jointly applied; when the average E_k of a group is higher than the corresponding E_p, bubble nucleation will occur. It is noteworthy that the E_p is less than zero, which reflects the restriction of interactions between atoms, and the value of E_k and E_p is affected.

The pressure profile change at the initial time of the non-equilibrium stage is shown in Figure 6 below. Analysis of this figure shows that, at the initial nucleation time, and the moment of bubble detachment, the initial positions of the solid-liquid interface are $z = 0$ Å. In the nonequilibrium phase, the density and energy in the corresponding liquid region basically remain uniform, with few fluctuations. Previous studies [22] have found that a small number of liquid atoms are always crystallized on the surface of hydrophilic substrate due to the interaction of solid and liquid atoms. In this case, the density of liquid atoms near the substrate is improved. In the analysis of boiling heat transfer mechanism, such liquid atoms are the focus of research.

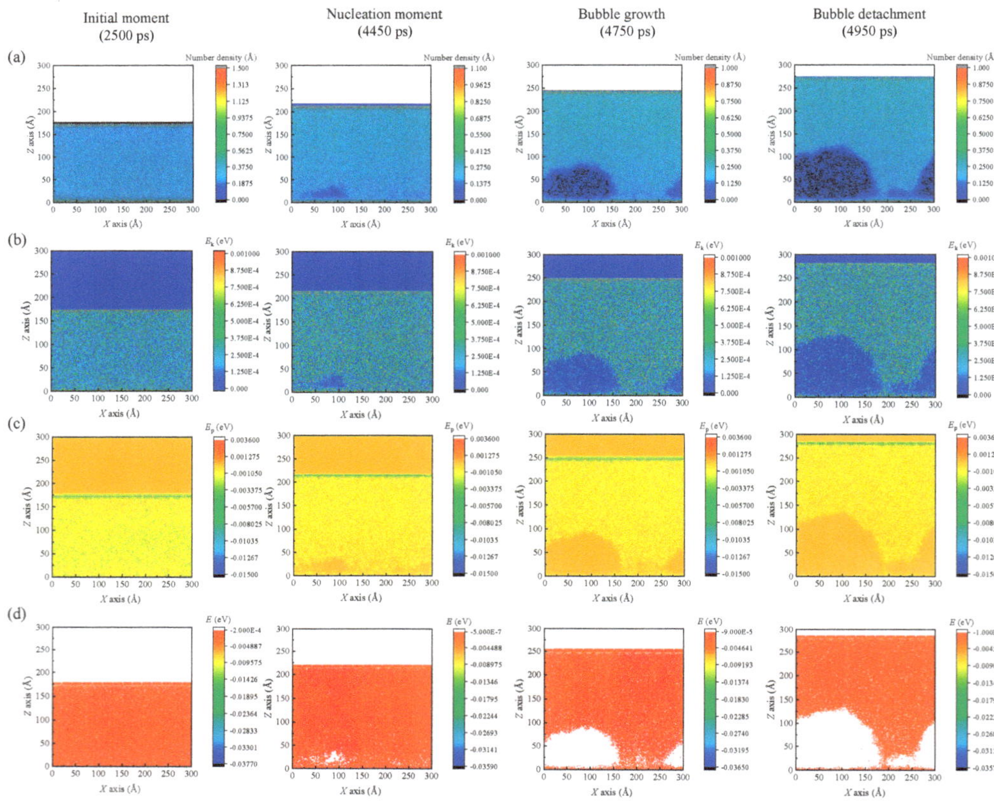

Figure 6. The contours of (**a**) number density, (**b**) kinetic energy (E_k), (**c**) potential energy (E_p), and (**d**) the total energy (E), at the initial moment (2500 ps), the nucleation moment (4450 ps), a certain bubble growth moment (4750 ps), and bubble detachment moment (4950 ps) of the pressure control case.

While time increases, the liquid film in the near wall area continuously absorbs heat. Some of the heat energy is converted into atomic kinetic energy, which will increase the average kinetic energy of the liquid atoms in this area, and some of the heat is converted into atomic potential energy, which will lead to the decrease of E_p, thus reducing the potential barrier of the horizontal potential energy. According to the above results, it can be judged that the average kinetic energy of liquid atoms in the near wall area will increase significantly during this process, which will lead to nucleating boiling phenomenon as shown in Figure 6d. The comparative analysis also shows that the bubble shape in the area with total energy greater than 0 eV is very similar to that in Figures 4 and 6a.

When the bubble nucleus appears in the near wall liquid, the potential energy barrier at the interface decreases significantly. In the process of continuous heat transfer, the kinetic energy of the liquid atoms at the interface of the bubble core continues to increase, along with the number of atoms overcoming the weak barrier, which also promotes the formation and growth of bubbles. After that, the bubbles enter a rapid growth stage and maintain certain thermodynamic stability. The heat transfer efficiency has a significant effect on the bubble growth rate. According to the above discussion, the mechanism of bubble nucleation and growth is closely related to energy, so it can be explained from an energy perspective.

3.4.2. Separate Energy Analyses in the Near-Wall Region

The atoms between the solid and liquid walls have a strong effect, which makes the atoms in them unable to evaporate due to a large force. Under the influence of this factor, the liquid layer is always confined to the substrate surface during the heating process. This also reflects the fact that solid-like liquid atoms do not play a role or participate in the bubble nucleation process. In this way, when analyzing the nucleation energy, the liquid film in the wall area (z = 0–20 Å) can be divided into two unrelated areas: a solid-like liquid layer (from z = 0–5 Å), and a bulk-liquid layer (z = 15–20 Å). See Figure 7a below for relevant information. Based on the analysis of the collected images, it is estimated that the thickness of such a liquid layer is 5 Å. The kinetic energy and total energy of these two regions as a function of time were further calculated separately and presented in Figure 7b–d.

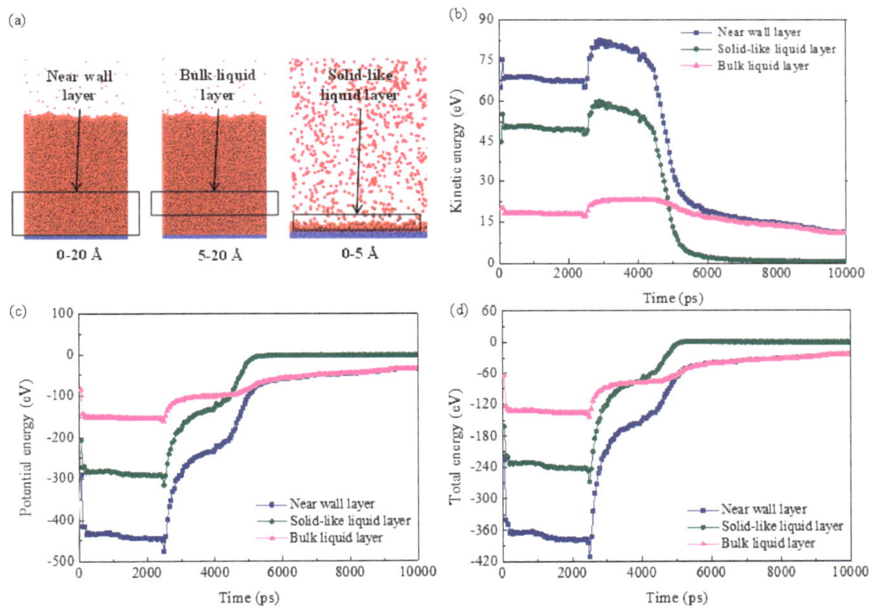

Figure 7. (**a**) The schematic of the liquid configuration in the near-wall region (from z = 0 Å to z = 20 Å), the bulk liquid layer (from z = 5 Å to z = 20 Å), and the solid-like liquid layer (from z = 0 Å to z = 5 Å). The time evolutions of (**b**) the kinetic energy, (**c**) the potential energy, and (**d**) the total energy in these three regions.

The body liquid layer near the wall is the key object of this analysis. Theoretical analysis shows that the heat transfer here is mainly related to the collision between solid and liquid atoms, and there is a positive correlation between the heat exchange efficiency and the collision probability. When the density of liquid atoms is large, the corresponding collision frequency also increases. There is a positive correlation between atomic density near substrate surface and collision frequency. According to the above discussion, it can be concluded that the solid liquid layer plays an important role in this interface heat transfer, which needs to be analyzed emphatically. The contrast analysis shows that the density of the solid liquid layer is higher, as shown in Figure 6a.

When the temperature of the bottom wall increases to the target value, the absorbed energy is mainly converted into atomic kinetic energy, which can increase the temperature of the liquid. However, during this process, the energy of the solid liquid layer has basically not changed. The liquid layer near the wall absorbs the most heat energy. It should also be noted that in various heating stages, the heat absorbed by the bulk liquid layer is mainly consumed to overcome the energy barrier, which also indicates that the near wall liquid layer should be the focus when analyzing the nucleation mechanism.

4. Conclusions

In summary, in this study, the molecular dynamics simulation method for nanoscale boiling heat transfer was modified by inducing a pressure-control wall. The pressure-control wall can move upwards or downwards according to the pressure of the simulation domain. Thus, the system pressure during the nanoscale phase-change process is able to be set and maintained at a given value. Moreover, the bubble nucleation and growth in nanoscale pool boiling were quantitatively investigated based on the pressure-control method. Mechanisms behind the nanoscale nucleate boiling were elaborated from the aspect of energies. This research is expected to improve current methods of pool boiling simulation. Moreover, the effects of the system pressure on bubble behavior and heat transfer of the nanoscale boiling can also be explored based on the proposed pressure-control method, which will be examined in future research.

Author Contributions: Methodology, L.G. and C.W.; writing—original draft preparation, Y.K. and Y.Y.; writing—review and editing, L.G. and Z.L.; funding acquisition, Z.L. All authors have read and agreed to the published version of the manuscript.

Funding: This work was financially supported by the National Natural Science Foundation of China under contract Nos. 52076113, and the Collaborative Innovation Project of Colleges in Jinan under Contract Nos. 2021GXRC059, and 2020GXRC045, and the International Cooperation Project of Science, Education, Industry Integration in Qilu University of Technology (No. 2022GH021).

Conflicts of Interest: The authors declare no conflict of interest.

Nomenclature

Ar	Argon atom
Cu	Copper atom
ε	The energy units, eV
σ	The distance units, Å
r	Inter-atomic distance, Å
r_c	The cut-off radius, Å
α	Energy parameter, eV
q	Heat flux, $W \cdot m^{-2} \cdot K^{-1}$
V	The volume of liquid computation domain, $Å^3$
N	The number of liquid argon atoms
v_i	The thermal velocity, Å/ps
e_i	The total energy of single atom, eV
F	The force acting on atom, eV/Å
F_{total}	The total force applied on the pressure-control wall, eV/Å
N_{up}	The number of atoms on the pressure-control wall
p	The pressure set in the simulation, bars
S_{xy}	The cross-sectional area in the x-y plane, $Å^2$
H	The height of the simulation box, Å
t_{in}	The inflection point of the bubble growth curve
E_k	The kinetic energy, eV
E_p	The potential energy, eV
E	The total energy, eV

References

1. Ma, L.; Zhang, T.; Zhang, X.L.; Wang, B.; Mei, S.W.; Wang, Z.F.; Xue, X.D. Optimization of parabolic though solar power plant operations with nonuniform and degraded collectors. *Sol. Energy* **2021**, *214*, 551–564. [CrossRef]

2. Manesh, M.H.K.; Aghdam, M.H.; Modabber, H.V.; Ghasemi, A.; Talkhoncheh, M.K. Techno-economic, environmental and energy analysis and optimization of integrated solar parabolic trough collector and multi effect distillation systems with a combined cycle power plant. *Energy* **2022**, *240*, 122499. [CrossRef]
3. Sonawane, C.R.; Panchal, H.N.; Hoseinzadeh, S.; Ghasemi, M.H.; Alrubaie, A.J.; Sohani, A. Bibliometric analysis of solar desalination systems powered by solar energy and CFD modelled. *Energies* **2022**, *15*, 5279. [CrossRef]
4. Sonawane, C.R.; Alrubaie, A.J.; Panchal, H.; Chamkha, A.J.; Jaber, M.M.; Oza, A.D.; Zahmatkesh, S.; Burduhos-Nergis, D.D.; Burduhos-Nergis, D.P. Investigation on the impact of different absorber materials in solar still using CFD simulations-Economic and environmental analysis. *Water* **2022**, *14*, 3031. [CrossRef]
5. Raul, A.; Jain, M.; Gaikwad, S.; Saha, S.K. Modelling and experimental study of latent heat thermal energy storage with encapsulated PCMs for solar thermal applications. *Appl. Therm. Eng.* **2018**, *143*, 415–428. [CrossRef]
6. Aftab, W.; Usman, A.; Shi, J.M.; Yuan, K.J.; Qin, M.L.; Zou, R.Q. Phase change material-integrated latent heat storage systems for sustainable energy solutions. *Energy Environ. Sci.* **2021**, *14*, 4268–4291. [CrossRef]
7. Wahile, G.S.; Malwe, P.D.; Aswalekar, U. Latent heat storage system by using phase change materials and their application. *Mater. Proc.* **2022**, *52*, 513–517. [CrossRef]
8. Chen, Y.J.; Zou, Y.; Wang, Y.; Han, D.X.; Yu, B. Bubble nucleation on various surfaces with inhomogeneous interface wettability based on molecular dynamics simulation. *Int. J. Heat Mass Transf.* **2018**, *98*, 135–142. [CrossRef]
9. Cui, Y.F.; Yu, H.Y.; Wang, H.J.; Wang, Z.Y.; Yan, X.L. The numerical modeling of the vapor bubble growth on the silicon substrate inside the flat plate heat pipe. *J. Heat Transfer.* **2020**, *147*, 118945. [CrossRef]
10. Datta, S.; Pillai, R.; Borg, M.K.; Sefiane, K. Acoustothermal nucleation of surface nanobubbles. *Nano Lett.* **2021**, *21*, 1267–1273. [CrossRef]
11. Dhillon, N.S.; Buongiorno, J.; Varanasi, K.K. Critical heat flux maxima during boiling crisis on textured surfaces. *Nat. Commun.* **2015**, *6*, 1–12. [CrossRef]
12. Han, H.X.; Merabia, S.; Muller-Plathe, F. Thermal transport at a solid-nanofluid interface: From increase of thermal resistance towards a shift of rapid boiling. *Nanoscale* **2017**, *9*, 8314–8320. [CrossRef]
13. Haramura, Y.; Katto, Y. A new hydrodynamic model of critical heat flux, applicable widely to both pool and forced convection boiling on submerged bodies in saturated liquids. *Int. J. Heat Mass Transf.* **1983**, *26*, 389–399. [CrossRef]
14. Jaikumar, A.; Kandlikar, S.G. Enhanced pool boiling heat transfer mechanisms for selectively sintered open microchannels. *Int. J. Heat Mass Transf.* **2015**, *88*, 652–661. [CrossRef]
15. Jaikumar, A.; Kandlikar, S.G. Ultra-high pool boiling performance and effect of channel width with selectively coated open microchannels. *Int. J. Heat Mass Transf.* **2016**, *95*, 795–805. [CrossRef]
16. Katto, Y.; Yokoya, S. Principal mechanism of boiling crisis in pool boiling. *Int. J. Heat Mass Transf.* **1968**, *11*, 993–1002. [CrossRef]
17. Kong, X.; Zhang, Y.H.; Wei, J.J. Experimental study of pool boiling heat transfer on novel bistructured surfaces based on micro-pin-finned structure. *Exp. Therm. Fluid Sci.* **2018**, *91*, 9–19. [CrossRef]
18. Kumar, G.U.; Suresh, S.; Thansekhar, M.R.; Babu, P.D. Effect of diameter of metal nanowires on pool boiling heat transfer with FC-72. *Appl. Surf. Sci.* **2017**, *423*, 509–520. [CrossRef]
19. Li, D.H.; Jiang, C.Y.; Cao, X.; Li, H.; Hekmatifar, M.; Sabetvand, R. Effect of cross-sectional area and number of Fe nanoparticles on the thermal behavior of pool boiling heat transfer of the water-based nanofluid: A molecular dynamics study. *Case Stud. Therm. Eng.* **2022**, *36*, 102242. [CrossRef]
20. Liang, G.T.; Mudawar, I. Review of pool boiling enhancement by surface modification. *Int. J. Heat Mass Transf.* **2019**, *128*, 892–933. [CrossRef]
21. Liu, B.; Yang, X.; Li, Q.; Chang, H.Z.; Qiu, Y. Enhanced pool boiling on composite microstructured surfaces with microcavities on micro-pin-fins. *Int. Commun. Heat Mass* **2022**, *138*, 106350. [CrossRef]
22. Liu, H.Q.; Ahmad, S.; Chen, J.T.; Zhao, J.Y. Molecular dynamics study of the nanoscale boiling heat transfer process on nanostructured surfaces. *Int. Commun. Heat Mass* **2020**, *119*, 104963. [CrossRef]
23. Lu, Y.W.; Kandlikar, S.G. Nanoscale surface modification techniques for pool boiling enhancement a critical review and future directions. *Heat Transf. Eng.* **2011**, *32*, 827–842. [CrossRef]
24. Nukiyama, S. Maximum and minimum values of heat q transmitted from metal to water under atmospheric pressure. *Int. J. Heat Mass Transf.* **1934**, *37*, 354–367. [CrossRef]
25. Patil, C.M.; Kandlikar, S.G. Pool boiling enhancement through microporous coatings selectively electrodeposited on fin tops of open microchannels. *Int. J. Heat Mass Transf.* **2014**, *79*, 816–828. [CrossRef]
26. Pham, Q.N.; Zhang, S.W.; Hao, S.; Montazeri, K.; Lin, C.H.; Lee, J.; Mohraz, A.; Won, Y. Boiling heat transfer with a well-ordered microporous architecture. *ACS Appl. Mater. Interfaces* **2020**, *12*, 19174–19183. [CrossRef]
27. Shi, L.; Hu, C.Z.; Yi, C.L.; Lyu, J.Z.; Bai, M.L.; Tang, D.W. A study of interface evolution-triggering different nucleate boiling heat transfer phenomenon on the structured surfaces. *Int. J. Heat Mass Transf.* **2022**, *190*, 122754. [CrossRef]
28. Shin, S.; Choi, G.; Rallabandi, B.; Lee, D.; Il Shim, D.; Kim, B.S.; Kim, K.M.; Cho, H.H. Enhanced boiling heat transfer using self-actuated nanobimorphs. *Nano Lett.* **2018**, *18*, 6392–6396. [CrossRef]
29. Song, Y.; Gong, S.; Vaartstra, G.; Wang, E.N. Microtube surfaces for the simultaneous enhancement of efficiency and critical heat flux during pool boiling. *ACS Appl. Mater. Interfaces* **2021**, *13*, 12629–12635. [CrossRef]

30. Van Erp, R.; Soleimanzadeh, R.; Nela, L.; Kampitsis, G.; Matioli, E. Co-designing electronics with microfluidics for more sustainable cooling. *Nature* **2020**, *585*, 211–216. [CrossRef]
31. Wang, Q.Y.; Chen, R.K. Ultrahigh flux thin film boiling heat transfer through nanoporous membranes. *Nano Lett.* **2018**, *18*, 3096–3103. [CrossRef]
32. Wang, Y.Q.; Lyu, S.S.; Luo, J.L.; Luo, Z.Y.; Fu, Y.X.; Heng, Y.; Zhang, J.H.; Mo, D.C. Copper vertical micro dendrite fin arrays and their superior boiling heat transfer capability. *Appl. Surf. Sci.* **2017**, *422*, 388–393. [CrossRef]
33. Wen, R.F.; Li, Q.; Wang, W.; Latour, B.; Li, C.H.; Li, C.; Lee, Y.C.; Yang, R.G. Enhanced bubble nucleation and liquid rewetting for highly efficient boiling heat transfer on two-level hierarchical surfaces with patterned copper nanowire arrays. *Nano Energy* **2017**, *38*, 59–65. [CrossRef]
34. Yi, C.L.; Hu, C.Z.; Shi, L.; Bai, M.L.; Lv, J.Z. Wettability of complex Long-Chain alkanes droplets on Pillar-type surfaces. *Appl. Surf. Sci.* **2021**, *566*, 150752. [CrossRef]
35. Yin, X.Y.; Hu, C.Z.; Bai, M.L.; Lv, J.Z. Effects of depositional nanoparticle wettability on explosive boiling heat transfer: A molecular dynamics study. *Int. Commun. Heat Mass* **2019**, *109*, 104390. [CrossRef]
36. Yin, X.Y.; Hu, C.Z.; Bai, M.L.; Lv, J.Z. Molecular dynamic simulation of rapid boiling of nanofluids on different wetting surfaces with depositional nanoparticles. *Int. J. Multiph. Flow* **2019**, *115*, 9–18. [CrossRef]
37. Yuan, B.; Liu, L.; Cui, C.Y.; Fang, J.B.; Zhang, Y.H.; Wei, J.J. Micro-pin-finned surfaces with fractal treelike hydrophilic networks for flow boiling enhancement. *ACS Appl. Mater. Interfaces* **2021**, *13*, 48189–48195. [CrossRef]
38. Zhang, S.W.; Chen, G.; Jiang, X.C.; Li, Y.J.; Shah, S.W.; Tang, Y.; Wang, Z.K.; Pan, C. Hierarchical gradient mesh surfaces for superior boiling heat transfer. *Appl. Therm. Eng.* **2023**, *219*, 119513. [CrossRef]
39. Zhou, W.B.; Fan, Q.X.; Zhang, Q.; Cai, L.; Li, K.W.; Gu, X.G.; Yang, F.; Zhang, N.; Wang, Y.C.; Liu, H.P.; et al. High-performance and compact-designed flexible thermoelectric modules enabled by a reticulate carbon nanotube architecture. *Nat. Commun.* **2017**, *8*, 1–9. [CrossRef]
40. Zhou, W.B.; Han, D.M.; Ma, H.L.; Hu, Y.K.; Xia, G.D. Molecular dynamics study on enhanced nucleate boiling heat transfer on nanostructured surfaces with rectangular cavities. *Int. J. Heat Mass Transf.* **2022**, *191*, 122814. [CrossRef]
41. Stukowski, A. Visualization and analysis of atomistic simulation data with Ovito-The open visualization tool. *Model Simul. Mater. Sci. Eng.* **2010**, *18*, 2154–2162. [CrossRef]
42. Zhang, L.Y.; Xu, J.L.; Liu, G.L.; Lei, J.P. Nucleate boiling on nanostructured surfaces using molecular dynamics simulations. *Int. J. Therm. Sci.* **2020**, *152*, 106325. [CrossRef]

energies

MDPI

Article

A New Two-Foci V-Trough Concentrator for Small-Scale Linear Fresnel Reflectors

Alberto Pardellas [1], Pedro Fortuny Ayuso [2], Luis Bayón [2,*] and Arsenio Barbón [1]

[1] Department of Electrical Engineering, University of Oviedo, 33003 Oviedo, Spain
[2] Department of Mathematics, University of Oviedo, 33003 Oviedo, Spain
* Correspondence: bayon@uniovi.es

Abstract: We present the design of an original secondary cavity for use in Small-Scale Fresnel Reflectors in photovoltaic applications. The cavity is similar to the classical V-trough, but the primary reflector system is configured so that there are two focal points on the aperture. The rays coming from each side of the primary system reach the opposite side of the cavity, producing a non-symmetrical distribution of the irradiance. This modifies the acceptance half-angle and allows us to break the maximum limit for the concentration ratio of ideal symmetric concentrators. Our study is analytic, and we provide formulas for any number of reflections. Numerical simulations with a ray-tracing program based on MATLAB are included. We provide a comparison of optical concentration ratio, height and cost parameter between our system and two classical designs with a single focal point: the V-trough and the Compound Parabolic concentrators. This way, we verify that our design yields better concentration ratios while keeping the ray acceptance rate at one. Our solution proves to be better than both the classical one-focus V-trough and the Compound Parabolic concentrator. Specifically, the proposed solution is significantly better than the classical one-focus V-trough in optical concentration ratio, with an increase between 15.02 and 35.95%. As regards the compound parabolic concentrator, the optical concentration ratio is always slightly better (around 4%). The height of the cavity, however, is notably less in this design (around 54.33%).

Keywords: V-trough concentrator; concentration ratio; small linear Fresnel reflector

1. Introduction

Human activity is the greatest source of greenhouse gas (GHG) emissions, as fossil fuels are the main energy source for these activities: hence, the emergence of the term Anthropocene to describe the human modification of the Earth's climate [1]. This has led to the organization of international meetings between representatives of most countries, the so-called Conferences of Parties (COP). At the last meeting (COP27), in Egypt in 2022, attended by 196 countries plus the European Union as a whole, stringent decisions to reduce global greenhouse gas emissions were supported [2].

Renewable energies are possibly the main solution to GHC emissions. In particular, decentralised energy systems with storage systems [3] give hope to meet the challenge [4].

Photovoltaic systems are one of the main solar energy technologies used to avoid climate change. A typical application of this technology is concentrator photovoltaic (CPV) systems. They can be grouped in three classes, according to what is called their geometric concentration ratio: the ratio between the area of the lens or primary mirror and the area of the PV cells. The classes are: low, medium and high concentration [5] systems. This work will focus on low concentration photovoltaic (LCPV) systems with a geometric concentration coefficient between 2 and 10 suns.

Among the types of concentrators used in the design of LCPVs [6], this work will focus on the small-scale linear Fresnel reflector (SSLFR). In addition to having a lower manufacturing cost than other solar concentrators, they showcase a well-proven technology that has been the subject of multiple studies (see [7–10], for instance). These reflectors concentrate the sunlight onto a secondary system by means of a row of longitudinal mirrors.

The geometric concentration ratio [11,12] of an SSLFR is a critical measure of its efficiency, especially when solar cells are used. Systems with a higher concentration ratio allow for less (or smaller) cells and prevent complications in the design of the primary field: the lower the concentration ratio, the thinner the mirrors must be, which makes the system either costlier or more difficult to maintain, or both (more mirrors are needed, hence more movable parts, etc.). Another important effect to avoid is a heterogeneous distribution of the flux of light [13], which causes inefficiencies and may lead to the appearance of hotspots, which can even lead to the total failure of the system as long as they persist.

Many concentration methods [14] have been proposed to improve the yield of solar systems. They can be divided in two large categories: nonimaging concentrators (which do not produce clearly defined images of the Sun on the absorber) and imaging ones (which form clear images). The former reflect all of the incident radiation towards the receiver as long as the incidence angle remains in a specific range (acceptance angle). The most important example of these systems is the compound parabolic concentrator (CPC). Non-imaging concentrators are usually the most appropriate for use in solar concentration (v.gr. concentrated photovoltaic systems). Imaging concentrators (such as parabolic reflectors or Fresnel lenses), for their part, provide wider acceptance angles, higher tolerances for imperfections and errors, higher solar concentrations, more uniform illumination of the receiver and greater design flexibility.

Madala and Boehm [15] provide a thorough review of solar concentrators, including the large family of CPCs (with different reflector and absorber geometries), Fresnel mirrors, V-trough concentrators, etc. Some conventional imaging type concentrators (such as parabolic troughs) are also covered in their study. More recent designs can be found in [16] (v.gr. concentrators with multi-surface and multielement combinations). As we propose a modification of the classical V-trough concentrator, we are only going to review this one.

Duffie [17] is one of the main references in this area: the acceptance angle (the angle such that any incident ray forming a lesser angle with the cavity gets to the absorber) is introduced in that work. He approximates the system using the tangents to a reference circle passing through one of the end points of the aperture. That reference [17] also contains the study of linear, two-dimensional, V-trough concentrators. He studies a system with two flat plate reflectors, an ideal concentrator perfectly aligned with the Sun and with a single reflection. V-trough concentrators have been considered the best in terms of uniformity of flux distribution [18]; their reduced complexity and lower manufacturing costs [19] also make them very convenient.

There is a good amount of literature on V-trough concentrators for different applications. Shaltout et al. [20] evaluated a V-trough concentrator on a photovoltaic system with two-axis solar tracing in a hot desert climate. Their system, which has a concetration ratio of 1.6, generates 40% more PV power than the same system without a concentrator. Different concentrator geometries, depending on the incidence angle of the solar irradiation as well as the effect of the wall angle of the trough, are considered by Solanki et al. [21], Maiti et al. [22], Chong et al. [23], Tina and Scandura [24], Singh et al. [19] and Al-Shohani et al. [25]. Recently, Al-Najideen et al. [26] proposed a new design by adding two additional elements to the Hollands concentrator, resulting in four symmetrical reflectors surrounding the PV cell. They call their design "Double V-trough Solar Concentrator". Our proposal follows their spirit, as we provide a new modification, specifically designed for SSLFR systems: a V-trough with two foci.

The optical behaviour of the cavity can be described using the method of images applied to V-trough linear concentrators by Duffie [17]. Other papers also present optimal designs of concentrators using analytical solutions of the equations describing the number of reflections of rays through the trough [27]. Fraidenraich [18,28,29] used the method of images, with an additional condition on the design (the illumination of the module's surface has to be uniform) to describe the optical behaviour of the class of V-trough cavities.

More recently, Tang [30] presented a detailed mathematical procedure for the design of a V-trough concentrator with attached solar cells. The solution is found using the method of images and determining the fraction of solar rays arriving at the cells after any number of reflections.

Another widely used approach uses ray-tracing techniques in order to design, simulate and optimize different types of V-trough concentrators. For instance, Chong et al. [23] use this technique implemented in Microsoft Visual C++. Maiti et al. [22] use Monte Carlo ray-tracing, as does Paul [31]. Narasimman and Selvarasan [32] use the software Trace-Pro to simulate. Finally, Hadavinia y Singh [33] use Comsol.

The concentration ratio of V-trough concentrators depends on the acceptance half-angle θ_c: the largest incidence angle for which no radiation is rejected. When the design of the system is symmetric and the distribution of irradiance going into the concentrator is uniform for any θ with $|\theta| \leq |\theta_c|$, it is well-known [17] that a two-dimensional (linear) concentrator (such as the V-trough) has a concentration ratio bounded by $C^{2D}_{ideal} = \sin^{-1}\theta_c$. In our design, we take advantage of the nature of an SSLFR in order to develop a configuration of the primary reflector system which gives rise to a non-uniform distribution of the irradiance on the cavity with two different focal points, so that each side of the SSLFR is focused on one of them.

As explained above, many authors have tried to determine the acceptance rate η_{ray} using both analytical and statistical methods. We have a different objective: to maximize the concentration ratio under the condition $\eta_{ray} = 1$ for all the incidence angles less than the acceptance half-angle θ_c. We do not use the method of images, only planar mirror geometry, and we compute the optimal design using closed-form analytic methods. In our study, we can analyze any number of reflections. The advantage over ray-tracing is obvious: we provide a universal method for computing the optimal design, regardless of simulations.

The inspiration for this new design comes from a previous study by some of the authors [34]. In that paper, dedicated to the application of SSLFRs to illumination by means of optic fiber, two non-symmetrical trapezes were joined along their vertical side, in order to construct what can be called a "half-trough concentrator". From the design and the specific functioning of the SSLFR, the reflected rays reach each of the two cavities from the corresponding side of the primary field. That way, light from a wider inlet aperture was concentrated into a narrower area (the absorber). What we have realized is that this design can be improved, removing the vertical wall but keeping the two different foci of the (previous) cavities. This reduces the number of reflections required for a ray to reach the absorber area (by a little less than one-half), and the amount of material required to build the concentrator. Thus, we both improve the efficiency and reduce costs. Furthermore, we carry out the whole theoretical study with analytical solutions in this paper, whereas in [34] only simulations with Maltab were performed.

Each focus of the receiver cavity is reached by rays satisfying the requirement $\theta \leq \theta_c$ on each side. These conditions are weaker than the global condition $|\theta| \leq |\theta_c|$, the one governing the usual one-focus half-trough design. The breakthrough is that this change of condition for the acceptance angle provides a concentration ratio greater than the one for symmetric concentrators with uniform irradiance distribution (one focal point)namely C^{2D}_{ideal}, and this happens for any acceptance half-angle. As far as we know, there is no such result in the literature.

All the previous literature covers the single-focus V-trough concentrator and the compound parabolic concentrator. We have not found any reference dealing with a double-foci V-trough concentrator.

Thus, the aim of this work is very specific. Starting from a two-focus configuration that causes a non-symmetric distribution of irradiance at the secondary reflector aperture, the cavity will be designed with the following property: it must maximise the geometrical concentration ratio under the condition that the ray acceptance rate is equal to one for all angles of incidence less than the half-angle of acceptance.We do not impose any restriction on the height of the cavity.

The specific contributions of this study can be summarized in the following proposals:

(i) We propose a secondary cavity prepared for low-concentration photovoltaic systems based on SSLFRs with a large concentration ratio and non-uniform solar irradiance distribution.
(ii) The design obtains a concentration ratio greater than the maximum possible for systems with uniform irradiance distribution [17].

The paper is organized as follows: Section 2 contains the basic notions about concentrators. A brief description of the SSLFR for concentration PV applications with the two-foci design and the geometric design of the two-foci V-trough concentrator is given in Section 3. Section 4 includes numerical results and validations of the proposed design, which is compared with other classical concentrators (CPC and V-trough) in Section 5. Finally, Section 6 contains a summary of the main conclusions.

2. Main Parameters of a Concentrator

Duffie [17] states the fundamental problem in the design of concentrators as follows: "How can radiation which is uniformly distributed over a range of angles $|\theta| \leq |\theta_c|$ and incident on an aperture of area A_a, be concentrated on a smaller absorber area A_{abs}, and what is the highest possible concentration". He is using the most common definition of concentration ratio, the area or geometric concentration ratio:

$$C_a = \frac{\text{aperture area}}{\text{absorber area}} = \frac{A_a}{A_{abs}} \tag{1}$$

Notice that in his statement, he assumes that the radiation is uniformly distributed (and uniformly reflected). From this hypothesis, it follows that this ratio has an upper limit which depends on whether the concentrator is three-dimensional (spherical) or a two-dimensional (linear) concentrator, such as our V-trough design. From the Second Law of Thermodynamics, Rabl concludes that the maximum possible concentration ratio for a given acceptance half angle θ_c is:

$$C^{2D}_{ideal} = \sin^{-1}\theta_c \tag{2}$$

For two-dimensional (trough-like) concentrators, as he remarks, a concentrator is ideal if and only if the exchange factor that measures the radiation going from absorber to the source is one. It is known that compound parabolic concentrators (CPC) actually reach this limit, so that they have been called "ideal concentrators" [17].

In addition to this index, there are other indices that measure the goodness of a concentrator, such as the flux concentration ratio. It is defined as the ratio of the average energy flux on the receiver to that on the aperture; the local flux concentration ratio which is the ratio of the flux at any point on the receiver to that on the aperture and which varies across the receiver. In order to avoid confusion, we will use the following notation (commonly used in the literature, e.g., in [24,25] or [33]):

$$C_{opt} = \frac{\text{flux at the receiver}}{\text{flux at the absorber}} = C_a \cdot \eta_{ray} \tag{3}$$

where C_{opt} is the optical concentration ratio, C_a is the area concentration ratio, and η_{ray} is the ray acceptance rate, defined as the fraction of incident light rays reaching the absorber.

Duffie [17] presented the following equation for the classical "single-focus" (so to say) V-concentrator, considering: ideal concentrator, perfectly aligned with the Sun, and with a single reflection:

$$C_a = 1 + 2\cos(2\Phi) \tag{4}$$

where Φ is the trough angle or half angle of the V-shaped cone. Equation (4) has also been used by [24]. Duffie [17] also used the concept of the half angle of the V-shaped cone, obtaining the following equation:

$$C_a = \frac{1}{\sin(\theta_c + \Phi)} \tag{5}$$

Equation (5) has also been used by [15].

Fraidenraich [28] presented a paper in which he showed that the optical concentration ratio can be approximated by a function of two parameters: the ray acceptance rate and the average number of reflections, n. In Fraidenraich [18], the main hypothesis is the condition of uniform illumination of the absorber's surface within an angular interval of light incidence. Using it, he provides analytical expressions relating C_a and Φ and also a cost analysis. Tang [30] presents a study where he considers Φ and C_a as independent parameters that determine the geometry of a V-trough concentrator. Rabl [35] shows that slight errors in the calculation of n are almost irrelevant on the final value of ρ_m^n.

Finally, thereflector-to-aperture area ratio parameter is defined:

$$R_a = \frac{\text{reflector area}}{\text{aperture area}} = \frac{A_r}{A_a} \tag{6}$$

It can be intuited that the height of the cavity plays a key role. The R_a parameter is necessary for cost analysis. For example, the high efficiency of ideal CPC concentrators has as a negative trade-off their high R_a.

3. Design of a Two-Foci V-Trough Concentrator

In this section, we provide a detailed description of the optimal design of the two-foci V-trough cavity, using analytical formulas exclusively.

3.1. Brief Description of the SSLFR with Two-Foci V-Trough

An SSLFR consists of a set of flat mirrors (the primary reflector system) concentrating the direct solar irradiance onto an element with much smaller area which, in our case, is a row of PV cells. The primary reflector system contains a set of parallel stretched mirrors mounted on a frame; in order to follow the Sun's motion, in our design, each mirror can rotate in the north–south axis. A secondary reflector system—a reflective cavity— is positioned so that the irradiance reflected from the primary system, which does not fall directly on the PV cells, is reflected again and directed towards them.

Figure 1 shows the schematics of the SSLFR: notice the symmetry of the system (except for the orientation of the mirrors). Its main constructive magnitudes are: mirror width (W_M), height to the receiver (f), separation between two consecutive mirrors (d), distance from the mirror centers to the center of SSLFR (L_i), width of the PV cells (b), aperture of the secondary reflector system (the V-trough cavity) (B) and number of mirrors on each side of the SSLFR ($N_r = N_l$, which we will call N, as we assume the same number of mirrors on each side). The secondary cavity is symmetric with respect to the central axis, but there are two different focal points for the optical system: F_1 and F_2, one on each side.

For each side of the SSLFR, the angle between the vertical line through the focal point and the line connecting this point with the center of the i-th mirror is:

$$\beta_i = \arctan\frac{L_i}{f};\ 1 \le i \le N \tag{7}$$

The maximum β_i on each side (that is, $\beta_{N_r} = \beta_{N_l}$) is the acceptance angle of the secondary cavity:

$$\theta_c = \beta_N \tag{8}$$

Finally, notice that we are not fixing (at all) the height of the secondary cavity (H, as we will see later): in fact, this is one of the most important variables in our design, as a large value implies a big concentrator, which is undesirable.

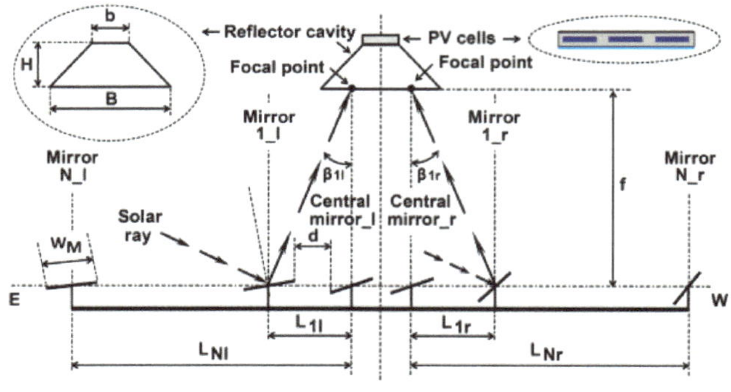

Figure 1. SSLFR with two-foci V-trough secondary concentrator.

3.2. *Two-Foci V-Trough Reflector*

Consider a classical V-trough cavity such as the one depicted in Figure 2. There are two linear side walls (PQ and $P'Q'$) which concentrate light from the wider inlet opening PP' towards the narrower absorber area QQ'. Four parameters are considered in this study: the incidence angle of each ray θ_i on the cavity aperture B, the height H of the cavity and the trough wall angle τ. Note that the upper width $QQ\prime$ of the cavity is not a free parameter but a constraint because it is equal to the width of the PV cells. Notice that our angle τ is the complement of Hollands' and Rabl's Φ. The central axis OR will be the reference axis for angles, and we will consider θ_i to be positive for rays coming from the left side and negative for those coming from the right. Using the notation of Figure 2:

$$\theta_i = \alpha_0 \tag{9}$$

For simplicity, we will denote the angle between the ray reaching the cavity and OR as α_0 (i.e., $\alpha_0 = \theta_i$), and each of its successive reflections will be α_j, for $j = 1, 2, \ldots$

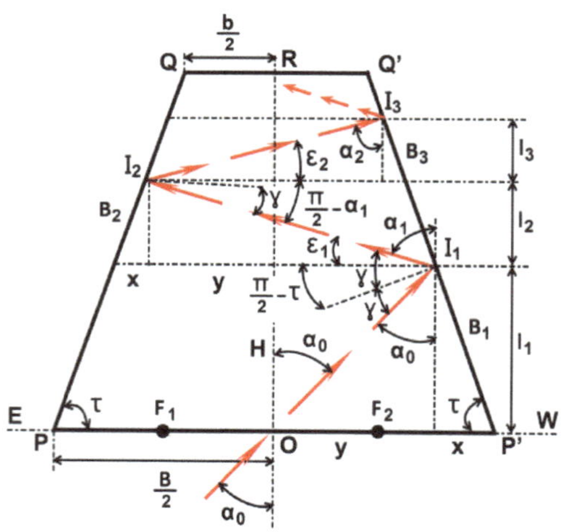

Figure 2. Two-foci V-trough (Case A).

As already stated, the key to our design is to assume that on each side of the concentrator only the rays coming from the same side of the primary reflector system arrive, each with an incidence angle θ_i. The position and orientation of the mirrors of the primary field create two focal points F_1 and F_2, one for each set of mirrors on each side of the field; the left F_1 and F_2 are on the midpoint of each of the half-bases of B. For simplicity, we will speak of the right and left sides of the cavity, separated by the axis OR despite there being no physical separation.

Of course, our design still aims at computing an acceptance angle θ_c such that the ray acceptance rate η_{ray} is one, and thus $C_{opt} = C_a$. However, we do not impose the classical condition $|\theta_i| \leq |\theta_c|$, but instead:

$$\begin{array}{ll} 0 \leq \theta_i \leq \theta_c & \text{on the left side of the cavity} \\ 0 \geq \theta_i \geq -\theta_c & \text{on the right side of the cavity} \end{array} \tag{10}$$

We will only state the left-side case (with rays coming from the right of the SSLFR focused on F_1). In these terms, the problem can be stated as: given b, in order to find the maximum C_a, we will maximize B under the restriction that all the rays reaching PP', after a number of reflections, get to the PV cell (whose width is b), that is $\eta_{ray} = 1$:

$$\max C_a = \max B; \; 0 \leq \theta_i \leq \theta_c \tag{11}$$

The following property is key to finding the optimal design.

Property 1. *In order to achieve* (11)*, the optimal solution of the most unfavorable case is that in which the vertical component of each reflection on the walls (if there are any) is largest and touches the base b either on Q or Q′.*

We make use of Property 1, forcing the reflected rays to be as high as possible and to reach the corners Q or Q' of the basis b. However, we are no longer in a symmetric geometry, and we have two different cases to consider (see Figures 2 and 3) :

Case A: ray passing through O with $\theta_i = \theta_c$;

Case B: ray falling on P with $\theta_i = 0$.

If we make these two rays reach b, any other ray will also, and usually with less reflections. From Figures 2 and 3, one can obtain the formulas for each number n of reflections.

Let us study each of the cases A and B separately.

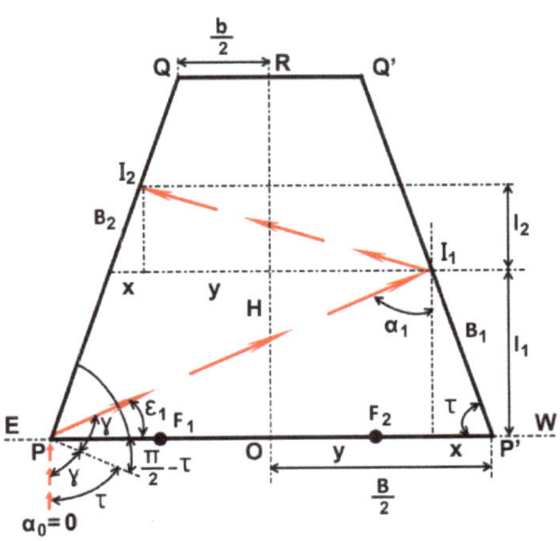

Figure 3. Two-foci V-trough (Case B).

3.2.1. Case A

Using the Law of Reflection, if n is the number of reflections required to reach the PV cells and starting at $\theta_i = \alpha_0$, the following equalities follow (see Figure 2):

$$\alpha_1 = (\pi - 2\tau) + \alpha_0;\ \alpha_2 = (\pi - 2\tau) + \alpha_1;\ \cdots;\ \alpha_n = (\pi - 2\tau) + \alpha_{n-1} \tag{12}$$

As for the angle between PP' and the $i-$th reflection, called ε_j:

$$\varepsilon_1 = (2\tau - \pi/2) - \alpha_0;\ \varepsilon_2 = (2\tau - \pi/2) - \alpha_1;\ \cdots;\ \varepsilon_n = (2\tau - \pi/2) - \alpha_{n-1} \tag{13}$$

Finally, the vertical lengths traveled by the reflected ray l_i after each reflection are given by the following equations:

For $i = 1$ we obtain:

$$l_1 = \frac{B/2}{\cot\tau + \tan\alpha_0} \tag{14}$$

For the following ones:

$$l_2 = \frac{B - 2(l_1)\cot\tau}{\cot\tau + \tan\alpha_1};\ l_3 = \frac{B - 2(l_1 + l_2)\cot\tau}{\cot\tau + \tan\alpha_2};\cdots$$

$$\cdots;\ l_n = \frac{B - 2\sum_{i=1}^{n-1}(l_i)\cot\tau}{\cot\tau + \tan\alpha_{n-1}} \tag{15}$$

We now have the tools required for describing the algorithm which gives the optimal design (11). The nested structure of the formulas leads to an easily implementable method.

Where n is the number of reflections on the lateral walls, the algorithm considers different cases A_n, depending on n. The height H_n of the cavity for n reflections is:

$$H_n(B) = \sum_{i=1}^{n} l_i \tag{16}$$

substituting into

$$B = b + 2H_n(B)\cot\tau \tag{17}$$

and solving for B, we obtain:

$$\begin{aligned}
&A_1 : B_1(\tau) = b + b\cot(\alpha_0)\cot(\tau)\\
&A_2 : B_2(\tau) = -b\cos(\alpha_0 - 3\tau)\csc(\alpha_0)\csc(\tau)\\
&A_3 : B_3(\tau) = b\cos(\alpha_0 - 5\tau)\csc(\alpha_0)\csc(\tau)\\
&\cdots\\
&A_n : B_n(\tau) = (-1)^{n-1}b\cos(\alpha_0 - (2n-1)\tau)\csc(\alpha_0)\csc(\tau)
\end{aligned} \tag{18}$$

The case $n = 1$ has no physical meaning, as $B \to \infty$ when $\tau \to 0$. The rest of the cases are possible, though. The algorithm finishes by maximizing, using numerical methods, the transcendental equations for $B_n(\tau)$, thus finding the optimal design angles τ_n^* which give the maximal C_a. Some qualitative properties can be deduced:

(i) The optimal value of τ_n^* increases with n, and $\tau^* \to 90^\circ$ as $n \to \infty$.
(ii) The optimal value for C_a is reached for $n = 3$ (A_2) and decreases afterwards asymptotically towards (2).

For each case A_n, the number of reflections is $n - 1$. We will see this in detail in the next section when we show the example.

3.2.2. Case B

In this case, we just need to set $\theta_i = \alpha_0 = 0$ in Formulas (12) and (13) to obtain (see Figure 3):

$$\alpha_1 = (\pi - 2\tau);\ \alpha_2 = (\pi - 2\tau) + \alpha_1;\ \cdots;\ \alpha_n = (\pi - 2\tau) + \alpha_{n-1} \tag{19}$$

$$\varepsilon_1 = (2\tau - \pi/2);\ \varepsilon_2 = (2\tau - \pi/2) - \alpha_1;\ \cdots;\ \varepsilon_n = (2\tau - \pi/2) - \alpha_{n-1} \tag{20}$$

However, the vertical lengths l_i of the i−th reflected ray are different from Equation (15). The vertical lengths traveled by the reflected ray l_i after each reflection are given by:

$$l_1 = \frac{B}{\cot\tau + \tan\alpha_1};\ l_2 = \frac{B - 2(l_1)\cot\tau}{\cot\tau + \tan\alpha_2};\ l_3 = \frac{B - 2(l_1 + l_2)\cot\tau}{\cot\tau + \tan\alpha_3};\ \cdots$$
$$\cdots;\ l_n = \frac{B - 2\sum_{i=1}^{n-1}(l_i)\cot\tau}{\cot\tau + \tan\alpha_n} \tag{21}$$

Reasoning as above, stating each case B_n and substituting the height H_n

$$H_n(B) = \sum_{i=1}^{n} l_i \tag{22}$$

into:

$$B = b + 2H_n(B)\cot\tau \tag{23}$$

and solving for B, we obtain:

$$\begin{aligned}
&B_1 : B_1(\tau) = b(1 - 2\cos(2\tau))\\
&B_2 : B_2(\tau) = b(1 - 2\cos(2\tau) + 2\cos(4\tau))\\
&B_3 : B_3(\tau) = b(1 - 2\cos(2\tau) + 2\cos(4\tau) - 2\cos(6\tau))\\
&\quad \cdots\\
&B_n : B_n(\tau) = b\Big(1 - 2\sum_{i=1}^{n}(-1)^i\cos(2i\tau)\Big)
\end{aligned} \tag{24}$$

First of all, notice that this family of functions does not depend on $\alpha_0 = \theta_c$. Secondly, and as the main result, when increasing the number of reflections n, we obtain a family of functions whose maximum (without physical meaning) is the asymptotic value:

$$\max B_n(\tau) = \lim_{\tau\to 90^o} B_n(\tau) \tag{25}$$

As a consequence, the concentration ratios C_a tend to

$$\lim C_a = 3, 5, 7, \ldots \tag{26}$$

as we increase the number of reflections $n = 1, 2, \ldots$ (obviously, allowing for $H_n \to \infty$).

We think it is remarkable that the classical formula of Hollands for an ideal concentrator which is perfectly aligned with the Sun and with a single reflection:

$$C_a = 1 + 2\cos(2\Phi) \tag{27}$$

is just a particular case of our family $B_n(\tau)$: specifically, $B_1(\tau)$, as one can verify readily because $\Phi = \pi/2 - \tau$. As a consequence, our study generalizes to any number of reflections the case of an ideal concentrator (perfectly aligned with the Sun).

From the above, it follows that the aim of this case B is no longer to maximize the function $B_n(\tau)$ but to choose the optimal solutions among the "candidate solutions" A_n.

To this end, we need to obtain the general expressions for the heights of the cavity in each case B_n. The simplest way is to use:

$$H_n = \frac{B_n - b}{2}\tan(\tau) \tag{28}$$

which gives:

$$\begin{aligned}
B_1 &: H_1(\tau) = -b\cos(2\tau)\tan(\tau) \\
B_2 &: H_2(\tau) = -b(\cos(2\tau) - \cos(4\tau))\tan(\tau) \\
B_3 &: H_3(\tau) = -b(\cos(2\tau) - \cos(4\tau) + \cos(6\tau))\tan(\tau) \\
&\dots \\
B_n &: H_n(\tau) = -b\tan(\tau)\textstyle\sum_{i=1}^{n}(-1)^{i-1}\cos(2i\tau)
\end{aligned} \tag{29}$$

As the value of b is fixed, using the envelope of this family of curves $H_n(\tau)$, for each value of τ we obtain the largest height under the condition $\eta_{ray} = 1$. This way it becomes easier to verify if the optimal candidate solutions (the values τ_n^* and $H(\tau_n^*)$) computed for the cases A_n also satisfy this case B. The example provided in Section 5 clarifies this step.

The process can be made as long as desired, and we can choose the optimal design depending on the number of reflections n. Qualitatively, the main result is that the larger n is, the larger B_n is, so that C_a increases as well. Actually, C_a tends asymptotically to the ideal value (2).

3.3. *Number of Reflections in the Two-Foci V-Trough*

Finally, in order to compute the approximate value of n, we will use the property proved by Rabl [35] that the average number of reflections in a V-trough is essentially the same as those in a CPC. Thus, we consider a truncated CPC with the same height as our two-foci V-trough, starting with a whole CPC designed for the specific value of θ_c.

We must not forget that the influence of n on the factor ρ_m^n is rather small because ρ_m is always very near to one.

4. Numerical Results and Validation

In this section, we present an example in order to clarify the method and also as a verification of our results. We set $b = 10$ (cm) and $\theta_c = \alpha_0 = 30°$ (the acceptance angle), a plausible value for the typical dimensions of an *SSLFR* [36]. All the computations have been carried out on a budget PC using the Mathematica™ Computer Algebra System.

We start by computing the candidates to the optimum of case A. Table 1 shows the concentrations C_{opt}, optimal angles τ^* and heights corresponding to each A_n. Recall that as our method ensures that $\eta_{ray} = 1$, we always have $C_{opt} = C_a$. The lack of influence of the longitudinal study implies that $B = b \cdot C_a$.

Table 1. Example of optimal design of a two-foci V-trough.

Case A	A_2	A_3	A_4	A_5	A_6	A_7	A_8
C_{opt}	2.146	2.047	2.023	2.014	2.009	2.007	2.005
τ^* (°)	67.36	77.49	81.25	83.25	84.50	85.36	85.98
H (cm)	13.74	23.59	33.23	42.82	52.40	61.97	71.53
Case B	B_1						

Figure 4 shows in a clearer way the evolution of the three parameters above. Notice how H is linear in n, but τ^* is (obviously) not, as it tends asymptotically to 90°.

The largest value of C_{opt} (these are computed numerically) corresponds to the first physically possible case, A_2. The values decrease with n and approach the ideal value (2) asymptotically from above (in this specific example, $\sin^{-1}\theta_C = 2$). This phenomenon happens for any value of θ_c and is quite relevant, as it implies the number of reflections n is less (so that ρ_m^n is greater) and also a lesser height H of the cavity (and, hence, less R_a).

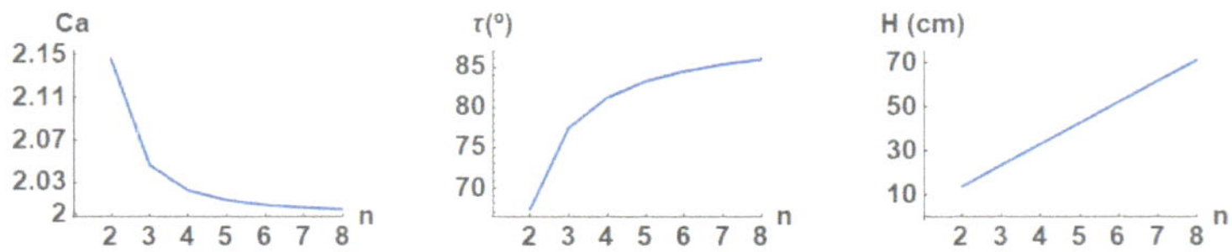

Figure 4. Optimal candidate solutions of case A.

Finally, we need to check whether the solutions above are also valid for some case B. To this end, we use the family of curves $H_n(\tau)$ given by (29) (Figure 5 contains the first four of these). For a fixed b, this sequence of functions is valid for any value θ_c, which simplifies the computations. Notice how as τ increases, the largest height (which is given by the envolvent of the curves) is reached for a greater number of reflections n. Even though the envolvent gives the maximum value of H, there may be cases B_n with less n which also satisfy the condition (which is good, as it means a lesser number of reflections).

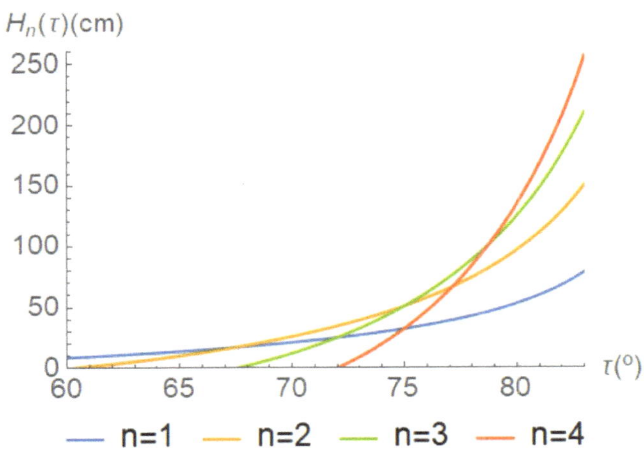

Figure 5. Choosing from the optimal candidates A_n using case B.

Thus, the last step of the optimization algorithm consists of taking the first optimal solution A_n (for decreasing values of C_{opt}) which also satisfies the condition of case B. In this example, for A_2, we have $\tau^* = 67.36$, $H = 13.74$, and

$$H_1(\tau^*) = 16.87 > H = 13.74 \tag{30}$$

So the best candidate A_2 is also valid because the conditions of case B_1 hold. However, this might not always be the case, as we will see later.

Ray Tracing Simulation and Verification

We verified our results using a Matlab™ray-tracing program which models solar power optical systems [37,38], using geometric optics. This program has already been used in other studies [38,39].

Figure 6 contains the simulation of our two-foci V-trough concentrator for $\alpha_0 = 30°, 40°$ and $50°$. Notice how for $\theta_i \leq \theta_c$ (the incidence angle), all the rays reaching the base B end up on the cells at b, as shown in Figure 6a. For $\theta_i > \theta_c$, part of the rays entering the cavity end up at b (Figure 6b), and finally, the worst case happens for $\theta_i \gg \theta_c$, when no ray entering the cavity reaches b.

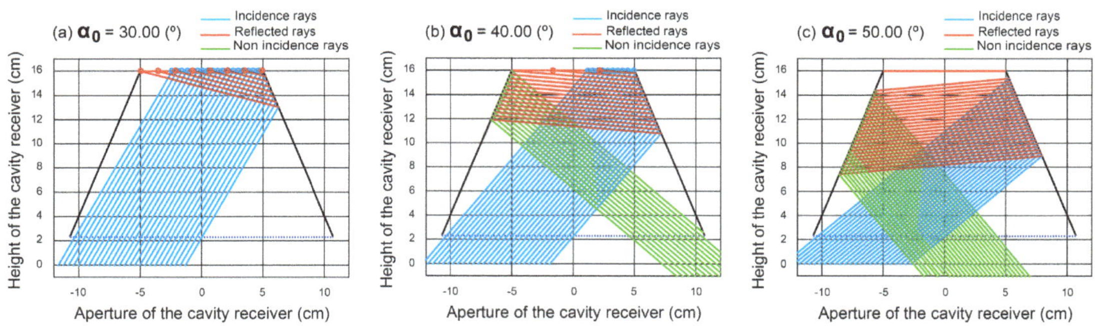

Figure 6. Two-foci V-trough: several examples of ray-tracing.

In Figure 7 we provide an analysis of the evolution of the ray acceptance rate η_{ray} for angles θ_i greater than θ_c. Recall that in ideal *CPCs*, this value goes from one to zero instantaneously [33]. In the classical V-trough [33], η_{ray} decreases depending strongly on the incidence ray and the design of the cavity. One can see that in our model, one goes from $\eta_{ray} = 1$ for $\theta_i \leq \theta_c$ to decreasing values in a progressive but not too sharp a way. This is, in our opinion, another strength of our proposed design.

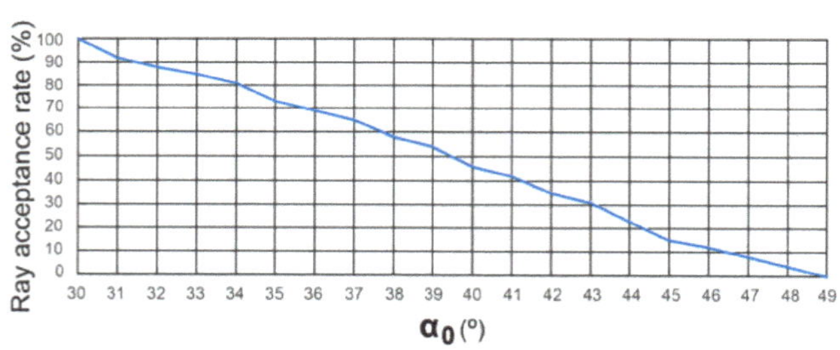

Figure 7. Plot of η_{ray} in two-foci V-trough for 30° = θ_c <θ_i < 50°.

As a cost analysis, we compute the reflector-to-aperture area ration R_a, disregarding the length, which has no relevance:

$$\left.\begin{array}{l} A_r = 2\dfrac{H}{\sin\tau^*} = 2.977 \\ A_a = 2.146 \end{array}\right\} \rightarrow R_a = \frac{A_r}{A_a} = 1.387 \tag{31}$$

5. Comparison with Other Classical Receivers

Finally, we compare our design with two different classical concentrators: the classical V-trough and the ideal CPC. We do this for several acceptance angles θ_c.

5.1. One-Focus V-Trough

For ease of comparison with the two-focus design proposed here, the classical one-focus V-trough design is shown. Using Property 1, we start the iterative algorithm stating a sequence of different cases C_n for an increasing number of reflections n. We will use, in each of them, the worst-case condition $\theta_i = \theta_c$. For each case C_n, the height of the cavity H_n can now be computed using (12), (13) and (21):

$$H_n(B) = \sum_{i=1}^{n} l_i \tag{32}$$

Now, we substitute the value of $H_n(B)$ into the formula relating b, B and H with τ:

$$B = b + 2H_n(B)\cot\tau \tag{33}$$

and, solving for B, after some easy computations, we obtain:

$$\begin{aligned}
C_1 &: B_1(\tau) = -b\cos(\alpha_0 - 3\tau)\sec(\alpha_0 - \tau) \\
C_2 &: B_2(\tau) = b\cos(\alpha_0 - 5\tau)\sec(\alpha_0 - \tau) \\
C_3 &: B_3(\tau) = -b\cos(\alpha_0 - 7\tau)\sec(\alpha_0 - \tau) \\
&\dots \\
C_n &: B_n(\tau) = (-1)^n b\cos(\alpha_0 - (2n+1)\tau)\sec(\alpha_0 - \tau)
\end{aligned} \tag{34}$$

This gives the functions $B_n(\tau)$ analytically in terms of b and α_0. In the last step of the algorithm, we need to compute the maximum of those $B_n(\tau)$ in order to obtain the optimum angles τ_n^* maximizing B and hence C_a.

The process can be made as long as desired, and we can choose the optimal design depending on the number of reflections n. From the qualitative point of view, the main result is that the larger n, the larger B_n, so that C_a increases as well. Actually, C_a tends asymptotically to the ideal value (2). In each case C_n, the number of reflections is n.

Recall that our design guarantees $\eta_{ray} = 1$. Table 2 and Figure 8 summarize our results for $b = 10$ and $\theta_c = 30°$.

Table 2. Example of optimal design of a V-trough.

Case C	C_1	C_2	C_3	C_4	C_5	C_6	C_7	C_8
C_{opt}	1.369	1.535	1.631	1.694	1.738	1.772	1.797	1.818
τ^* (°)	76.44	80.75	82.96	84.31	85.22	85.88	86.38	86.77
H (cm)	7.66	16.43	25.55	34.83	44.18	53.59	63.03	72.49

Notice how C_n approaches the ideal value $C_{opt} = \sin^{-1}\theta_C = 2$ but forcing the walls to become practically vertical ($\tau \to 90°, \Phi \to 0°$). This follows easily from (5) (see [35]):

$$C_a = \frac{1}{\sin(\theta_C + \Phi)} \tag{35}$$

Quoting [35], "We see that a V-trough can, at least in principle, approach the ideal concentration if it is very narrow, that is if $\Phi \to 0°$. In that limit, however, the number of reflections as well as the aperture to reflector ratio become very unfavorable". Furthermore, as n increases, H does so much faster, and the factor ρ_m^n decreases the power reaching the PV cells. Table 3 contains the summary of this study for $\theta_c \in [10, 45]$ in steps of 5°.

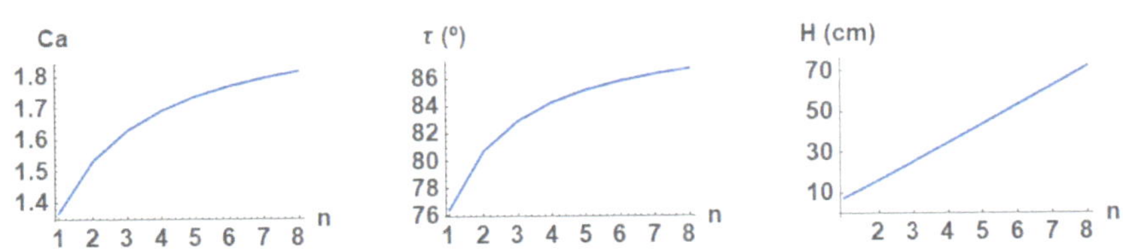

Figure 8. Sequence of optimal solutions of case C.

Table 3. Comparative table for several θ_c.

10°	V-2F	CPC	V-1F	30°	V-2F	CPC	V-1F
Case	A_5-B_4		C_6	Case	A_2-B_1		C_2
C_{opt}	5.829	5.759	3.733	C_{opt}	2.146	2.000	1.535
τ	80.99		85.09	τ	67.36		80.75
H	152.5	191.6	158.9	H	13.74	25.98	16.43
n	0.977	1.043	0.989	n	0.468	0.674	0.522
R_a	5.295	6.774	8.55	R_a	1.387	2.674	2.169
15°	V-2F	CPC	V-1F	35°	V-2F	CPC	V-1F
Case	A_4-B_2		C_4	Case	A_2-B_1		C_2
C_{opt}	3.934	3.864	2.616	C_{opt}	1.849	1.743	1.418
τ	79.06		83.41	τ	69.27		81.37
H	75.89	90.76	69.95	H	11.22	19.59	13.78
n	0.847	0.902	0.822	n	0.439	0.621	0.502
R_a	3.930	4.813	5.38	R_a	1.298	2.308	1.966
20°	V-2F	CPC	V-1F	40°	V-2F	CPC	V-1F
Case	A_3-B_2		C_3	Case	A_2-B_1		C_2
C_{opt}	3.017	2.924	2.067	C_{opt}	1.633	1.555	1.325
τ	75.40		82.16	τ	71.17		82.04
H	38.73	53.90	38.76	H	9.28	15.23	11.63
n	0.701	0.807	0.701	n	0.411	0.572	0.480
R_a	2.653	3.792	3.786	R_a	1.201	2.006	1.773
25°	V-2F	CPC	V-1F	45°	V-2F	CPC	V-1F
Case	A_3-B_1		C_3	Case	A_2-B_1		C_2
C_{opt}	2.431	2.366	1.817	C_{opt}	1.471	1.414	1.250
τ	76.44		82.52	τ	73.06		82.76
H	29.68	36.09	31.12	H	7.73	12.07	9.84
n	0.669	0.734	0.684	n	0.382	0.525	0.456
R_a	2.512	3.141	3.455	R_a	1.099	1.743	1.587

5.2. CPC

The design of a CPC is basically a pair of skewed parabolas whose length is such that at the extremes the parabolic arcs are parallel to the axis of the concentrator. The angle between the axis of the CPC and the line connecting the focus of one of the parabolas with the opposite edge of the aperture is the acceptance half-angle θ_c. When the reflector is ideal, any radiation entering the aperture at angles between $\pm\theta_c$ will be reflected to the base of the concentrator. However, CPCs must be very high to achieve great concentrations, and they are usually truncated in order to cut them down from h to an acceptable height h_T. This truncation is convenient for the reflector-to-area ratio, and the decrease in performance (acceptance angle and concentration ratio) is low. See Appendix A for the relevant formulas.

5.3. Results

Table 3 contains the comparison for $\theta_c \in [10, 45]$ (°) between our design and the other two concentrators. We provide the values of three parameters: optical concentration ratio C_{opt}, optimum angle τ^* and height H, and the cost parameter R_a. We also show the combinations of cases A and B yielding our optimal design. The values for the CPC correspond to the ideal case (in the formulas of the Appendix A, $\phi_T = 2\theta_C$, $a_T = a$, and $h_T = h$). To compute the average number of reflections n, we used the formulas for the truncated CPC with the same height as our two-foci V-trough. In order to give a meaningful comparison with the classical V-trough, we have chosen the case C_n with the same height as our design.

We remark the following:

1. Our solution is clearly better than the classical one-focus V-trough in optical concentration ratio C_{opt}, with an increase between 15.02 and 35.95%. It is also better from a cost-analysis point of view, as our R_a is generally 46.63% better.
2. As regards the CPC, our C_{opt} is always slightly better (around 4%). The height H is notably less in our design (around 54.33%), which leads to a much more compact element, and R_a is generally 57.63% less. Notice that our design is much easier to build than the CPC, obviously.

6. Conclusions

The design of the secondary cavity of a small-scale linear Fresnel reflector is key to maximizing the concentration ratio, which allows for a decrease in the number of photovoltaic cells required and for an increase in the width of the mirrors of the primary field, both of which lower the final cost.

In this work, we have computed analytically, the optimal design of a cavity which, using a non-symmetric distribution of the irradiance reaching its opening, has a concentration ratio greater than those of classical designs. Our analytic approach provides formulas for any number of reflections, which are easily implemented as an iterative algorithm. Furthermore, we prevent the combinatorial explosion inherent in ray-tracing.

We use a two-foci configuration in which rays from each side of the small-scale linear Fresnel reflector reach the other side of the secondary cavity, so that the distribution of irradiance cannot be assumed uniform. We show that our design produces an optical concentration above the ideal value for classical concentrators with uniform distributions. The values for the reflector-to-aperture area ratio are also better, and the design is both more compact and easier to build. Finally, our proposal always yields a value of $\eta_{ray} = 1$, as the classical compound parabolic concentrator, but for $\theta_i > \theta_c$, the values of η_{ray} decrease progressively but slowly.

Future research might include the possibility of modifying the design to have two secondary cavities instead of just one, one on each side of the small-scale linear Fresnel reflector. This would halve the acceptance ratio while notably increasing the concentration. However, there would probably be a cost increment which should be taken into account. This study can be applied to daylighting systems using fibre optics.

Author Contributions: Conceptualization, A.B. and L.B.; methodology, A.B. and L.B.; software, P.F.A.; validation, A.P.; writing—original draft preparation, A.P. and P.F.A.; visualization, A.B.; supervision, A.B. and L.B. All authors have read and agreed to the published version of the manuscript.

Funding: This research received no external funding.

Data Availability Statement: Not applicable.

Conflicts of Interest: The authors declare no conflict of interest.

Nomenclature

A_a	Aperture area (m^2)
A_{abs}	Absorber area (m^2)
A_r	Reflector area (m^2)
a	Aperture of the secondary (CPC) cavity (m)
B	Aperture of the secondary V-trough cavity (m)
b	width of the PV cells (m)
C_a	Area concentration ratio (dimensionless)
C_{opt}	Optical concentration ratio (dimensionless)
d	Separation between two consecutive mirrors (m)
f	Vertical coordinate of the receiver (m)
L_i	Position of the i-th mirror (m)
l_i	Vertical length of the i-th reflection (m)

H	Height of the cavity (m)
h	Height of the CPC cavity (m)
N	Number of mirrors on each side of the SSLFR
n	Mean number of reflections
R_a	Reflector-to-aperture area ratio (dimensionless)
W_M	Mirror width (m)
β_i	Angle between the horizontal and the line from the focal point to the midpoint of mirror i (°)
$\alpha_{0,1,\ldots,n}$	Angle between reflected rays and vertical axis (°)
ε_i	Angle between the i-th reflection and the horiz. axis (°)
η_{rai}	Ray acceptance rate (dimensionless)
θ_c	Acceptance angle (°)
θ_i	Angle between the normal to the mirror and the incidence direction of the Sun rays (°)
ρ_m	Reflectivity of the mirror (dimensionless)
τ	Trough wall angle (°)
Φ	Trough angle or half angle of the V-shaped cone (°)
CPC	Compound parabolic concentrator
CPV	Concentrator photovoltaic
LCP	Low concentration photovoltaic
$SSLFR$	Small-scale linear Fresnel reflector

Appendix A. Formulas for the CPC

The focal distance of the parabola f, the total length h, the aperture length a and the concentration C_a are [17]:

$$f = a'(1 + \sin\theta_C);\ h = \frac{f\cos\theta_C}{\sin^2\theta_C};\ a = \frac{a'}{\sin\theta_C} \rightarrow C_a = \frac{a}{a'} = \frac{1}{\sin\theta_C} \quad \text{(A1)}$$

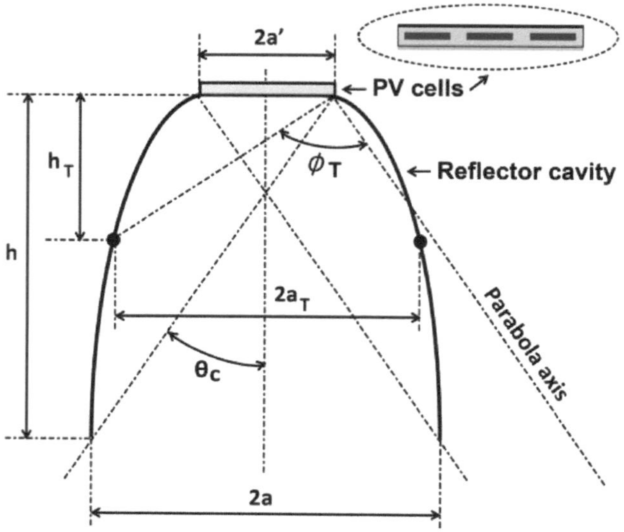

Figure A1. CPC.

When the CPC is truncated to reduce its height from h to h_T, we obtain:

$$h_T = \frac{f\cos(\phi_T - \theta_C)}{\sin^2(\phi_T/2)};\ a_T = \frac{f\sin(\phi_T - \theta_C)}{\sin^2(\phi_T/2)} - a' \rightarrow C_a = \frac{a_T}{a'} \quad \text{(A2)}$$

where a_T is the aperture area of the truncated system and ϕ_T the polar angle at which the parabola is truncated. For the reflector-to-aperture area ratio R_a and the average number of reflections n, we have (after fixing some minor errata in [17]):

$$R_a = \frac{f}{a_T}\left[\frac{\cos(\phi/2)}{\sin^2(\phi/2)} + \ln\cot\frac{\phi}{4}\right]_{\theta_C+\pi/2}^{\phi_T} \tag{A3}$$

$$n = \max\left[C_a\frac{R_a}{2} - \frac{x^2-\cos^2\theta_C}{2(1+\sin\theta_C)}, 1-\frac{1}{C}\right] \tag{A4}$$

$$x = \left(\frac{1+\sin\theta_C}{\cos\theta_C}\right)\left(-\sin\theta_C + \sqrt{1+\frac{h_T}{h}\cot^2\theta_C}\right) \tag{A5}$$

References

1. Kotsopoulos, D. Organizational Energy Conservation Matters in the Anthropocene. *Energies* **2022**, *15*, 8214. [CrossRef]
2. In Proceedings of the Sharm El-Sheikh Climate Change Conference, Sharm el-Sheikh, Egypt, 6–20 November 2022. Available online: https://unfccc.int/cop27 (accessed on 23 January 2023).
3. Shukhobodskiy, A.A.; Zaitcev, A.; Pogarskaia, T.; Colantuono, G. RED WoLF hybrid storage system: Comparison of CO_2 and price targets. *J. Clean. Prod.* **2021**, *321*, 128926. [CrossRef]
4. Grosspietsch, D.; Saenger, M.; Girod, B. Matching decentralized energy production and local consumption: A review of renewable energy systems with conversion and storage technologies. *Wiley Interdiscip. Rev. Energy Environ.* **2019**, *8*, e336. [CrossRef]
5. Alzahrani, M.; Shanks, K.; Mallick, T.K. Advances and limitations of increasing solar irradiance for concentrating photovoltaics thermal system. *Renew. Sustain. Energy Rev.* **2021**, *138*, 110517. [CrossRef]
6. Barbón, A.; Pardellas, A.; Fernández-Rubiera, J.A.; Barbón, N. New daylight fluctuation control in an optical fiber-based daylighting system. *Build. Environ.* **2019**, *153*, 35–45. [CrossRef]
7. Wang, G.; Shen, F.; Wang, F.; Chen, Z. Design and experimental study of a solar CPV system using CLFR concentrator. *Sustain. Technol. Assess.* **2020**, *40*, 100751. [CrossRef]
8. Wang, G.; Wang, F.; Shen, F.; Jiang, T.; Chen, Z.; Hu, P. Experimental and optical performances of a solar CPV device using a linear Fresnel reflector concentrator. *Renew. Energy* **2020**, *146*, 2351-2361. [CrossRef]
9. Liu, Y.; Hu, P.; Zhang, Q.; Chen, Z. Thermodynamic and optical analysis for a CPV/T hybrid system with beam splitter and fully tracked linear Fresnel reflector concentrator utilizing sloped panels. *Solar Energy* **2014**, *103*, 191–199. [CrossRef]
10. Ghodbane, M.; Bellos, E.; Said, Z.; Boumeddane, B.; Kadhim Hussein, A.; Kolsi, L. Evaluating energy efciency and economic efect of heat transfer in copper tube for small solar linear Fresnel refector. *J. Therm. Anal. Calorim.* **2021**, *143*, 4197–4215. [CrossRef]
11. Zubi, G.; Bernal-Agustín, J.L.; Fracastoro, G.V. High concentration photovoltaic systems applying III–V cells. *Renew. Sustain. Energy Rev.* **2009**, *139*, 2645–2652. [CrossRef]
12. Sánchez-González, A.; Gómez-Hernández, J. Beam-down linear Fresnel reflector: BDLFR. *Renew. Energy* **2020**, *146*, 802–815. [CrossRef]
13. Barbón, A.; Fortuny Ayuso, P.; Bayón, L.; Fernández-Rubiera, J.A. Non-uniform illumination in low concentration photovoltaic systems based on smallscale linear Fresnel reflectors. *Energy* **2022**, *239*, 122217. [CrossRef]
14. Ullah, I. Development of Fresnel-based concentrated photovoltaic (CPV) system with uniform irradiance. *J. Daylighting* **2014**, *1*, 2–7. [CrossRef]
15. Madala, S.; Boehm, R.F. A review of nonimaging solar concentrators for stationary and passive tracking applications. *Renew. Sustain. Energy Rev.* **2017**, *71*, 309–322. [CrossRef]
16. Ma, X.; Zheng, H.; Liu, S. A Review on Solar Concentrators with Multi-surface and Multi-element Combinations. *J. Daylighting* **2019**, *6*, 80–96. [CrossRef]
17. Duffie, J.A.; Beckman, W.A. *Solar Engineering of Thermal Processes*, 4th ed.; John Wiley & Sons: Hoboken, NJ, USA, 2013.
18. Fraidenraich, N. Design procedure of V-trough cavities for photovoltaic systems. *Prog. Photovolt. Res. Appl.* **1998**, *6*, 43–54. [CrossRef]
19. Singh, H.; Sabry, M.; Redpath, D.A.G. Experimental investigations into low concentrating line axis solar concentrators for CPV applications. *Sol. Energy* **2016**, *136*, 421–427. [CrossRef]
20. Shaltout, M.A.M.; Ghettas, A.; Sabry, M. V-trough cencentrator on a photovoltaic full tracking system in a hot desert climate. *Renew. Energy* **1995**, *6*, 527–532. [CrossRef]
21. Solanki, C.S.; Sangani, C.S.; Gunashekar, D.; Antony, G. Enhanced heat dissipation of V-trough PV modules for better performance. *Sol. Energy Mater. Sol. Cells* **2008**, *92*, 1634–1638. [CrossRef]
22. Maiti, S.; Banerjee, S.; Vyas, K.; Patel, P.; Ghosh, P.K. Self regulation of photovoltaic module temperature in V-trough using a metal–wax composite phase change matrix. *Sol. Energy* **2011**, *85*, 1805–1816. [CrossRef]
23. Chong, K.K.; Chay, K.G.; Chin, K.H. Study of a solar water heater using stationary V-trough collector. *Renew. Energy* **2012**, *39*, 207–215. [CrossRef]

24. Tina, G.M.; Scandura, P.F. Case study of a grid connected with a battery photovoltaic system: V-trough concentration vs. single-axis tracking. *Energy Convers. Manag.* **2012**, *64*, 569–578. [CrossRef]
25. Al-Shohani, W.A.; Al-Dadah, R.; Mahmoud, S.; Algareu, A. Optimum design of V-trough concentrator for photovoltaic applications. *Sol. Energy* **2016**, *140*, 241–254. [CrossRef]
26. Al-Najideen, M.; Al-Shidhani, M.; Min, G. Optimum design of V-trough solar concentrator for photovoltaic applications. In Proceedings of the AIP Conference Proceedings, Jodhpur, India, 18–22 December 2019; AIP Publishing LLC.: Melville, NY, USA, 2019; Volume 2149, p. 030001.
27. Ejaz, A.; Babar, H.; Ali, H.M.; Jamil, F. Concentrated photovoltaics as light harvesters: Outlook, recent progress, and challenges. *Sustain. Energy Technol. Assess.* **2021**, *46*, 101199. [CrossRef]
28. Fraidenraich, N.; Almeida, G.J. Optical properties of V-trough concentrators. *Sol. Energy* **1991**, *47*, 147–155. [CrossRef]
29. Fraidenraich, N. Analytic solutions for the optical properties of V-trough concentrators. *Appl. Opt.* **1992**, *31*, 131–139. [CrossRef]
30. Tang, R.; Liu, X. Optical performance and design optimization of V-trough concentrators for photovoltaic applications. *Sol. Energy* **2011**, *85*, 2154–2166. [CrossRef]
31. Paul, D.I. Theoretical and experimental optical evaluation and comparison of symmetric 2D CPC and V-trough collector for photovoltaic applications. *Int. J. Photoenergy* **2015**, *2015*, 693463. [CrossRef]
32. Narasimman, K.; Selvarasan, I. Design construction and analysis of solar ridge concentrator photovoltaic (PV) system to improve battery charging performance. *Ecotoxicol. Environ. Saf.* **2016**, *127*, 187–192. [CrossRef]
33. Hadavinia, H.; Harjit, S. Modelling and experimental analysis of low concentrating solar panels for use in building integrated and applied photovoltaic (BIPV/BAPV) systems. *Renew. Energy* **2019**, *139*, 815–829. [CrossRef]
34. Barbon, A.; Sanchez-Rodriguez, J.A.; Bayon, L.; Barbon, N. Development of a fiber daylighting system based on a small scale linear Fresnel reflector: Theoretical elements. *Appl. Energy* **2018**, *212*, 733–745. [CrossRef]
35. Rabl, A.; Goodman, N.B.; Winston, R. Practical design considerations for CPC solar collectors. *Sol. Energy* **1979**, *22*, 373–381. [CrossRef]
36. Barbón, A.; Bayón, L.; Bayón-Cueli, C.; Barbón, N. A study of the effect of the longitudinal movement on the performance of small scale linear Fresnel reflectors. *Renew. Energy* **2019**, *138*, 128–138. [CrossRef]
37. Ma, J.; Wang, C.-L.; Zhou, Y.; Wang, R.-D. Optimized design of a linear Fresnel collector with a compound parabolic secondary reflector. *Renew. Energy* **2021**, *171*, 141–148. [CrossRef]
38. Barbón, A.; Bayón-Cueli, C.; Bayón, L.; Fortuny Ayuso, P.Influence of solar tracking error on the performance of a small-scale linear Fresnel reflector. *Renew. Energy* **2020**, *162*, 43–54. [CrossRef]
39. Barbón, A.; Barbón, N.; Bayón, L.; Sánchez-Rodríguez, J.A. Optimization of the distribution of small-scale linear Fresnel reflectors on roofs of urban buildings. *Appl. Math. Model.* **2018**, *59*, 233–250. [CrossRef]

energies

MDPI

Article

Structure Optimization of Longitudinal Rectangular Fins to Improve the Melting Performance of Phase Change Materials through Genetic Algorithm

Yang Xu, Hang Yin, Chen He, Yong Wei, Ming Cui and Zhang-Jing Zheng *

School of Low-carbon Energy and Power Engineering, China University of Mining and Technology, Xuzhou 221116, China

* Correspondence: zhengzj@cumt.edu.cn

Abstract: In this paper, the structural parameters of longitudinal rectangular fins used in a horizontal shell-and-tube latent heat storage unit (LHSU) are optimized to increase the melting rate of phase-change materials. The influence of natural convection on the melting process is considered. Due to the extremely nonlinear and expensive computational cost of the phase-change heat-transfer-optimization problem, a new coupling algorithm between genetic algorithm and computational fluid dynamics is developed. The effects of the thermal conductivity of fins; the filling rate of fins; and the number of fins on the optimal structure parameters, including the length, width, and position of each fin, are discussed. The results show that when a single fin is inserted in the half-ring region, the optimal dimensionless fin angle is about 0.2, and the optimal dimensionless fin length is about 0.96. The use of optimal single fin can shorten the dimensionless total melting time by 68% compared with the case of no fin, and 61.3% compared with uniformly arranged single fin. When the number of fins exceeds one, each fin should have a specific length (L), thickness (Δ), and position (ψ) instead of uniform distribution. The advantage of the optimized fins decreases as the number of fins increases. When the number of fins is four, the optimized fin distribution is almost uniform, and the dimensionless total melting time is only 15.9% less than that of the absolutely uniform fin. The number of fins is a more sensitive parameter affecting the optimal position and structure of fins than the filling rate and thermal conductivity of fins.

Keywords: phase-change material (PCM); melting; heat-transfer enhancement; fin; computational fluid dynamics (CFD); genetic algorithm (GA)

1. Introduction

Thermal energy storage (TES) technology has a wide application in solar thermal utilization and waste heat recovery systems. Latent heat TES (LHTES) is one of the promising TES technologies, storing heat in liquid–solid phase-change materials (PCMs) in the form of latent heat. In the LHTES system, the PCM is encapsulated in the storage unit called LHTES unit. Heat is carried into and out of the LHTES unit by heat-transfer fluid (HTF). The shell-and-tube LHTES unit is a commonly used LHTES unit since it can be easily manufactured [1]. In addition, the shell-and-tube LHTES unit gives a better charging and discharging performance than rectangular LHTES unit [2]. Numerous research studies related to the shell-and-tube LHTES unit have focused on the heat-transfer enhancement to promote the charging and discharging efficiency [3–6], such as adding heat pipe [7], metal foam [8], nanoparticles [9], graphite [10], etc.

The utilization of fin is a cheap and easy-to-process strategy for heat-transfer enhancement [11,12]. In a shell-and-tube LHTES unit, HTF usually flows in the tubes, and the PCM is encapsulated in the shell. During the heat-transfer process between the HTF and PCM, the maximum thermal resistance regularly occurs in the PCM, so fins are usually fixed on the outside of the tube to enhance heat transfer in the PCM. There are two types of fins that can be attached to the outside of the tube: longitudinal and transversal. Generally, the rectangular fin is longitudinal, and the ring fin is transversal. This paper concerns the longitudinal rectangular fins, and the research progress is summarized as follows.

The research of the longitudinal fin with a rectangular shape can be divided into two stages according to the arrangement of fins. In the first stage, the fins are evenly arranged. Some research of the first stage focused on verifying the thermal strengthening effect of the rectangular fins and revealing the melting and solidification characteristics under the influence of the fins. Rabienataj Darzi et al. [13] compared the melting and solidification performances of a double-tube LHTES unit enhanced by adding longitudinal rectangular fins and using an elliptical inner tube instead of a round inner tube through the numerical method. Twelve longitudinal rectangular fins with fixed structural parameters were uniformly attached to the outside of the inner tube. The results showed that the use of longitudinal fins can shorten the melting and solidification time compared to the use of the elliptical tube. Agyenim et al. [14] experimentally compared the melting and solidification performances of a shell-and-tube LHTES unit enhanced by longitudinal and transversal fins. Eight longitudinal rectangular fins with fixed structural parameters were evenly fixed outside the inner tube. The results indicated that the use of longitudinal fins can achieve a faster melting and solidification rate than the use of transversal fins. Li et al. [15] numerically studied the melting and solidification performances of a horizontal shell-and-tube LHTES unit enhanced by longitudinal fins. Six rectangular fins were evenly fixed outside the inner tube. The results indicated that the total melting/solidification time can be shortened by more than 14%. Rozenfeld et al. [16] experimentally investigated the close-contact melting in a horizontal shell-and-tube LHTES unit enhanced by longitudinal fins. Three full-length rectangular fins were evenly fixed outside the inner tube. The results indicated that close-contact melting shortens the melting time by 2.5 times.

The parameter analysis on the added fins is also a research topic in the first stage. Padmanabhan et al. [17] numerically investigated the melting and solidification characteristics of the PCM filled in a double-tube LHTES unit with evenly arranged longitudinal fins. Conduction was considered to be the only mode of heat transfer. In other words, natural convection was ignored. The results indicated that the melting/solidification rate is a function of the number of fins, fin width, and fin length. Ismail et al. [18] studied the solidification of the PCM around a vertical axially finned isothermal cylinder through numerical and experimental methods. The results showed that the number of fins and fin length have greater influence on the solidification rate of the PCM than fin width. Zhang et al. [19] numerically studied the solidification behaviors of a horizontal shell–tube LHTES unit with several longitudinal fins evenly fixed on the inner tube. The Taguchi method was utilized to optimize the fin geometry. The results suggested that the number and length of fins should be increased to obtain the best solidification behavior. Hosseini et al. [20] studied the melting behaviors of a horizontal double-tube LHTES unit with eight longitudinal fins evenly fixed on the inner tube through experimental and numerical methods. The results indicated that there is an optimal value for the fin length if the melting rate, storage capacity, and cost are considered comprehensively. Solomon et al. [21] experimentally studied the solidification behaviors of a vertical shell–tube LHTES unit with eight longitudinal fins evenly fixed on the inner tube. The results showed that setting the fin length to 60% of the annular gap is sufficient to achieve the maximum heat-transfer enhancement. Yuan et al. [22] numerically investigated the effect of the installation angle of longitudinal fins on the melting characteristics of a horizontal double-tube LHTES unit. Two longitudinal fins were evenly fixed on the outside of the inner tube. The results showed that installing one fin directly below the inner tube and another fin directly above the inner tube leads to the maximum melting rate.

Some other studies in the first stage focus on revealing the different effects of longitudinal fins on the melting and solidification processes. Abhat et al. [23] experimentally studied the melting and solidification behaviors of a horizontal double-tube LHTES unit with twelve longitudinal fins evenly fixed on the inner tube. The results showed that even if the annular space is separated into 12 small pieces, the uneven melting phenomenon is still obvious. On the contrary, the solidification of the PCM is more uniform. This conclusion can confirm that the effect of natural convection on the melting process is more significant than on the solidification process. Bathelt et al. [24] experimentally investigated the melting and solidification around a horizontal tube. Three longitudinal fins were evenly fixed on the outside of tube. The results showed that the orientation of fins has more influence on the melting process than on the solidification process. Sparrow et al. [25] experimentally investigated the freezing around a vertical tube. Four longitudinal fins were evenly fixed on the outside of the tube. It was found that the liquid–solid interface when conduction is dominant is different from the liquid–solid interface when natural convection is dominant. Rathod et al. [26] experimentally investigated the thermal performance enhancement of a vertical shell-and-tube LHTES unit with longitudinal rectangular fins. Three fins with the same fixed structural parameters were evenly arranged. The results indicated that using fins can shorten the total melting time and solidification time by 25 and 44%, respectively. The inconsistent effects of fins on the melting and solidification process are due to the different heat-transfer mechanisms.

Natural convection leads to non-uniform melting and solidification, so recently some researchers have claimed that the fins should be arranged unevenly in order to take advantage of convection, which marks the beginning of the second stage. Since convection has more significant influence on the melting process than solidification, most of the research in the second stage focused on the melting process [27]. Wang et al. [28] numerically investigated the melting performance of a horizontal shell-and-tube LHTES unit with longitudinal rectangular fins. Three equidistant fins with the same structural parameters were placed in the bottom of the annular region. The results indicated that there is an optimal angle between adjacent fins to reduce the total melting time. Mahood et al. [29] numerically investigated the effect of fin design on the melting of the PCM in a horizontal shell-and-tube LHTES unit. Five equidistant fins with the same structural parameters were designed. The results showed that the small angle between adjacent fins and the placement of the fins in the bottom of the annular region can optimize thermal performance. Tao et al. [30] numerically investigated the effect of the number of longitudinal fins on the melting performance of a horizontal LHTES unit. Three, five, and seven equidistant fins with the same structural parameters were designed, and all the fins were place in the bottom of the annular region. The results indicated that the local enhanced fin tube can improve the uniformity of the PCM melting process. Deng et al. [31] numerically investigated the melting performance of a horizontal shell-and-tube LHTES unit with two longitudinal rectangular fins. The effects of the location of fins and the angle between adjacent fins on the melting characteristics of the PCM were studied. The results showed that both fins should be placed in the bottom of annular region, and the best angle between the two fins should be 120°. Deng et al. [32] further numerically investigated the effect of the number of fins on the optimal layout of fins. The results showed that when the number of fins is less than six, all the fins should be placed in the bottom of the annular region; otherwise, the fins should be fixed evenly on the outside of the inner tube. Kumar et al. [33] investigated the effect of longitudinal finned tube eccentric configuration on melting behavior. They proposed a special arrangement of three fins to accommodate the eccentric structure. One long fin is arranged in the top region, and two short fins are arranged symmetrically in the bottom. The optimal angle between the two bottom fins was found to be 60°. Yagci et al. [34] studied the melting (charge) characteristics of a phase change material (PCM) in a vertical shell-and-tube LHTES unit. Four equidistant fins with the same structural parameters were fixed on the inner tube. Since the natural convection in the top is stronger than in the bottom, the fins were designed to be wedge-shaped. That is, the bottom of the fin is longer than the top. It was found that decreasing the fin edge lengths' ratio could significantly shorten the melting time, but no optimal value was proposed.

To sum up, many research studies have focused on the structural parameters optimization of longitudinal rectangular fins used in shell-and-tube LHTES unit. In addition, many research studies focus on optimizing the structures of other types of fins [35–37]. However, most of the research is based on univariate analysis. Actually, univariate analysis is not suitable for the optimization of fin structure, because the number of structural parameters of a single fin exceeds two especially when natural convection is considered. Not to mention, the number of structural parameters will increase with the increase of the number of fins. It has been preliminary found that, to utilize the convection, the fins should not only be arranged unevenly, but also have different structural parameters. So far, few research studies have proposed fins with different structural parameters due to the expensive computational cost of phase-change optimization problems and lack of a smarter optimization method for the complex dynamic nonlinear characteristic [38]. Therefore, in this study, a modified genetic algorithm coupled with computational fluid dynamics (CFD) was developed to optimize the structural parameters of longitudinal rectangular fins in a shell-and-tube LHTES unit. The influence of natural convection on the optimal structure of fins was considered. The length, width, and position of each fin were selected as variables, and the filling rate of the fins was taken as the constraint. In this paper, the effects of the thermal conductivity of fins, the filling rate of fins, and the number of fins on the optimal structure of fins are discussed. The optimization method proposed by this paper can be used to optimize the structure of other types of fins.

2. Presentation of the Optimization Problem

2.1. Physical Model

This paper focuses on optimizing the design parameters of the longitudinal rectangular fins inserted in a horizontal single-pass shell-and-tube LHTES unit so as to accelerate the melting rate of the PCM. As shown in Figure 1, the PCM is encapsulated within the annular space between the inner and outer tube. The radii of outer and inner tube are denoted by r_o and r_i, respectively. Heat-transfer fluid (HTF) with a temperature higher than the melting point of the PCM flows in the inner tube. Heat is transferred to the PCM region through the wall of the inner tube. Several longitudinal rectangular fins are fixed on the outer surface of the inner tube to enhance heat transfer. The design parameters of fins are selected as the optimization variables, including fin length, l; fin thickness, w; and fin position (represented by angle θ). The definitions of l, w, and θ refer to Figure 1.

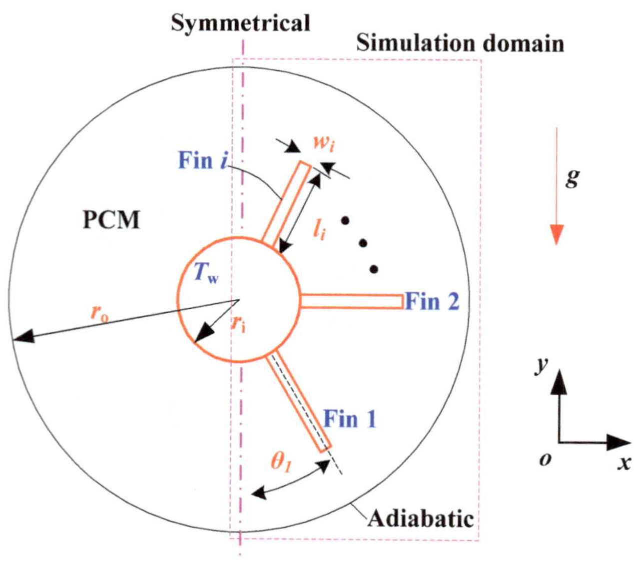

Figure 1. Physical model for present problem.

For the convenience of optimization, l, w, and θ are replaced by three dimensionless parameters, L, Δ and ψ, respectively. The definitions of these three dimensionless parameters are expressed in Equation (1). During the optimization, the three dimensionless parameters all range from 0 to 1.

$$L = \frac{l}{r_o - r_i} \quad \Delta = \frac{w}{2r_i} \quad \psi = \frac{\theta}{\pi} \tag{1}$$

The melting performance of the whole fin-enhanced LHTES unit is evaluated through studying the typical cross-section of the LHTES unit. In addition, due to the symmetry, half of the cross-section is selected as the simulation domain (see Figure 1). The wall of the inner tube is usually made of metal alloy material with high thermal conductivity, and it is far thinner than the simulation domain size, so the thickness of the wall can be ignored without having effects on the simulation results [8,39–44]. The inner tube maintains a constant temperature of T_w, and the outer tube is insulated from the environment [44]. Paraffin is selected as the PCM, and its thermo-physical properties can be found in Table 1. The values of T_w, r_o, and r_i are 420 K, 20 mm, and 10 mm, respectively. The thermo-physical properties of the fin satisfy that the density ratio of the fin to the PCM is equal to 10.95 (ρ_{fin}/ρ_{PCM} = 10.95), the specific heat ratio of the fin to the PCM equals 0.15 ($c_{p,fin}/c_{p,PCM}$ = 0.15), and the thermal conductivity ratio of the fin to the PCM varies from 83.6 to 1987.7 ($\lambda_{fin}/\lambda_{PCM}$ = 83.6 − 1987.7).

Table 1. Physical properties of paraffin [45].

Physical Properties	Paraffin
ρ (kg m^{-3})	820
c_p (J kg^{-1} K^{-1})	2500
λ (W m^{-1} K^{-1})	0.195
μ (kg s m^{-1})	0.205
γ (K^{-1})	0.0006
L (J kg^{-1})	210,000
T_{m1} (K)	320.66
T_{m2} (K)	321.66

2.2. Governing Equations

Numerical methods are employed to solve the optimization problem of the fin structure. In the traditional numerical simulation process, the parameters of the fins need to be determined first. The grids of the fin area and the phase-change material area are drawn according to these parameters. Each specific parameter corresponds to a specific group of grids. However, in the optimization process, the parameters of the fins are unknown and obtained through the optimization process. If the traditional method is used, a large number of grids need to be drawn, and this is time-consuming. Therefore, in order to improve the computing efficiency, the unified governing equations are used in this study to describe the phase-change heat transfer of the PCM and the pure heat conduction of the fins. The enthalpy–porosity model is employed to express the solid–liquid phase change process, and the porous model is used to differentiate the fin region from the PCM region. The governing equations, namely the energy equation, continuity equation, and momentum equation, are listed as follows.

Energy equation [42]:

$$\begin{aligned}&\left[(1-\varepsilon)\left(\rho c_p\right)_{\text{fin}}+\varepsilon\left(\rho c_p\right)_{\text{PCM}}\right]\frac{\partial T}{\partial t}+\varepsilon\rho_{\text{PCM}}\Delta h\frac{\partial f}{\partial t}+\left(\rho c_p\right)_{\text{PCM}}\left(u\frac{\partial T}{\partial x}+v\frac{\partial T}{\partial y}\right)\\&=\left[\varepsilon\lambda_{\text{PCM}}+(1-\varepsilon)\lambda_{\text{fin}}\right]\left(\frac{\partial^2 T}{\partial x^2}+\frac{\partial^2 T}{\partial y^2}\right)\end{aligned} \tag{2}$$

where ε is an integer used to distinguish the fin region from the PCM region. For the fin region and the PCM region, ε is equal to 0 and 1, respectively; and f in the second term on the left side of Equation (2) is the liquid fraction of the PCM, whose definition is given in Equation (3). Solidification temperature, T_s, is 1 K lower than the melting temperature, T_m.

$$\begin{cases} f = 1 & T > T_m \\ f = \frac{T - T_s}{T_m - T_s} & T_s < T < T_m \\ f = 0 & T < T_s \end{cases} \tag{3}$$

Momentum equations [41]:
x-direction:

$$\rho\left(\frac{\partial u}{\partial t} + u\frac{\partial u}{\partial x} + v\frac{\partial u}{\partial y}\right) = \mu\left(\frac{\partial^2 u}{\partial x^2} + \frac{\partial^2 u}{\partial y^2}\right) - \frac{\partial p}{\partial x} - \frac{\mu}{K}u - A_{mush}\frac{(1-f)^2}{f^3 + \delta}u \tag{4}$$

y-direction:

$$\rho\left(\frac{\partial v}{\partial t} + u\frac{\partial v}{\partial x} + v\frac{\partial v}{\partial y}\right) = \mu\left(\frac{\partial^2 v}{\partial x^2} + \frac{\partial^2 v}{\partial y^2}\right) - \frac{\partial p}{\partial y} - \frac{\mu}{K}v - A_{mush}\frac{(1-f)^2}{f^3+\delta}v + \rho_{ref} g\beta(T - T_{ref}) \tag{5}$$

The third terms on the right side of Equations (4) and (5) are the viscous loss terms from the porous model. In the fin region, the permeability (K) is set to an extremely small value (10^{-20} m^2), which means no flow in the fin region, while in the PCM region, $1/K$ is set to zero. A_{mush} is the mushy zone constant with a value of 10^5 kg m^{-3} s^{-1} [40], and δ is a small constant (0.001) to prevent the denominator from being equal to zero. The fifth term on the right side of Equation (5) is used to estimate the effect of natural convection on the velocity field.

Continuity equation:

$$\frac{\partial u}{\partial x} + \frac{\partial v}{\partial y} = 0 \tag{6}$$

2.3. Numerical Method

The governing Equations (2)–(6) are solved numerically through the FVM (finite volume method) because they have no analytical solutions. The unified governing equations make it feasible to construct fins with different structural parameters in a fixed grid system, and the detailed method can be found in our previous articles [42,43,46–49]. For specific settings of the FVM, see Ref. [50].

To determine the appropriate mesh size, a single-fin case is studied. As shown in Figure 2, the dimensionless length, L; thickness, Δ; and angle, ψ, of the fin are set to 0.5, 0.05, and 0.5, respectively. The thermal conductivity ratio of the fin to the PCM (denoted as $\lambda_{fin}/\lambda_{PCM}$) is set to 1987.7. Four grid sizes are designed, and the total number of grids in the whole simulation domain is 819, 2765, 12,346, and 48,269, respectively. In addition, the cases with different grid numbers have different time steps so as to determine the appropriate time step. Figure 2 shows the variation of dimensionless total melting time, Fo_{tot}, with the number of grids. The dimensionless melting time is expressed by the Fourier number, Fo, and its definition is given in Equation (7). Fo_{tot} can be calculated by substituting the total melting time, t_{tot}, for t in Equation (7). It can be seen from Figure 2 that, for the grid numbers 48,269 and 12,346, the deviation of Fo_{tot} is only 0.02%.

Fourier number:

$$Fo = \frac{t\alpha}{D_e{}^2}, D_e = D_o - D_i \tag{7}$$

Figure 3 shows the effect of the number of grids on the distribution of dimensionless thermal conductivity. The dimensionless thermal conductivity is defined as the thermal conductivity ratio of the local grid to the PCM. The outline of the fin can be seen from the distribution graph of the dimensionless thermal conductivity. It can be seen from Figure 3 that the more grids there are, the clearer the outline of the fin. The fin structure is very close to the real one when the number of grids is 48,269. The comparison results shown in Figures 2 and 3 can prove that the fourth grid system with a grid number of 48,269 and a time step of 0.1 s is suitable for studying the optimization problem presented in this paper.

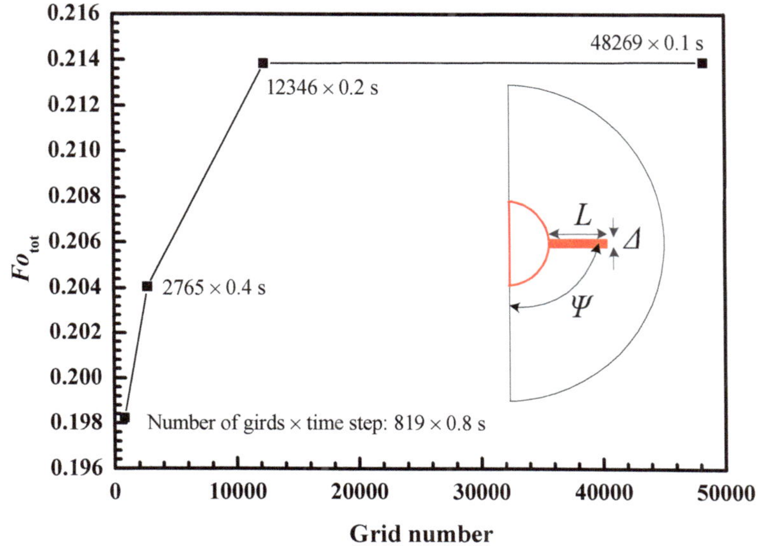

Figure 2. Variation of dimensionless total melting time with grid number.

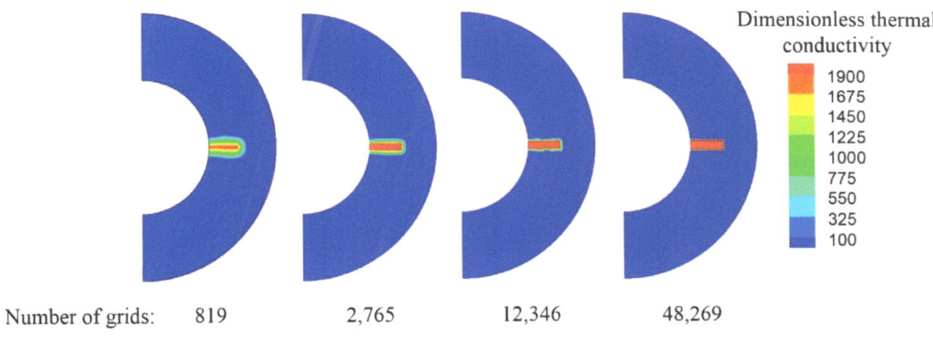

Figure 3. Effect of the number of grids on the distribution of dimensionless thermal conductivity.

The mathematical model presented in Section 2.3 has been widely used to investigate the solid–liquid phase-change problem. Our previous articles have verified the accuracy of this model through comparing the numerical results with the experimental data and analytical solutions. The verification of the model can be found in Refs. [40–43,50] and is not repeated here.

2.4. Optimization Method

The optimization problem that concerns this paper is listed in Table 2. The design parameters of the fins under different numbers and total areas of fins (L_i, Δ_i, θ_i) are selected as variables. The total melting time, t_{tot}, is taken as the optimization objective. The number of variables depends on the number of fins. For example, there are two variables for one fin and five variables for two fins.

Table 2. Description of the optimization problem.

Optimization function:	$\text{Min}(t_{\text{tot}}) = \text{Fun}\{L_i, \Delta_i, \theta_i\}$
Constraint:	The dimensionless fin area (A) is fixed

The genetic algorithm (GA) was chosen as the optimization method because it is suitable for solving nonlinear optimization problems [51]. Due to the expensive computational cost of phase-change optimization problems, a new coupling algorithm between GA and CFD is proposed. The flowchart of this new coupling algorithm is shown in Figure 4. The characteristics of this new coupling algorithm are summarized as follows.

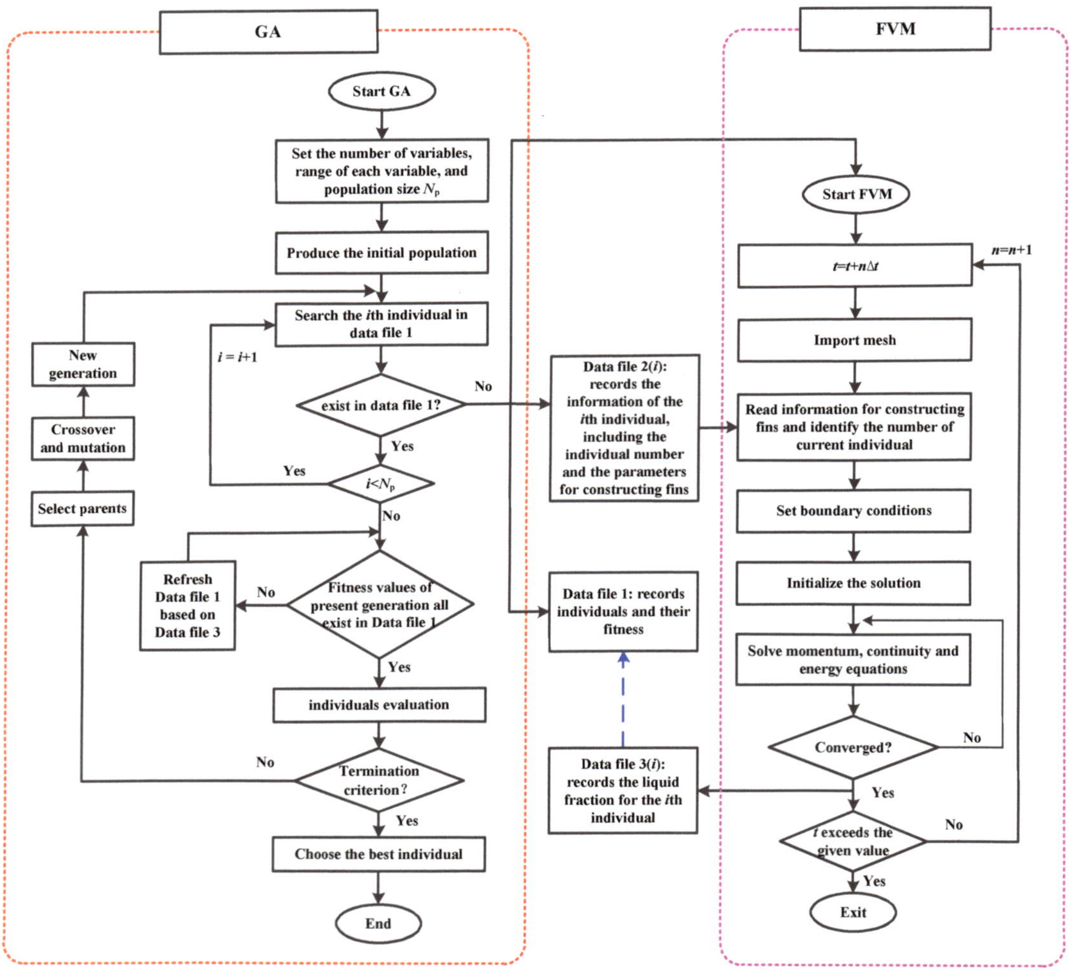

Figure 4. The flowchart of the optimization process.

The first characteristic is that Data File 1 is used to record all individuals generated by GA and their fitness values. Each individual is composed of independent variables with fixed values, and the number of variables of each individual is equal to the number of variables of the optimization problem. Data File 1 is used for two reasons: one reason is that some excellent individuals with low fitness values will be preserved from the present generation to the next generation. Since the new algorithm searches Data File 1 for the fitness evaluation, Data File 1 can ensure that the GA needs to call the CFD only once for the same individual. Most of the optimization time is spent on the iterations of the CFD, so reducing the number of CFD calls can greatly reduce the time consumption. Another reason is that Data File 1 is very useful in dealing with the unexpected interruption of the optimization process caused by power failure or computer restart.

The second characteristic of the new algorithm is that the CFD can be called in parallel to estimate the fitness of all individuals in the same generation. Calling the CFD in parallel is the most effective way to reduce the time cost. Nevertheless, the next generation is allowed to be created only after the fitness assessment of all individuals in the current generation has been completed. The third characteristic of the new algorithm is that the series of data files named Data File 3 (i) is created to record the real-time evaluation progress of each individual. Data File 3 can be used to refresh Data File 1 and monitor the running CFD cases.

Optimization starts with GA. GA randomly creates the first generation of individuals and searches for each individual in Data File 1. If an individual does not exist, FVM will be called to calculate the fitness value of that individual. After checking all individuals, if the fitness value of some individuals has not been finished, the GA will pause. A loop is used to pause the GA and refresh the fitness value recorded in Data File 1. If the fitness calculation of the first generation is completed, the GA enters the loop of the second generation. More details on the combination between the GA and FVM can be found in Ref. [52].

3. Results and Discussion

The purpose of this paper is to optimize the structural parameters of fins with a given fin area and improve the melting rate of a shell-and-tube latent heat storage unit. In this section, some novel fin structures are presented, and the effects of fin thermal conductivity, fin area, and fin number on the optimization results are discussed.

3.1. Optimization Results of a Single Fin in Half Ring Region

The structural parameters of a single fin with a given fin area are optimized in this section. The dimensionless fin area (the ratio of fin area to simulation-domain area) is fixed to 0.02 (A = 0.02). The thermal conductivity ratio of the fin to the PCM is fixed at 1987.7 ($\lambda_{fin}/\lambda_{PCM}$ = 1987.7). In this case, an individual consists of two variables: the dimensionless fin length and fin angle (L and ψ). The definitions of L and ψ refer to Equation (1). Both the L and ψ can vary from 0 to 1. The fin width can be calculated if L and ψ have specific values.

For genetic algorithms, the population size is a key parameter to determine the accuracy of the optimization results. Figures 5 and 6 give the fitness values of dimensionless total melting time (Fo_{tot}) for different L and ψ values generated by the genetic algorithm (GA) with different population sizes. Theoretically, the larger the population size is, the better the optimization result is, because that larger population size will lead to more individuals generated by GA. However, from the results shown in Figures 5 and 6, it is the population size of 10 rather than 20 or 40 that produces a minimum Fo_{tot}. This is due to the randomness of the GA. Despite the minor difference, the optimization results for the three population sizes are consistent. As shown in Figure 6, Fo_{tot} first decreases and then increases with ψ. The optimal individuals with population size of 10, 20, and 40 are named Optimal Case 1, 2, and 3, respectively. The L and ψ of the Optimal Case 1 are 0.964 and 0.199, respectively. Compared with the Optimal Cases 1, the L of the Optimal Cases 2 and 3 has only a deviation of 2% and 0.9%, and the ψ of the Optimal Cases 2 and 3 has only a deviation of 3.7% and 1.6%. From the view of the fitness value (Fo_{tot}), the deviations between Optimal Cases 1 and 2 and between Optimal Cases 1 and 3 are as low as 1.3% and 0.4%, respectively.

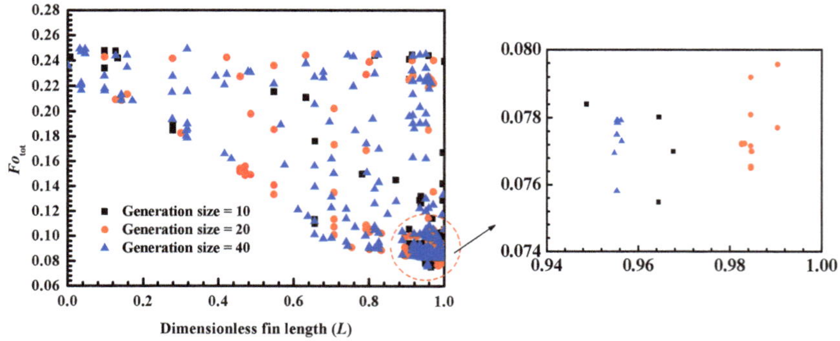

Figure 5. Different dimensionless fin lengths generated by genetic algorithm (GA) with different population sizes.

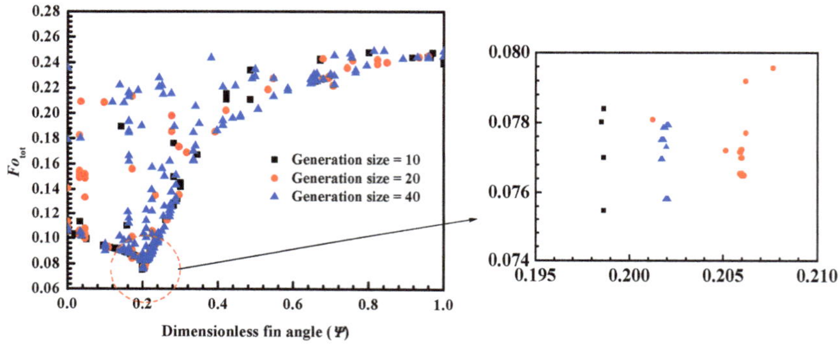

Figure 6. Different dimensionless fin angles generated by genetic algorithm (GA) with different population sizes.

Increasing population size will lead to a huge surge in optimization time and CPU assumption, because many more CFD cases will be awakened. When population sizes are 10, 20, and 40, the number of calls to CFD is 140, 239, and 474, respectively. To control the time cost and guarantee the accuracy, the population size of 20 is selected for all the following cases.

To reveal why using the structural parameters optimized by GA can obtain the highest melting rate, the melting characteristic of the Optimal Case 1 is compared with three special cases. Special Case 1 has no fin. In Special Case 2, a single fin is placed in the bottom of the domain. In other words, the L and ψ of Special Case 2 are 1.0 and 0, respectively. Special Case 3 has the same ψ as Optimal Case 1, but its L is 1.0.

Figure 7 shows the distribution of liquid fraction and velocity magnitude $(u^2+v^2)^{0.5}$ for Special Cases 1–3 and Optimal Case 1 at three moments. In each annular region, the left half is the distribution of liquid fraction, and the right is the distribution of velocity magnitude. In the left half, the blue region denotes the non-melted region (solid PCM), and the red region represents the fully melted region (liquid PCM). It can be observed from the contour map of Special Case 1 that natural convection results in non-uniform melting. The melting rate of the PCM in the top region is higher than in the bottom region. Inserting the fin in the bottom, like in Special Case 2, is helpful to decrease the Fo_{tot}. However, a large area of the solid PCM can still be found in the bottom region when Fo = 0.071. This phenomenon indicates that inserting the fin vertically under the inner tube is not the best solution. The fin used in Special Case 3 is longer and thinner than that in Optimal Case 1 since the fin area is fixed. From the contour graphs of Special Case 3, it can be seen that the fin is helpful to accelerate the melting of the PCM in the lower region. There is more solid PCM in the upper part of the fin than in the lower part when $Fo = 0.071$. This non-uniform melting phenomenon leads to an increase in the total melting time. On the contrary, a more uniform melting can be found in the contour graphs of Optimal Case 1. Therefore, a well-designed fin should ensure that the PCM in the different regions tends to have a consistent melting rate. Figure 8 gives the dimensionless total melting time, Fo_{tot}, for Special Cases 1–3 and Optimal Case 1. It can be seen that Optimal Case 1 has the minimum Fo_{tot}. Compared with the case of no fin (Special Case 1), The use of optimal fins can shorten the dimensionless total melting time by 68%.

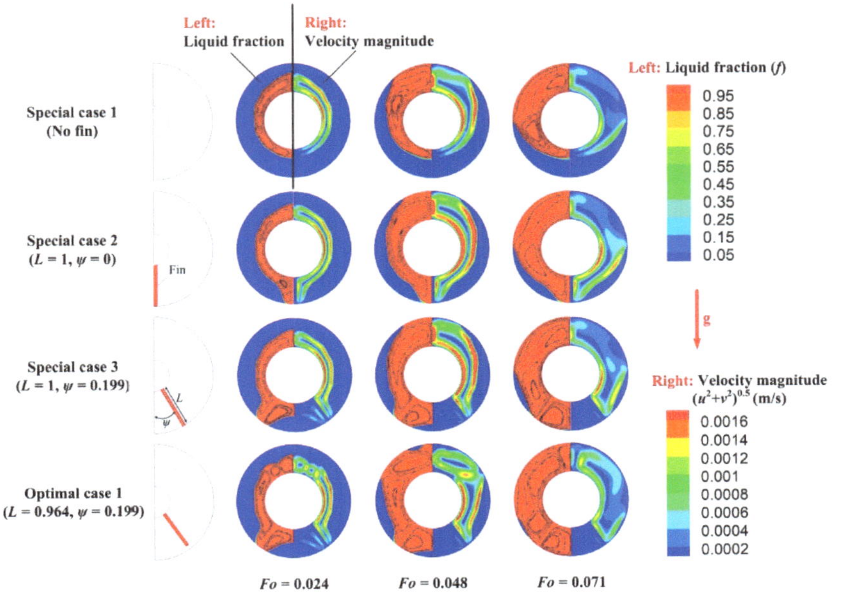

Figure 7. Contour maps of liquid fraction and velocity magnitude for Special Cases 1–3 and Optimal Case 1.

Based on the optimal results of GA, the optimal value of ψ is about 0.2, which means that the fin is still in the lower half of the annular region. Actually, the optimal value of ψ depends on the magnitude of convection relative to conduction. When convection plays a more significant role in heat transfer than conduction, the value of ψ should be reduced. On the contrary, if the conduction is dominant in the heat transfer, the value of ψ should be increased. For an extreme case of neglecting natural convection, the best value of ψ should be 0.5. That is, the fin should be placed in the middle region between the upper and lower parts of the annular space. Therefore, natural convection is a sensitive parameter that affects the optimal position and structure of the fin.

3.2. *Effect of Fin Thermal Conductivity on the Optimal Structural Parameters of Fin*

This section discusses the effect of the thermal conductivity ratio of the fin to the PCM (denoted as $\lambda_{fin}/\lambda_{PCM}$) on the optimal structure parameters of the fin. The single fin with dimensionless area of 0.02 ($A = 0.02$) is considered. Figure 9 shows the variation of the optimal structure parameters (dimensionless fin length, L, and dimensionless fin angle, ψ) with $\lambda_{fin}/\lambda_{PCM}$ when the population size is 20. The $\lambda_{fin}/\lambda_{PCM}$ changes from 83.6 to 1987.7. It can be seen from Figure 9 that the optimal values of L and ψ both increase with the increase of $\lambda_{fin}/\lambda_{PCM}$. Increasing the $\lambda_{fin}/\lambda_{PCM}$ leads to the increase of heat conduction through the fin. In other words, heat conduction plays a more important role in total heat transfer. Therefore, the value of ψ should be increased to move the fin from the bottom toward the middle. As for L, its value should be carefully designed to balance the melting progress in the upper and lower regions of the fin. It also can be found from Figure 9 that, although the value of $\lambda_{fin}/\lambda_{PCM}$ is increased by nearly 24 times, the optimal values of L and ψ are only increased by 14.7% and 11%, respectively. This is because most of the heat absorbed by the PCM is not through the fin since the fill rate of the fin is low. In addition, increasing $\lambda_{fin}/\lambda_{PCM}$ does not significantly affect the role of natural convection in total heat transfer. Therefore, $\lambda_{fin}/\lambda_{PCM}$ is not a sensitive parameter that affects the optimal position and structure of the fin.

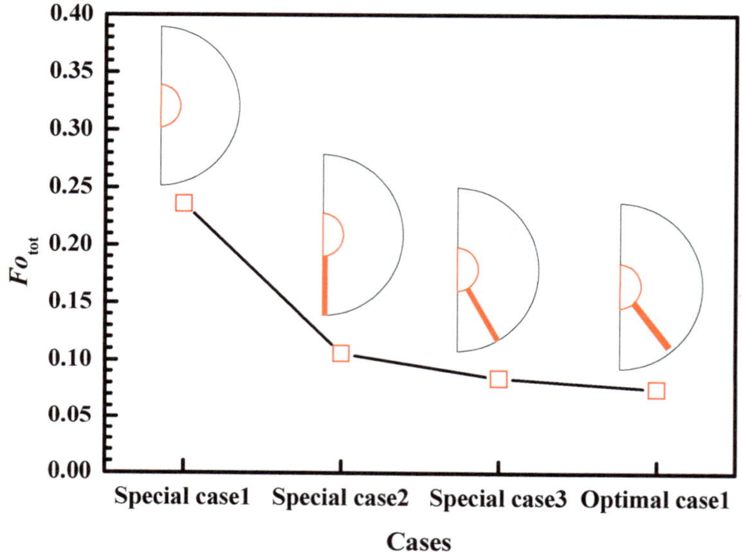

Figure 8. Dimensionless total melting time, Fo_{tot}, for Special Cases 1–3 and Optimal Case 1.

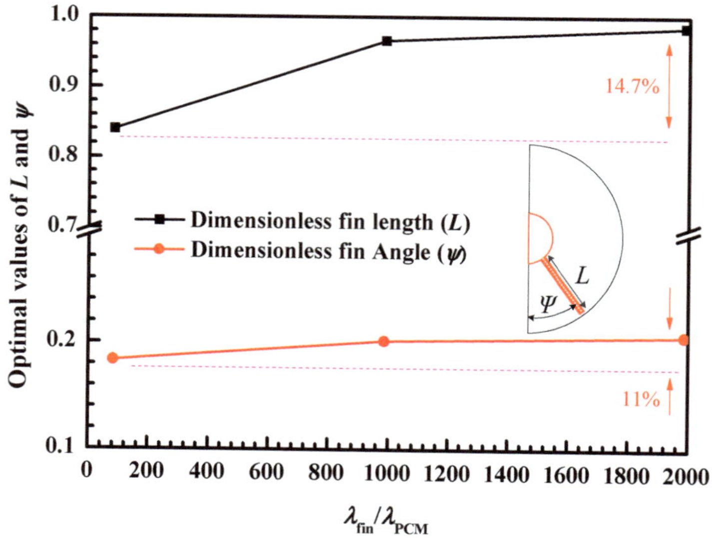

Figure 9. Variation of the optimal dimensionless fin length, L, and dimensionless fin angle, ψ, with the thermal conductivity ratio of fin to PCM ($\lambda_{fin}/\lambda_{PCM}$).

Figure 10 shows the distribution of liquid fraction and dimensionless temperature for Optimal Cases 2, 4, and 5 at three moments. Optimal Case 2 is mentioned in Section 3.1, which uses the fin optimized for $\lambda_{fin}/\lambda_{PCM}$ = 1987.7. Optimal Cases 4 and 5 use the fins optimized for $\lambda_{fin}/\lambda_{PCM}$ = 1037.9 and $\lambda_{fin}/\lambda_{PCM}$ = 83.6, respectively. It can be seen from Figure 10 that the melting progress of these three optimal cases is almost the same.

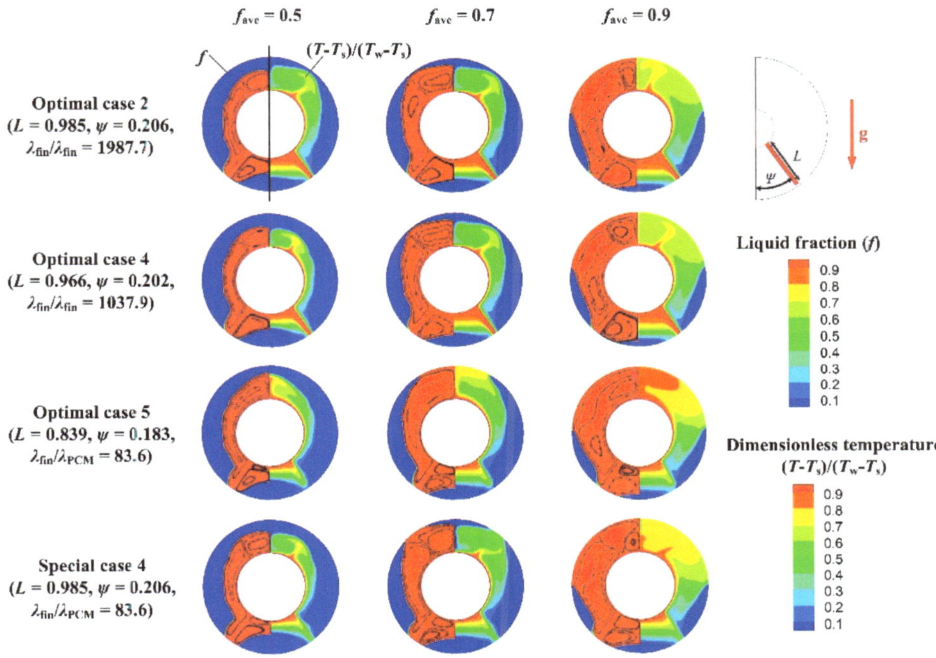

Figure 10. Contour maps of liquid fraction and dimensionless temperature for Optimal Cases 2, 4 and 5 and Special Case 4.

To further prove different fin designs for different $\lambda_{fin}/\lambda_{PCM}$ values, a special case called Special Case 4 was designed. Special Case 4 uses the same fin as Optimal Case 2, but the $\lambda_{fin}/\lambda_{PCM}$ is set to 83.6. The distribution of liquid fraction and dimensionless temperature for Special Case 4 is also shown in Figure 10. It can be seen that, for the same $\lambda_{fin}/\lambda_{PCM}$, the melting rate of the PCM under the fin in Special Case 4 is slower than that in Optimal Case 5. Figure 11 gives the variation of the liquid fraction (f) with the dimensionless melting time (Fo) for Optimal Case 5 and Special Case 4. It can be seen that Optimal Case 5 has a faster later-stage melting rate than Special Case 4. When Fo = 0.095, the PCM in Optimal Case 5 has been completely melted (f = 1.0), while the solid PCM can still be found in the bottom of the annular region in Special Case 4.

3.3. Effect of Fin Area on the Optimal Structural Parameters of Fin

In this section, the effect of fin area on the optimal structure parameters of fin is studied. The dimensionless fin area, A, varies from 0.02 to 0.1, and the number of fins is fixed at 1. The definition of A refers to Section 3.1. The thermal conductivity ratio of the fin to the PCM is fixed at 1987.7 ($\lambda_{fin}/\lambda_{PCM}$ =1987.7). Figure 12 shows the variation of the optimal dimensionless fin length (L) and dimensionless fin angle (ψ) with A when the population size is 20. It can be seen that ψ increases with the increase of A. The reason is that the role of heat conduction in total heat transfer becomes more important with the increase of A. Although increasing $\lambda_{fin}/\lambda_{PCM}$ and fin area both increase the role of heat conduction, they have different effects on the optimal L. It can be found from Figures 9 and 12 that the optimal L increases with the increase of $\lambda_{fin}/\lambda_{PCM}$, while it decreases with the increase of the fin area. This result indicates that, unlike the optimal ψ, the optimal L does not simply change with the relative size of heat conduction and natural convection.

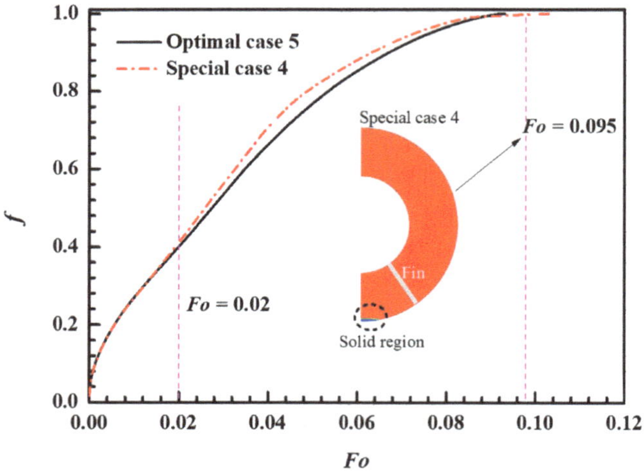

Figure 11. Variation of the liquid fraction, f, with dimensionless melting time, Fo, for Optimal Case 5 and Special Case 4.

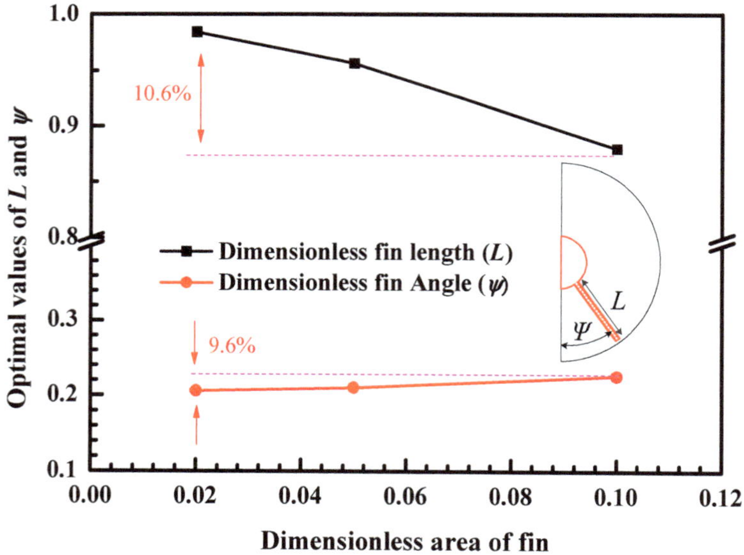

Figure 12. Variation of the optimal dimensionless fin length, L, and dimensionless fin angle, ψ, with the dimensionless fin area, A.

Figure 13 shows the distribution of liquid fraction and dimensionless temperature for Optimal Cases 6 and 7 at three moments. Optimal Cases 6 and 7 use the fins optimized for $A = 0.05$ and $A = 0.1$, respectively. For the optimal design of fin for $A = 0.02$, see Optimal Case 2 mentioned in Section 3.1. The contour maps of Optimal Case 2 can be found in Figure 10, so it is not repeated in Figure 13. It can be seen that these three optimal cases have almost the same melting process, and the PCM in the upper and lower region of the fin has a similar melting ratio.

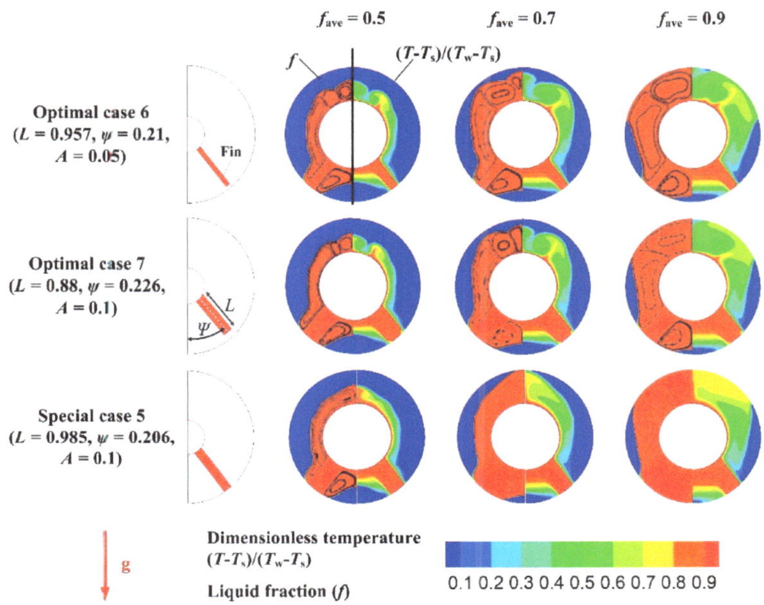

Figure 13. Contour maps of liquid fraction and dimensionless temperature for Optimal Cases 6 and 7 and Special Case 5.

It also can be found from Figure 12 that although the value of A is increased by four times, the optimal values of L and ψ only change by 10.6% and 9.6%, respectively. It means that the fin area or fin fill rate is not a sensitive parameter for the optimal position and structure of the fin. In fact, in order to maintain the heat storage capacity, reduce the investment cost, and control the total mass of the device, the fin area is not allowed to increase excessively.

To further prove that there should be different fin designs for different A values, a special case called Special Case 5 was designed. The values of L and ψ of Special Case 5 are the same as Optimal Case 2, but the A of Special Case 5 is set to 0.1. That is to say, the fin used in Special Case 5 is thicker than the fin used in Optimal Case 2. The distribution of liquid fraction and dimensionless temperature for Special Cases 5 is also shown in Figure 13. It can be observed that, compared with the fin design of Optimal Case 7, the fin design of Special Case 5 results in a greater melting rate of the PCM under the fin. Figure 14 gives the variation of the liquid fraction, f, with the dimensionless melting time, Fo, for Optimal Case 7 and Special Case 5. It can be seen that when the dimensionless melting time exceeds 0.02 ($Fo > 0.02$), Optimal Case 7 has a faster melting rate than Special Case 5. When $Fo = 0.076$, the PCM in Optimal Case 7 has been completely melted ($f = 1.0$), while the solid PCM can still be found in the upper region of Special Case 5. From Figures 13 and 14, it can be concluded that the optimal fin can make the melting more uniform, thus shortening the melting time.

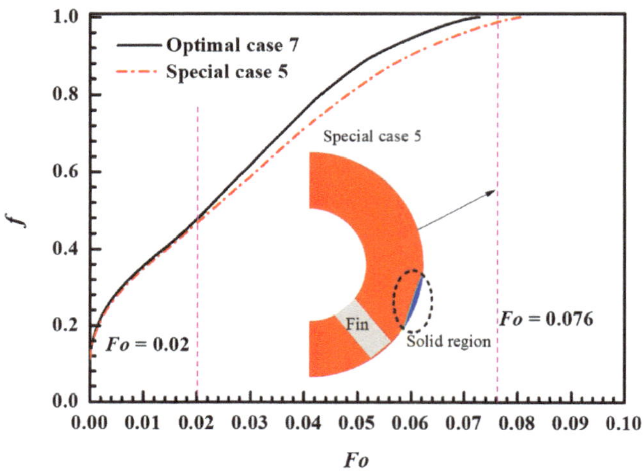

Figure 14. Variation of liquid fraction, f, with the dimensionless melting time, Fo, for Optimal Case 7 and Special Case 5.

3.4. Effect of the Number of Fins on the Optimal Structural Parameters of Fins

In this section, the effect of the number of fins on the optimization results is discussed. The dimensionless fin area is fixed at 0.05, and the thermal conductivity ratio of the fin to the PCM is 1987.7 ($\lambda_{fin}/\lambda_{PCM}$ = 1987.7). The structure and position of single fin, double fins, and four fins are optimized. For the structure and position of the single fin, see Optimal Case 2, which is given in Figure 10. The optimal structure and position of the double fins and four fins are given in Figure 15. It can be seen that if the number of fins is one or two, all fins should be placed in the lower half, indicating that natural convection is still dominant. However, when the number of fins reaches four, the fins are arranged almost uniformly. This signifies that conduction is dominant. Figure 16 gives the contour maps of liquid fraction and velocity magnitude for Optimal Cases 8 and 9. Optimal Cases 8 and 9 use the optimized two fins and four fins, respectively. It can be seen that the velocity magnitude $(u^2+v^2)^{0.5}$ of Optimal Case 9 is obviously slighter than that of Optimal Case 8 when Fo = 0.36. This result means that, in Optimal Case 9, which has four fins, natural convection is relatively weak, so the effects of heat conduction play a bigger role.

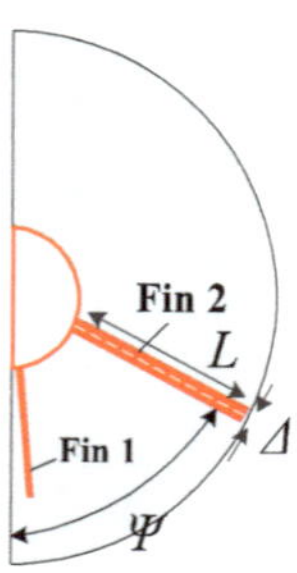

	L	Δ	Ψ
Fin 1	0.663	0.036	0.034
Fin 2	0.998	0.094	0.350

Two fins

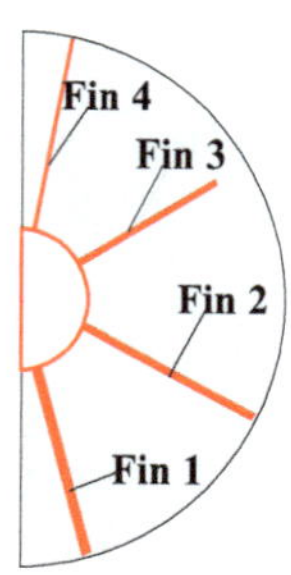

	L	Δ	Ψ
Fin 1	0.970	0.055	0.078
Fin 2	0.984	0.031	0.348
Fin 3	0.813	0.030	0.671
Fin 4	0.996	0.010	0.941

Four fins

Figure 15. Optimal structure and position of the double fins and four fins.

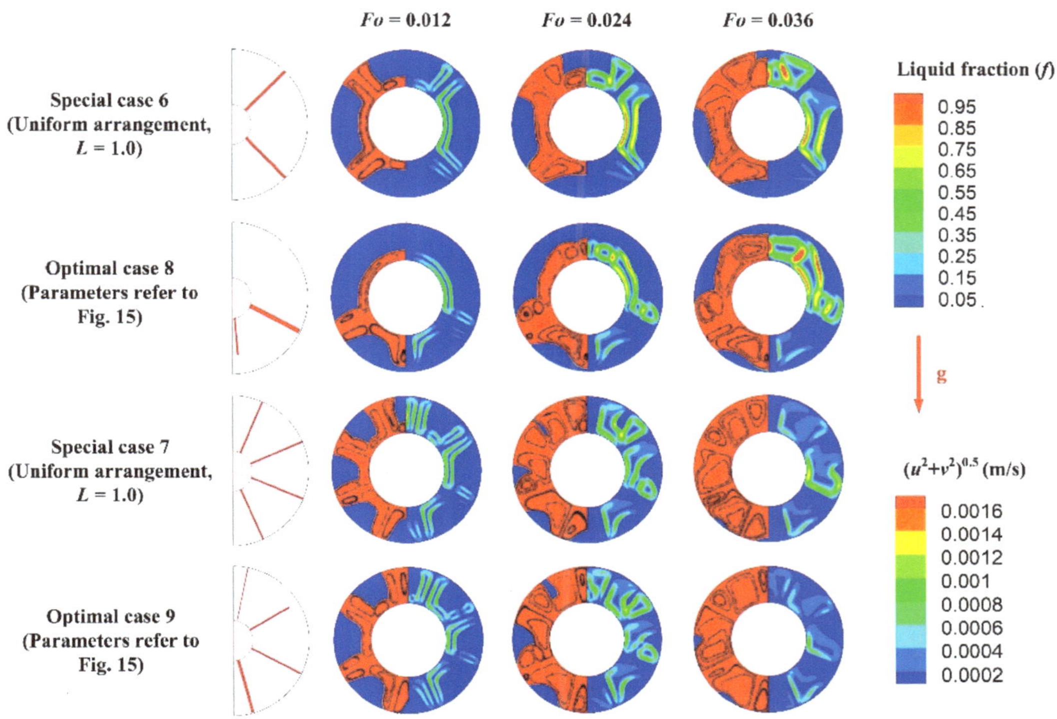

Figure 16. Contour maps of liquid fraction and velocity magnitude for Special Cases 6 and 7 and Optimal Cases 8 and 9.

Figure 15 also shows that when the number of fins exceeds one, different fins have different structures and locations after the optimization. To prove this, two special cases named Special Cases 6 and 7 were designed. Case 6 has two fins, and Case 7 has four fins. All the fins in Special Cases 6 and 7 have a full length ($L = 1.0$) and are uniformly arranged. The contour maps of liquid fraction and velocity magnitude for Special Cases 6 and 7 are also given in Figure 16. It can be seen that, compared with the optimized fins, the uniformly arranged fins lead to a more solid PCM in the bottom region. This limits the later melting rate because the PCM in the bottom is difficult to melt. Figure 17 shows the comparison of dimensionless total melting time, Fo_{tot}, with optimized fins and uniformly arranged fins. It can be seen that using optimized fins can obtain a smaller Fo_{tot} compared with using uniformly arranged fins. The highest reduction of Fo_{tot} is 61.3% when one fin is used. As the number of fins increases, the reduction of Fo_{tot} decreases. This is because, as the number of fins increases, the role of conduction in total heat transfer becomes more important. When the number of fins is four, Fo_{tot} can still be reduced by 15.9%, which indicates that natural convection still has a significant effect on heat transfer. The results of this section indicate that the number of fins is a sensitive parameter that affects the optimal position and structure of the fins.

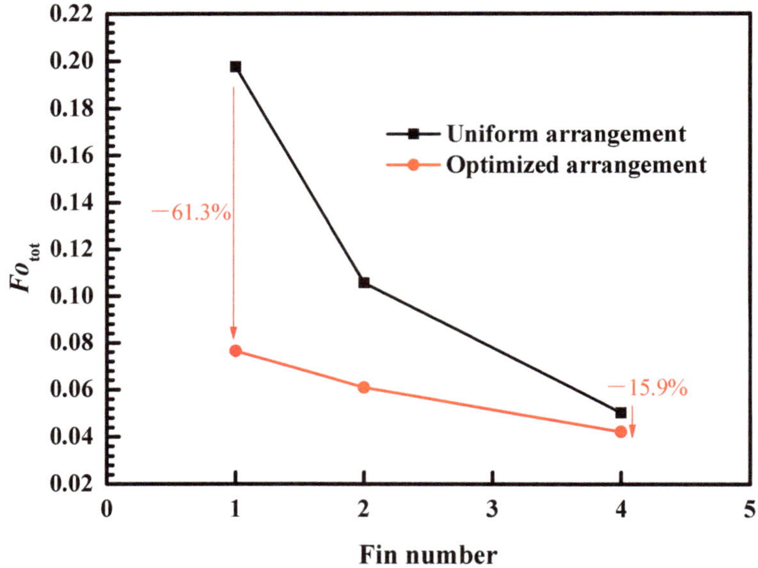

Figure 17. Variation of dimensionless total melting time, Fo_{tot}, with the number of fins.

4. Conclusions

Longitudinal rectangular fins are commonly applied in the shell-and-tube latent heat storage unit for melting-performance enhancement. In this study, a new coupling algorithm between genetic algorithm (GA) and computational fluid dynamics (CFD) was developed to optimize the structural parameters of the fins. The total melting time is selected as the objective; the length, width, and position of each fin as variables; and the area of fins as constraint. The effects of the thermal conductivity of fins, the area of fins, and the number of fins on the optimal structure of fins were discussed. The following conclusions can be made:

(1) A single fin inserted in the half-ring region should be placed in the lower half due to the effects of natural convection. The optimal dimensionless fin angle, ψ, is about 0.2, and the optimal dimensionless fin length, L, is about 0.96. The use of optimal fins can shorten the dimensionless total melting time, Fo_{tot}, by 68% compared with the case of no fin, and 61.3% compared with uniformly arranged single fin.

(2) Enlarging fin area will increase the optimal ψ and decrease the optimal L when the fin area is added. Nevertheless, the fin area is not a sensitive parameter that affects the optimal position and structure of the fin. When the value of A is increased by four times, the optimal values of L and ψ only change by 10.6% and 9.6%, respectively.

(3) Thermal conductivity is also not a sensitive parameter for the optimal position and structure of the fin since the area of the fin is limited. As the value of $\lambda_{fin}/\lambda_{PCM}$ increases from 83.6 to 1987.7, the optimal values of L and ψ are only increased by 14.7% and 11%, respectively.

(4) The number of fins sensitively affects the optimal position and structure of the fins. When the number of fins exceeds one, fins should have respective structural parameters, as well as non-uniform arrangements for effective utilization of natural convection. The advantage of non-uniform fins decreases as the number of fins increases. When the number of fins is four, using the optimized fins can only reduce the Fo_{tot} by 15.9% compared with using the uniformly arranged fins.

Author Contributions: Conceptualization, Z.-J.Z.; data curation, C.H.; formal analysis, C.H.; investigation, Y.X.; methodology, H.Y.; project administration, Z.-J.Z.; software, H.Y.; validation, Y.W.; visualization, M.C.; writing—original draft, Y.X. All authors have read and agreed to the published version of the manuscript.

Funding: This research was funded by the National Natural Science Foundation of China (No. 52176215).

Conflicts of Interest: The authors declare no conflict of interest.

Nomenclature

A	dimensionless fin area
A_{mush}	mushy zone constant (kg m^{-3} s)
c_p	specific heat (J kg^{-1} K^{-1})
D	diameter (m)
D_e	hydraulic diameter (m)
f	melting fraction
Fo	Fourier number
g	acceleration of gravity (m s^{-2})
Δh	latent heat of fusion (kJ kg^{-1})
K	permeability (m^2)
l	fin length (m)
L	dimensionless fin length
p	pressure (Pa)
r	radius (m)
Ra	Rayleigh number
Ste	Stefan number
t	time (s)
T	temperature (K)
u, v	x and r velocity components (m s^{-1})
w	fin width (m)
x, y	Cartesian coordinates (m)
Abbreviation	
CFD	computational fluid dynamics
FVM	finite volume method
GA	genetic algorithm
HTF	heat-transfer fluid
LHTES	latent heat thermal energy storage
PCM	phase-change material
TES	thermal energy storage
Greek symbols	
α	thermal diffusivity (m^2 s^{-1})
β	thermal expansion coefficient (K^{-1})
δ	small constant number with a value of 0.001
Δ	dimensionless fin width
ε	integer with a value of 0 or 1
θ	angle (rad)
λ	thermal conductivity (W m^{-1} K^{-1})
μ	kinetic viscosity (kg m^{-1} s^{-1})
π	ratio of the circumference to the diameter of a circle
ρ	density (kg m^{-3})
Ψ	dimensionless angle

subscripts

ave	average
i	inner tube
m	melting
o	outer tube
ref	reference
s	solidification
tot	total
w	wall

References

1. Guo, J.; Du, Z.; Liu, G.; Yang, X.; Li, M.-J. Compression effect of metal foam on melting phase change in a shell-and-tube unit. *Appl. Therm. Eng.* **2022**, *206*, 118124. [CrossRef]
2. Mosaffa, A.H.; Talati, F.; Basirat Tabrizi, H.; Rosen, M.A. Analytical modeling of PCM solidification in a shell and tube finned thermal storage for air conditioning systems. *Energy Build.* **2012**, *49*, 356–361. [CrossRef]
3. Huang, X.; Yao, S.; Yang, X.; Zhou, R.; Luo, J.; Shen, X. Comparison of solidification performance enhancement strategies for a triplex-tube thermal energy storage system. *Appl. Therm. Eng.* **2022**, *204*, 117997. [CrossRef]
4. Chen, D.; Riaz, A.; Aute, V.C.; Radermacher, R. A solid–liquid model based on lattice Boltzmann method for phase change material melting with porous media in cylindrical heat exchangers. *Appl. Therm. Eng.* **2022**, *207*, 118080. [CrossRef]
5. Liu, Z.; Wang, Z.; Li, P.; Qin, H.; Heng, W. Multi-parameter heat transfer analysis of molten PCM in an inclined enclosure. *Appl. Therm. Eng.* **2022**, *208*, 118209. [CrossRef]
6. Zhang, S.; Pu, L.; Xu, L.; Dai, M. Study on dominant heat transfer mechanism in vertical smooth/finned-tube thermal energy storage during charging process. *Appl. Therm. Eng.* **2022**, *204*, 117935. [CrossRef]
7. Behi, H.; Behi, M.; Ghanbarpour, A.; Karimi, D.; Azad, A.; Ghanbarpour, M.; Behnia, M. Enhancement of the Thermal Energy Storage Using Heat-Pipe-Assisted Phase Change Material. *Energies* **2021**, *14*, 6176. [CrossRef]
8. Yang, C.; Xu, Y.; Cai, X.; Zheng, Z.-J. Effect of the circumferential and radial graded metal foam on horizontal shell-and-tube latent heat thermal energy storage unit. *Sol. Energy* **2021**, *226*, 225–235. [CrossRef]
9. Tselepi, M.; Prouskas, C.; Papageorgiou, D.G.; Lagaris, I.E.; Evangelakis, G.A. Graphene-Based Phase Change Composite Nano-Materials for Thermal Storage Applications. *Energies* **2022**, *15*, 1192. [CrossRef]
10. Karimi, D.; Behi, H.; Van Mierlo, J.; Berecibar, M. An Experimental Study on Thermal Performance of Graphite-Based Phase-Change Materials for High-Power Batteries. *Energies* **2022**, *15*, 2515. [CrossRef]
11. Eslamnezhad, H.; Rahimi, A.B. Enhance heat transfer for phase-change materials in triplex tube heat exchanger with selected arrangements of fins. *Appl. Therm. Eng.* **2017**, *113*, 813–821. [CrossRef]
12. Kalhori, B.; Ramadhyani, S. Studies on heat transfer from a vertical cylinder, with or without fins, embedded in a solid phase change medium. *J. Heat Transf.* **1985**, *107*, 44–51. [CrossRef]
13. Rabienataj Darzi, A.A.; Jourabian, M.; Farhadi, M. Melting and solidification of PCM enhanced by radial conductive fins and nanoparticles in cylindrical annulus. *Energy Convers. Manag.* **2016**, *118*, 253–263. [CrossRef]
14. Agyenim, F.; Eames, P.; Smyth, M. A comparison of heat transfer enhancement in a medium temperature thermal energy storage heat exchanger using fins. *Sol. Energy* **2009**, *83*, 1509–1520. [CrossRef]
15. Li, Z.; Wu, Z.-G. Analysis of HTFs, PCMs and fins effects on the thermal performance of shell–tube thermal energy storage units. *Sol. Energy* **2015**, *122*, 382–395. [CrossRef]
16. Rozenfeld, T.; Kozak, Y.; Hayat, R.; Ziskind, G. Close-contact melting in a horizontal cylindrical enclosure with longitudinal plate fins: Demonstration, modeling and application to thermal storage. *Int. J. Heat Mass Transf.* **2015**, *86*, 465–477. [CrossRef]
17. Padmanabhan, P.V.; Krishna Murthy, M.V. Outward phase change in a cylindrical annulus with axial fins on the inner tube. *Int. J. Heat Mass Transf.* **1986**, *29*, 1855–1868. [CrossRef]
18. Ismail, K.A.R.; Alves, C.L.F.; Modesto, M.S. Numerical and experimental study on the solidification of PCM around a vertical axially finned isothermal cylinder. *Appl. Therm. Eng.* **2001**, *21*, 53–77. [CrossRef]
19. Zhang, C.; Sun, Q.; Chen, Y. Solidification behaviors and parametric optimization of finned shell-tube ice storage units. *Int. J. Heat Mass Transf.* **2020**, *146*, 118836. [CrossRef]
20. Hosseini, M.J.; Ranjbar, A.A.; Rahimi, M.; Bahrampoury, R. Experimental and numerical evaluation of longitudinally finned latent heat thermal storage systems. *Energy Build.* **2015**, *99*, 263–272. [CrossRef]
21. Solomon, G.R.; Velraj, R. Analysis of the heat transfer mechanisms during energy storage in a Phase Change Material filled vertical finned cylindrical unit for free cooling application. *Energy Convers. Manag.* **2013**, *75*, 466–473. [CrossRef]
22. Yuan, Y.; Cao, X.; Xiang, B.; Du, Y. Effect of installation angle of fins on melting characteristics of annular unit for latent heat thermal energy storage. *Sol. Energy* **2016**, *136*, 365–378. [CrossRef]
23. Abhat, A.; Aboul-Enein, S.; Malatidis, N.A. Heat-of-fusion storage systems for solar heating applications. In *Thermal Storage of Solar Energy*; Springer: Dordrecht, The Netherlands, 1981; pp. 157–171.
24. Bathelt, A.G.; Viskanta, R. Heat transfer and interface motion during melting and solidification around a finned heat source/sink. *J. Heat Transf.* **1981**, *103*, 720–726. [CrossRef]

25. Sparrow, E.M.; Larson, E.D.; Ramsey, J.W. Freezing on a finned tube for either conduction-controlled or natural-convection-controlled heat transfer. *Int. J. Heat Mass Transf.* **1981**, *24*, 273–284. [CrossRef]
26. Rathod, M.K.; Banerjee, J. Thermal performance enhancement of shell and tube Latent Heat Storage Unit using longitudinal fins. *Appl. Therm. Eng.* **2015**, *75*, 1084–1092. [CrossRef]
27. Dhaidan, N.S.; Khodadadi, J.M. Improved performance of latent heat energy storage systems utilizing high thermal conductivity fins: A review. *J. Renew. Sustain. Energy* **2017**, *9*, 034103. [CrossRef]
28. Wang, P.; Yao, H.; Lan, Z.; Peng, Z.; Huang, Y.; Ding, Y. Numerical investigation of PCM melting process in sleeve tube with internal fins. *Energy Convers. Manag.* **2016**, *110*, 428–435. [CrossRef]
29. Mahood, H.B.; Mahdi, M.S.; Monjezi, A.A.; Khadom, A.A.; Campbell, A.N. Numerical investigation on the effect of fin design on the melting of phase change material in a horizontal shell and tube thermal energy storage. *J. Energy Storage* **2020**, *29*, 101331. [CrossRef]
30. Tao, Y.B.; He, Y.L. Effects of natural convection on latent heat storage performance of salt in a horizontal concentric tube. *Appl. Energy* **2015**, *143*, 38–46. [CrossRef]
31. Deng, S.; Nie, C.; Wei, G.; Ye, W.-B. Improving the melting performance of a horizontal shell-tube latent-heat thermal energy storage unit using local enhanced finned tube. *Energy Build.* **2019**, *183*, 161–173. [CrossRef]
32. Deng, S.; Nie, C.; Jiang, H.; Ye, W.-B. Evaluation and optimization of thermal performance for a finned double tube latent heat thermal energy storage. *Int. J. Heat Mass Transf.* **2019**, *130*, 532–544. [CrossRef]
33. Kumar, R.; Verma, P. An experimental and numerical study on effect of longitudinal finned tube eccentric configuration on melting behaviour of lauric acid in a horizontal tube-in-shell storage unit. *J. Energy Storage* **2020**, *30*, 101396. [CrossRef]
34. Yagci, O.K.; Avci, M.; Aydin, O. Melting and solidification of PCM in a tube-in-shell unit: Effect of fin edge lengths' ratio. *J. Energy Storage* **2019**, *24*, 100802. [CrossRef]
35. Mao, Q.; Hu, X.; Zhu, Y. Numerical Investigation of Heat Transfer Performance and Structural Optimization of Fan-Shaped Finned Tube Heat Exchanger. *Energies* **2022**, *15*, 5682. [CrossRef]
36. Sun, X.; Mahdi, J.M.; Mohammed, H.I.; Majdi, H.S.; Zixiong, W.; Talebizadehsardari, P. Solidification Enhancement in a Triple-Tube Latent Heat Energy Storage System Using Twisted Fins. *Energies* **2021**, *14*, 7179. [CrossRef]
37. Chen, Q.; Wu, J.; Sun, K.; Zhang, Y. Numerical Study of Heat Transfer Enhancement by Arc-Shaped Fins in a Shell-Tube Thermal Energy Storage Unit. *Energies* **2022**, *15*, 7799. [CrossRef]
38. Abdulateef, A.M.; Mat, S.; Abdulateef, J.; Sopian, K.; Al-Abidi, A.A. Geometric and design parameters of fins employed for enhancing thermal energy storage systems: A review. *Renew. Sustain. Energy Rev.* **2018**, *82 Pt 1*, 1620–1635. [CrossRef]
39. Cai, X.; Zheng, Z.-J.; Yang, C.; Xu, Y. Improving the solidification performance of a shell-and-tube latent-heat thermal energy storage unit using a connected-Y-shaped fin. *Int. J. Energy Res.* **2022**, *46*, 12758–12771. [CrossRef]
40. Xu, Y.; Li, M.J.; Zheng, Z.J.; Xue, X.D. Melting performance enhancement of phase change material by a limited amount of metal foam: Configurational optimization and economic assessment. *Appl. Energy* **2018**, *212*, 868–880. [CrossRef]
41. Xu, Y.; Ren, Q.L.; Zheng, Z.J.; He, Y.L. Evaluation and optimization of melting performance for a latent heat thermal energy storage unit partially filled with porous media. *Appl. Energy* **2017**, *193*, 84–95. [CrossRef]
42. Zheng, Z.J.; Xu, Y.; Li, M.J. Eccentricity optimization of a horizontal shell-and-tube latent-heat thermal energy storage unit based on melting and melting-solidifying performance. *Appl. Energy* **2018**, *220*, 447–454. [CrossRef]
43. Xu, Y.; Zheng, Z.-J.; Chen, S.; Cai, X.; Yang, C. Parameter analysis and fast prediction of the optimum eccentricity for a latent heat thermal energy storage unit with phase change material enhanced by porous medium. *Appl. Therm. Eng.* **2021**, *186*, 116485. [CrossRef]
44. Zheng, Z.-J.; Cai, X.; Yang, C.; Xu, Y. Improving the solidification performance of a latent heat thermal energy storage unit using arrow-shaped fins obtained by an innovative fast optimization algorithm. *Renew. Energy* **2022**, *195*, 566–577. [CrossRef]
45. Zheng, Z.-J.; Yang, C.; Xu, Y.; Cai, X. Effect of metal foam with two-dimensional porosity gradient on melting behavior in a rectangular cavity. *Renew. Energy* **2021**, *172*, 802–815. [CrossRef]
46. Zheng, Z.J.; Li, M.J.; He, Y.L. Optimization of porous insert configurations for heat transfer enhancement in tubes based on genetic algorithm and CFD. *Int. J. Heat Mass Transf.* **2015**, *87*, 376–379. [CrossRef]
47. Zheng, Z.J.; He, Y.; He, Y.L.; Wang, K. Numerical optimization of catalyst configurations in a solar parabolic trough receiver–reactor with non-uniform heat flux. *Sol. Energy* **2015**, *122*, 113–125. [CrossRef]
48. Zheng, Z.J.; Li, M.J.; He, Y.L. Thermal analysis of solar central receiver tube with porous inserts and non-uniform heat flux. *Appl. Energy* **2017**, *185*, 1152–1161. [CrossRef]
49. Zheng, Z.J.; Xu, Y.; He, Y.L. Study on the performance of a shell-and-tube latent-heat storage unit enhanced by porous medium with graded porosity (in chinese). *J. Eng. Thermophys.* **2019**, *40*, 129–135.
50. Xu, Y.; Zheng, Z.J.; Li, M.J. A half-analytical correlation of total melting time for shell-and-tube latent-heat thermal energy storage unit. *Appl. Therm. Eng.* **2019**, *161*, 114176. [CrossRef]
51. Xu, Y.; Zheng, Z.J.; Yang, C.; Cai, X. Intelligent optimization of horizontal fins to improve the melting performance of phase change materials in a square cavity with isothermal vertical wall. *J. Energy Storage* **2021**, *44*, 103334. [CrossRef]
52. Zheng, Z.J.; Xu, Y.; He, Y.L. Thermal analysis of a solar parabolic trough receiver tube with porous insert optimized by coupling genetic algorithm and CFD. *Sci. China Technol. Sci.* **2016**, *59*, 1475–1485. [CrossRef]

energies

Article

Integration of a Linear Cavity Receiver in an Asymmetric Compound Parabolic Collector

Dimitrios N. Korres [1], Evangelos Bellos [1,2,*] and Christos Tzivanidis [1]

[1] Department of Thermal Engineering, National Technical University of Athens, Zografou, 157 80 Athens, Greece
[2] Department of Mechanical Engineering Educators, School of Pedagogical and Technological Education (ASPETE), 151 22 Amarousion, Greece
* Correspondence: bellose@central.ntua.gr

Abstract: The objective of this work is the integration of a linear cavity receiver in an asymmetric compound parabolic collector. Two different numerical models were developed; one for the conventional geometry and one for the cavity configuration. Both models were examined for inlet temperatures from 20 °C up to 80 °C, considering water as the operating fluid with a typical volume flow rate of 15 lt/h. Emphasis was given to the comparison of the thermal and optical performance between the designs, as well as in the temperature levels of the fluids and the receiver. The geometry of the integrated cavity receiver was optimized according to two independent parameters and two possible optimum designs were finally revealed. The optimization took place regarding the optical performance of the collector with the cavity receiver. The simulation results indicated that the cavity design leads to enhancements of up to 4.40% and 4.00% in the optical and thermal efficiency respectively, while the minimum possible enhancement was above 2.20%. The mean enhancements in optical and thermal performance were found to be 2.90% and 2.92% respectively. Moreover, an analytical solution was developed for verifying the numerical results and the maximum deviations were found to be less than 5% in all the compared parameters. Especially, in thermal efficiency verification, the maximum deviation took a value of less than 0.5%. The design and the simulations in the present study were conducted with the SolidWorks Flow Simulation tool.

Keywords: compound parabolic concentrator; asymmetric reflector; computational fluid dynamics; solar concentrating power; cavity receiver

1. Introduction

Concentrating Solar Thermal Collectors (CSTC) are able to serve a great range of solar thermal applications and are able to be integrated into domestic hot water systems, desalination and dehumidification set-ups, absorption chiller set-ups (solar cooling systems), and power production applications [1–5]. There is a wide agenda that has been developed around the optimization of such systems as concerns the optical and thermal efficiency, with a variety of optimization technics. The most well-known performance enhancement methods are related to the improvement of the convective heat transfer from the receiver to the thermal fluid and they are applied to linear concentrators such as Compound Parabolic Collectors (CPCs) and Parabolic Trough Collectors (PTCs). This enhancement could be achieved with the use of special working mediums such as nanofluids, with sophisticated flow tube geometries (ribbed and corrugated tubes), and with the integration of flow inserts (metal foams, twisted tapes, fins, etc.) along the flow of the working fluid.

There are many studies in which such methods have been proposed for CPCs and PTCs. Korres et al. [6] investigated a nanofluid-based compound parabolic collector (CPC) and they compared it with the case where only the base fluid is applied in laminar flow conditions. The results were positive since the thermal efficiency was improved sufficiently with the enhancement reaching 2.8%. Rehan et al. [7] studied two different nanofluids (Fe_2O_3/H_2O and Al_2O_3/H_2O) as working mediums in a PTC at several different concentrations. The efficiency of the collector was increased and the increment was more significant in higher concentrations and especially in the case where Al_2O_3/H_2O is applied.

A PTC with a hybrid nanofluid and a double twisted tape inside the flow tube was examined by Alnaqi et al. [8]. It was found that there was a specific orientation of the swirling direction where the thermal efficiency was maximized. As regards the hybrid nanofluid, a higher concentration leads to lower thermal losses. It is important to mention that many studies have examined nanofluid integrations numerically, with high particle concentrations and important enhancements came out. However, it should be taken into account that in experimental set-ups has been proved that agglomeration problem occurs and it has a negative impact in such enhancement when the concentrations take very high values [6,9,10]. Liu et al. [11] conducted experiments in outdoor conditions regarding a CPC and a significant enhancement was found with the use of water/CuO nanofluid. In another work, Bellos et al. [12] investigated the integration of a wavy-walled flow tube in a PTC, using a nanofluid as the thermal fluid. The simulation indicated a slight enhancement in thermal efficiency compared with a typical flow tube. Lu et al. [13] used a water-based nanofluid with CuO nanoparticles in a CPC by performing indoor experiments and a very good enhancement was revealed with the use of the nanofluid. Mwesigye et al. [14] used perforated plates parallel to each other inside a PTC flow tube receiver. The positioning of the plates was optimized in order to increase the convective heat transfer and thus the thermal efficiency was enhanced sufficiently. Liu et al. [15] inserted two twisted tapes in a PTC and they investigated the flow regime. It was revealed that the swirling of the fluid inside the tube led to the increment of the convective heat transfer coefficient. In the study of Ref. [16], a PTC with a wavy-type metal strip inserted in the flow tube was investigated. The research showed that there is an important drop in the heat losses which was greater than 15% and thus a significant enhancement was achieved in the thermal efficiency. Similar methods have been applied in studies [17–20].

An equally important role in the thermal output, also, is played by the receiver design and positioning as well as the reflector's geometry. As far as the enhancement technics in this field several studies have been conducted, especially with linear cavity receivers, to be an alternative solution. Most of these studies are referred to as PTCs. Korres and Tzivanidis [21] are the first who studied the effect of the angular aperture of circular cavity receivers in the absorbed solar irradiation and they developed two semi-empirical relationships in order to calculate the equivalent absorptance of such receivers, considering the entrapment of solar irradiation inside the cavity. In addition, the study of Korres and Tzivanidis [22] examined the operation of a PTC with an integrated single cavity receiver inside an evacuated glass envelope. This proposal, which appears for the first time in scientific literature, was found to ensure a great thermal performance enhancement of about 12.2% compared to a conventional PTC. Moreover, the same authors investigated the optical and thermal performance of a PTC with a double circular cavity receiver in a vacuum environment [23] and the thermal performance was enhanced by approximately 16%. It should be mentioned that in studies [22,23], the cavity receivers were first optimized. A PTC with a partially-evacuated circular cavity receiver was examined by Avargani et al. [24] and higher thermal performances were achieved compared to the conventional design. More proposals regarding cavity receivers and special reflector designs could be found in many literature studies [25–37].

Considering the lack of literature regarding enhancement methods in CPCs and especially in cavity receivers, the present study is dedicated to this field. In particular, an innovative linear single cavity receiver with a single glass evacuated tube is integrated into a novel asymmetric compound parabolic collector (ACPC) design which has been developed by the authors in a previous study [25]. The solar collector that appears in the study [25] has a maximum optical efficiency of 77.68% and it was found to exceed by much, three other similar geometries from literature as far as optical efficiency is concerned. In this study, two different numerical models were developed; one for the conventional geometry of study [25] and one for the cavity configuration. Both models were examined for inlet temperatures from 20 °C up to 80 °C as in studies [3,25], considering water as the working fluid with a typical volume flow rate of 15 lt/h [3]. Emphasis was given to the comparison of the thermal and optical performance as well as in the fluid's outlet and the absorber mean temperature. The geometry of the integrated cavity receiver was optimized according to two independent parameters and two possible optimum designs were revealed. The optimization aims to the maximization of the collector's optical performance. The global simulation results indicated an enhancement of up to 4.4% and 4% in the optical and thermal efficiency respectively. An analytical solution was developed for verifying the numerical results and the maximum deviations were found to be less than 1% in all the compared parameters. The design and the simulations were performed with SolidWorks which is a proper tool for performing both optical and thermal studies. To our knowledge, there is no other study in literature where such receiver designs have been proposed for CPCs. Hence, the proposed solar collector appears for the first time in the international scientific literature. In particular, it combines a linear cavity receiver, enclosed in a single glass evacuated tube, with an asymmetrical concentrator and it comes to substitute conventional systems, providing higher optical and thermal efficiency. This combination is originally unique and its investigation constitutes a significant contribution to global research on the solar thermal systems field.

2. Material and Methods

2.1. Examined Collectors

An innovative CPC with an asymmetric reflector and a single cavity receiver (SC-ACPC) is examined in the present study and compared with a similar geometry taken from literature with a conventional receiver (CNV-ACPC) [25]. The geometry of each collector is depicted in Figure 1. According to this Figure, the only difference between the two collectors is the receiver geometry and the evacuated tube type. More particularly, in the SC-ACPC there is a single cavity receiver with an integrated flow tube placed at the interior of a single glass evacuated tube. In the case of the CNV-ACPC, a typical double glass evacuated tube with a U-pipe hydroskeleton is applied. The reflector obtains the same design as in Ref. [25], while the placement of the glass tube with respect to the reflector is identical to the one that was followed in the study of Ref. [25]. The dimensions of the collectors can be found in Table 1. It should be mentioned that in the double glass case, the inner glass cover plays the role of the receiver. Hence, in this case, the copper fin's outer diameter is identical to the inner diameter of the receiver.

Table 1. Dimensions of the examined CPCs.

Dimension	SC-ACPC	CNV-ACPC
Mirror length	1440 mm	1440 mm
Mirror width	153 mm	153 mm
Cover outer diameter	47 mm	47 mm
Cover thickness	1.5 mm	1.5 mm
Copper fin outer diameter	-	30 mm
Copper fin thickness	-	0.6 mm
Receiver outer diameter	33 mm	33 mm
Receiver thickness	0.6 mm	1.5 mm
Flow pipe external diameter	10 mm	9.52 mm
Flow pipe thickness	1 mm	0.65 mm

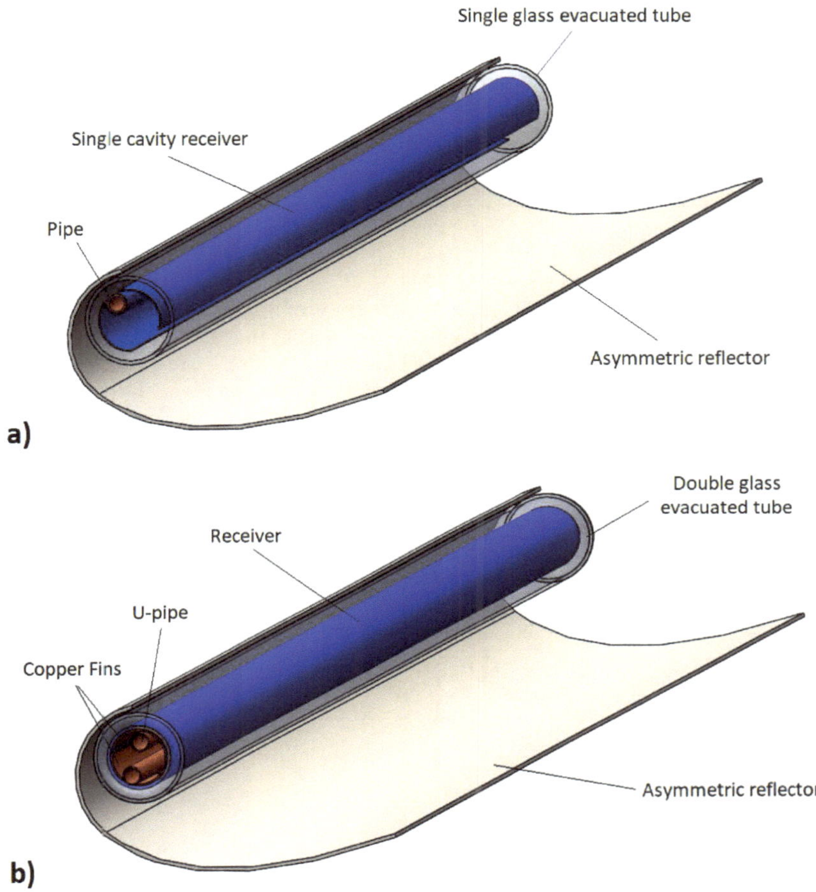

Figure 1. Design of the examined CPCs: (**a**) SC-ACPC (novel design) and (**b**) CNV-ACPC ([25]). Reprinted with permission from Ref. [25]. 2022, Elsevier.

The flow pipes and the fins were considered to be made of copper. The same was assumed for the receiver in the SC-ACPC case. All the considered materials of each component are given in Table 2, while the respective thermal and optical characteristics are provided in Table 3. The values in Table 3 are typical ones and were taken from studies [3,25]. It should be mentioned that the material of all the components in SC-ACPC was considered to be the same as the respective of the CNV-ACPC and thus with the experimental set-up in the study [25]. In addition, all the properties were taken according to the same study.

Table 2. CPCs components materials.

Object	Material
	SC-ACPC and CNV-ACPC
Flow tube	Copper
Receiver/Fins	Copper
Evacuated tube	Borosilicate glass
Concentrator	Stainless Steel with mirror finish

Table 3. Optical and thermal properties [3,25].

Parameter	Symbol	Value
Absorbance	α	0.92
Receiver emittance	ε_r	0.08
Transmittance	τ	0.915
Glass emittance	ε_g	0.88
Reflectance	ρ	0.902

2.2. Methodology and Implementation

2.2.1. Optical Optimization of the Cavity Receiver

The cavity receiver in the SC-ACPC geometry was optimized aiming to achieve maximum optical efficiency. It is essential to state that in the SC-ACPC case both the inner and the outer surfaces were considered to be coated with the same selective film of 0.92 absorptance and 0.08 emittance, as in the CNV-ACPC configuration. The optimization was conducted by modifying two independent parameters. More specifically, the cavity's angular aperture (φ) and the cavity rotation angle (ω) were taken to several different values (see Figure 2) and the optical performance of the collector was found each time.

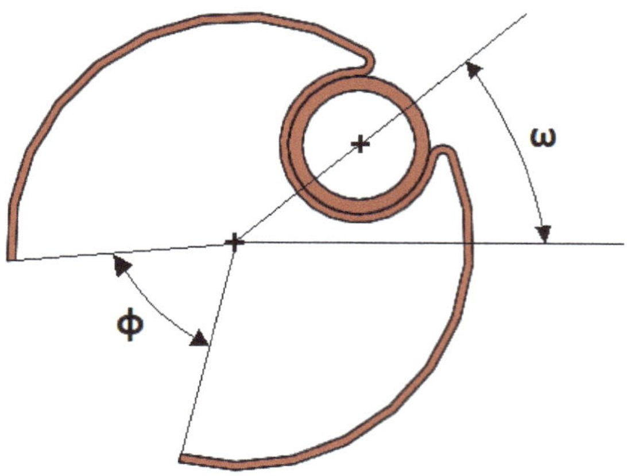

Figure 2. Parameters of optimization.

For the ray tracing process, solar rays were considered to be parallel to the collector's width, while several different transversal angles of incidence (θ_{tr}) were examined. The details of the ray tracing parameters are available in Table 4. There were 16 incident angles, 7 cavity rotation angles, and 11 angular apertures which were combined with each other with 1232 total simulation scenarios conducted.

Table 4. Optical simulation parameters.

Parameter	Value
Effective solar irradiation intensity ($G_{eff,n}$)	800 W/m^2
Simulated solar rays	10^6
Transversal incident angle (θ_T)	0°–60°
Cavity's angular aperture (φ)	40°–140° per 10°
Cavity's rotation angle (ω)	20°–80° per 10°

2.2.2. Operating Conditions in Thermal Analysis

In this section, a detailed thermal analysis is put forward for the calculation of the thermal efficiency, the thermal losses, and the temperature regime in both collectors. The analysis was conducted for a volume flow rate of 15 lt/h, which is a typical one in such applications [3], while the inlet temperature took values in the range of 20 °C up to 80 °C, as in other studies [3,25]. Water was assumed to be the working medium of the analysis. The ambient temperature was considered to be 20 °C and the heat transfer coefficient took the value of 10 W/m^2/K. The effective solar irradiation for the thermal analysis part was considered to be 800 W/m^2, while the simulations were conducted for a transversal incident angle of 10° and the consideration that the solar rays are falling parallel to the width of the collector. The operating conditions are given in Table 5.

Table 5. Thermal analysis operating conditions.

Parameter	Value
Effective solar irradiation intensity	800 W/m^2
Simulated rays	10^6
Transversal incident angle	10°
Environment temperature	20 °C
Wind heat transfer coefficient	10 W/m^2/K
Inlet temperature range	20 °C–80 °C
Volume flow rate	15 lt/h

2.2.3. Numerical Simulation

The software of the present simulation study is SolidWorks Flow Simulation [38]. A wide range of solar thermal systems have been examined with the particular software and the results have been verified via analytical and experimental results. It is critical to note that with the use of specific software, fluid dynamics, thermal simulation, and ray tracing can be conducted at the same time. More details regarding the present simulation tool and the assumptions which are adopted could be found in previous literature studies [3,6,19,21–23,25,26,39–42].

As regards the boundary conditions, several different conditions were introduced in the program. The inlet temperature and the inlet volume flow rate were the first two parameters that were defined at the inlets of the collectors. The ambient temperature and the wind heat transfer coefficient were defined later at the outer glass surfaces for simulating the convective thermal losses from the cover to the environment. The sky temperature was considered to be equal to the ambient one as in studies [22,43] since the glass envelope is very close to the asymmetric reflector in both cases. The radiative surface conditions were defined according to Table 3, for the proper simulation of the thermal radiation exchange between the components and losses to the ambient. The ray tracing was conducted by assuming a real concentrating consideration, where the reflector acts as a real mirror. Thus, the concentration on the receiver is not considered uniform but is dependent on according to the reflector's shape and the incident angle.

As far as the mesh grid is concerned, great emphasis was given to the fluid regions and the fluid-to-solid interfaces. More specifically, several different refinements were conducted in fluid and partial cells by developing local grids inside the U-pipes and on the interface between the tube walls and the fluid. Great attention was paid to the formation of the receiver mesh with several refinements applied on the body and its absorbing surfaces of it. This was performed for ensuring that the conduction heat transfer at the interior of the receiver's body is being performed well and for achieving a proper ray tracing in order to take into consideration all the incident solar irradiation on the receiver. Another important factor for obtaining an accurate ray tracing is the reflector's surface the grid of which was, also, refined enough.

2.3. Mathematical Formulation

The equations which were used for the analysis of the present models are given in the specific part of the manuscript. Equation (1) is referred to the thermal performance calculation using the useful thermal output and the available solar power, while Equation (2) corresponds to the optical efficiency calculation dividing the absorbed solar power by the available one.

$$\eta_{th} = \frac{Q_u}{Q_s} = \frac{Q_u}{A_a G_{eff,T}} \tag{1}$$

$$\eta_{opt} = \frac{Q_{abs}}{Q_s} \tag{2}$$

The useful thermal output is given by Equation (3) and it can be calculated with two different expressions. Equation (4) gives the mean temperature of the working fluid.

$$Q_u = \dot{V} \cdot \rho \cdot C_p \cdot (T_o - T_i) = h_f \cdot A_s \cdot \left(T_s - T_{f,m}\right) \tag{3}$$

$$T_{f,m} = \frac{T_o + T_i}{2} \tag{4}$$

The overall thermal losses are provided via Equation (5) with three different expressions to be available. The first consideration regards the heat exchange between the outer cover surface and the ambient while the second one calculates the losses considering the heat transfer between the cover and the receiver.

$$Q_L = \left[h_w \cdot (T_g - T_\alpha) + \varepsilon_g \cdot \sigma \cdot \left(T_g^4 - T_\alpha^4\right)\right] \cdot A_{g,o} = \frac{A_{p,o} \cdot \sigma \cdot \left(T_p^4 - T_g^4\right)}{\left(\frac{1}{\varepsilon_p} + \frac{1-\varepsilon_g}{\varepsilon_g} \cdot \frac{D_{p,o}}{D_{g,i}}\right)} = Q_{abs} - Q_u \tag{5}$$

Regarding the flow regime in the water tube, Equation (6) gives the Darcy friction factor, and Equation (7) provides the Reynolds number. Equation (8) calculates the critical values of the Reynolds number and enables the characterization of the flow as turbulent when the Reynolds number exceeds the respective critical limit and as laminar in the opposite case.

$$\lambda_t = 2 \cdot \frac{D_{t,i}}{L_t} \cdot \frac{\Delta p_t}{{u_t}^2 \cdot \rho_f} \tag{6}$$

$$Re_{D,t} = \frac{u_t \cdot D_{t,i}}{\nu} \tag{7}$$

$$Re_{cr,t} = 140 \cdot \sqrt{8/\lambda_t} \tag{8}$$

As far as the heat transfer equations from the flow tube to the fluid are concerned, Equation (9) provides us with the heat transfer coefficient expression arising by solving Equation (3) for (h_f).

$$h_f = \frac{Q_u}{A_s \cdot \left(T_s - T_{f,m}\right)} \tag{9}$$

Through the heat transfer coefficient, it is possible to calculate the Nusselt number as Equation (10) suggests.

$$Nu = \frac{h_f \cdot D_{t,i}}{k_f} \tag{10}$$

For the development of the analytical model, which will be introduced later in the manuscript, it is essential to use the theoretical approaches of the Nusselt number as these are described in Equations (11) and (12) for turbulent and laminar flow regimes [6,7,44–47].

$$Nu_{lam} = \frac{3.66 + 0.0668 \cdot Re_{D_t} \cdot Pr \cdot D_t/L_t}{\left(1 + 0.04 \cdot (Re_{D_t} \cdot Pr \cdot D_t/L_t)^{2/3}\right)} \tag{11}$$

$$Nu_{tur} = 0.023 \cdot {Re_{D_{1,i}}}^{0.8} \cdot Pr^{0.4} \tag{12}$$

Thus, the theoretical value for the heat transfer coefficient in the water tube could be calculated by solving Equation (10) as Equation (13) shows.

$$h_f = \frac{Nu \cdot k_f}{D_{t,i}} \tag{13}$$

2.4. Analytical Approach

In this part of the manuscript, the method around the analytical approach which was developed is described and the main assumptions are provided.

The main assumptions regarding the analytical solution are listed down:

- The solar energy which is absorbed from the receiver (Q_{abs}) in every case is available through ray tracing simulation.
- The solar energy which is available on the aperture of the collector (Q_s) was calculated as the product between the aperture area of the collector and the effective solar irradiation (see Table 5).
- The mean temperature of the working medium was defined as the average value between the inlet and the outlet temperature.
- The temperature of the flow tube walls was considered to be equal to the respective of the receiver.

The last assumption appears, also in studies [3,22], in which similar methods were developed. A flow chart of the analytical method is given in Figure 3 and the steps that were followed are presented clearly. First of all, a random value was given for the outlet temperature and the useful power was calculated according to the first part of Equation (3). Then the mean fluid temperature was calculated via Equation (4) and the thermal losses were defined through the third part of Equation (5). By using the first part of Equation (5) it was possible to calculate the glass temperature iterationally. More specifically, several (T_g) values were given until the first and the last part of Equation (5) come to a convergence. After that, the receiver temperature was found with the contribution of the second part of Equation (5). Given the mean fluid temperature, it was feasible to calculate the theoretical values of the Nusselt number (Equations (11) and (12)) and thus the theoretical heat convection coefficient can be calculated (Equation (13)). By having calculated all the above, the next step is the calculation of the new useful power value through the second part of Equation (3). Last but not least, a new value for the outlet temperature was calculated by solving properly the first part of Equation (3). The described procedure was applied in all the operating points and it was repeated, each time, until the old and the new outlet temperature value converge with each other with an absolute error lower than 10^{-4}.

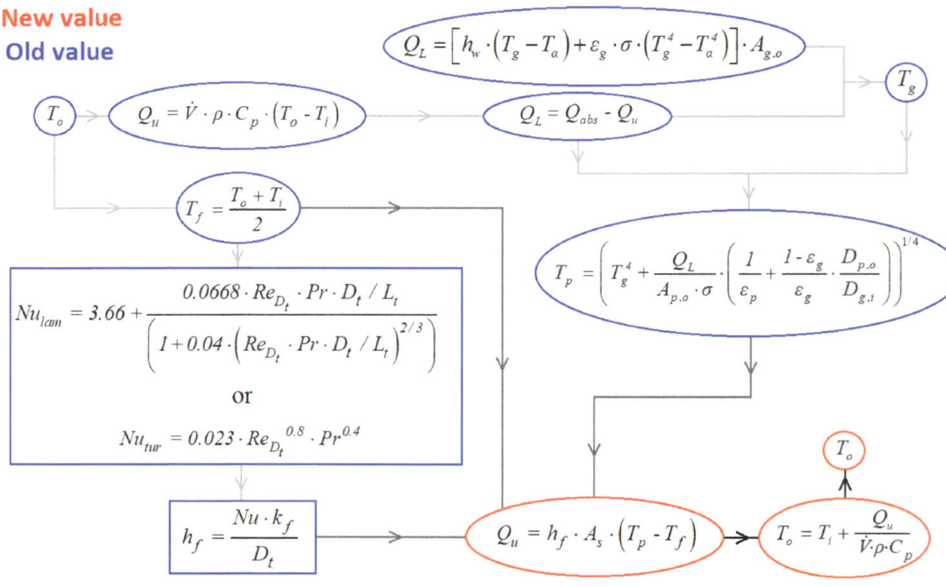

Figure 3. Analytical approach flow chart.

3. Results

The optical and thermal analysis results arising from the simulation are available in the present part of the manuscript. The comparison between the numerical and the analytical results is, also, presented for ensuring the reliability of the present simulation.

3.1. Verification of the Numerical Model

The verification of the simulation results is presented in this particular section. This verification was conducted as regards conventional design for the thermal efficiency, the thermal losses, and the temperature deviation between the receiver and the inlet of the fluid and between the glass and the ambient respectively. The particular deviation was calculated as $\left|\frac{\text{numerical result}}{\text{analytical result}} - 1\right|$. The temperature differences were selected in order to evaluate in a proper manner the results. Figures 4–7 present the results of the verification procedure and it can be said that there is a satisfying agreement between the numerical and the analytical results.

More specifically, Figure 4 declares that the maximum deviation in the thermal efficiency comparison was found to be less than 0.5% which is a relatively low value that indicates the high accuracy of the model. Figure 5 shows that the maximum deviation of the thermal losses is up to 5% which is an acceptable limit. Figure 6 indicates that the maximum deviation regarding the receiver-to-inlet fluid temperature difference does not exceed 1.0%, while Figure 7 shows that the glass-to-ambient temperature difference has a maximum deviation of up to 5%. The aforementioned values prove that the present model is a verified one and so it can further be used in the present study.

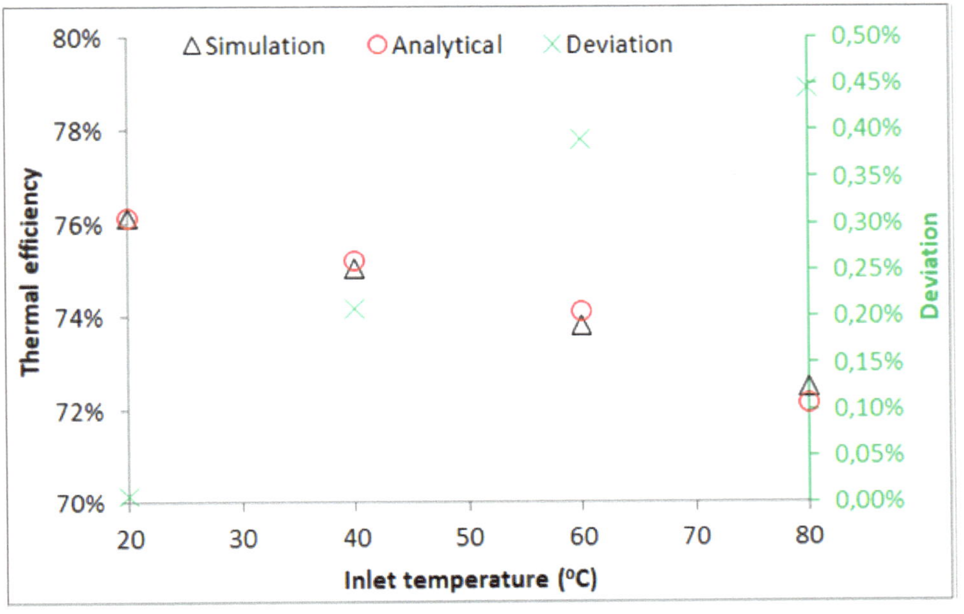

Figure 4. Verification regarding thermal efficiency.

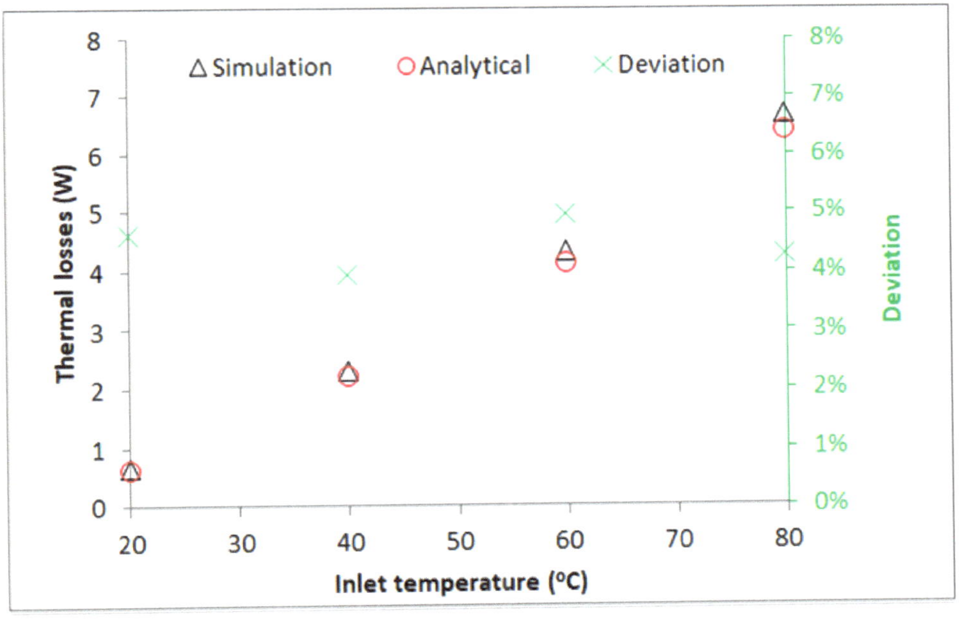

Figure 5. Verification regarding the thermal losses.

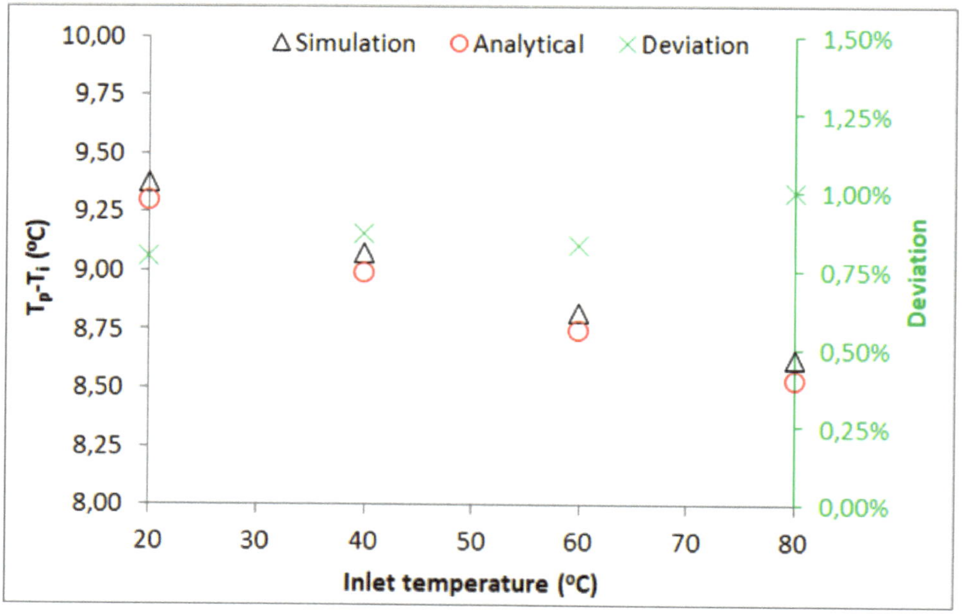

Figure 6. Verification regarding the temperature difference between the absorber and the fluid in the inlet.

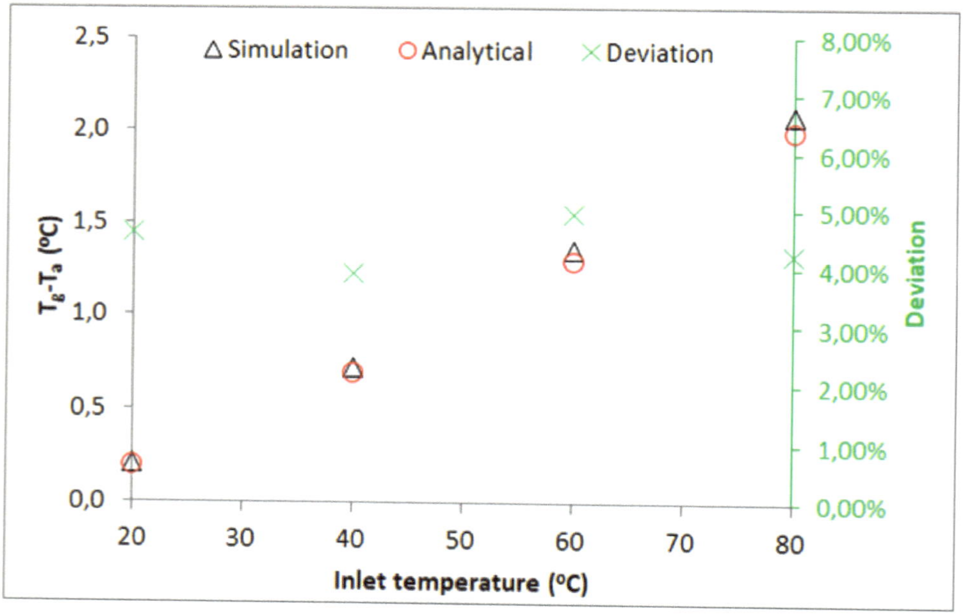

Figure 7. Verification regarding the temperature difference between the cover and the fluid in the inlet.

3.2. Ray Tracing Results

In this section, the SC-ACPC configuration is set under a detailed optical optimization procedure according to the details referred to previously in the manuscript (see Section 2.2.1).

At this point, it is important to provide some extra information regarding the ray tracing tool and the method that was followed. SolidWorks software has been successfully applied as a ray tracing tool in numerous other studies where solar collectors have been investigated. Indicatively, studies [19,22,23] use SolidWorks software for ray tracing, while study [39] verifies it with TracePro software. It is, also, significant to state that the used simulation tool adopts the Monte-Carlo ray tracing model as far as the optical analysis is concerned. For the present ray tracing analysis, the number of solar rays was selected after a detailed independency procedure and took the optimum value of 10^6. Another significant point that should be re-mentioned is that during the ray tracing process the longitudinal incident angle was considered to be zero.

Via the optimization process, two different cavity geometries SC-ACPC 1 ($\omega = 50°$, $\varphi = 90°$) and SC-ACPC 2 ($\omega = 40°$, $\varphi = 70°$) were picked out for being compared with the CNV-ACPC. These geometries were selected since they ensure maximum optical performances in wide incident angle ranges of (2° … 25°) and (2° … 15°) respectively. Figure 8 illustrates the ray tracing results. More specifically, Figure 8a shows that the proposed designs ensure greater optical efficiency than the conventional ACPC in wide ranges of solar incident angles. Figure 8b shows clearly the enhancement percentage in the range from 2° to 25°.

In particular, SC-ACPC 1 ensures a mean enhancement of 2.90% for an incident angle range of 23°, with the maximum possible enhancement reaching approximately 3.90%. SC-ACPC 2 configuration seems not to contribute significantly to an incident angle greater than 15°. However, it provides greater enhancements than the SC-ACPC 1 for incident angles lower than 15°, with the maximum value being 4.35% at 10°. The mean enhancement, in this case, corresponds to 3.48% for the range of (2° … 15°). The operation of the proposed geometries seems not to diversify significantly from the previous study model for incident angles greater than 30°.

For the determination of the optimum geometry between the two proposals, an efficiency index (EI) should be introduced. The product between the incident angle range and the mean optical efficiency enhancement in the respective range is a proper EI that could be used for comparing the two proposed geometries. The main criterion is the maximization of the mean optical efficiency enhancement in combination with the maximization, as much as possible, of the incident angle range. Hence, the EI could take the following form:

$$EI = enh_{opt,m}(\Delta\theta_T) \cdot \frac{\Delta\theta_T}{1^o} \tag{14}$$

where $enh_{opt,m}$ $(\Delta\theta_T)$ is the mean optical efficiency enhancement in $\Delta\theta_T$ range and $\Delta\theta_T$ is the transversal incident angle range. Table 6 gives EI values for the two cases. More specifically, Table 6 indicates that although the mean optical efficiency is greater in the case of SC-ACPC 2, the EI is higher in the case of SC-ACPC 1. This happens because the incident angle range is much wider in this case. Hence, SC-ACPC 1 geometry seems to be more efficient for a wide range of incident angles.

Table 6. Efficiency Index values for SC-ACPC 1 and SC-ACPC 2.

Geometry	SC-ACPC 1	SC-ACPC 2
$enh_{opt,m}$ $(\Delta\theta_T)$	2.90%	3.48%
$\Delta\theta_T$	23°	13°
EI	0.667	0.452

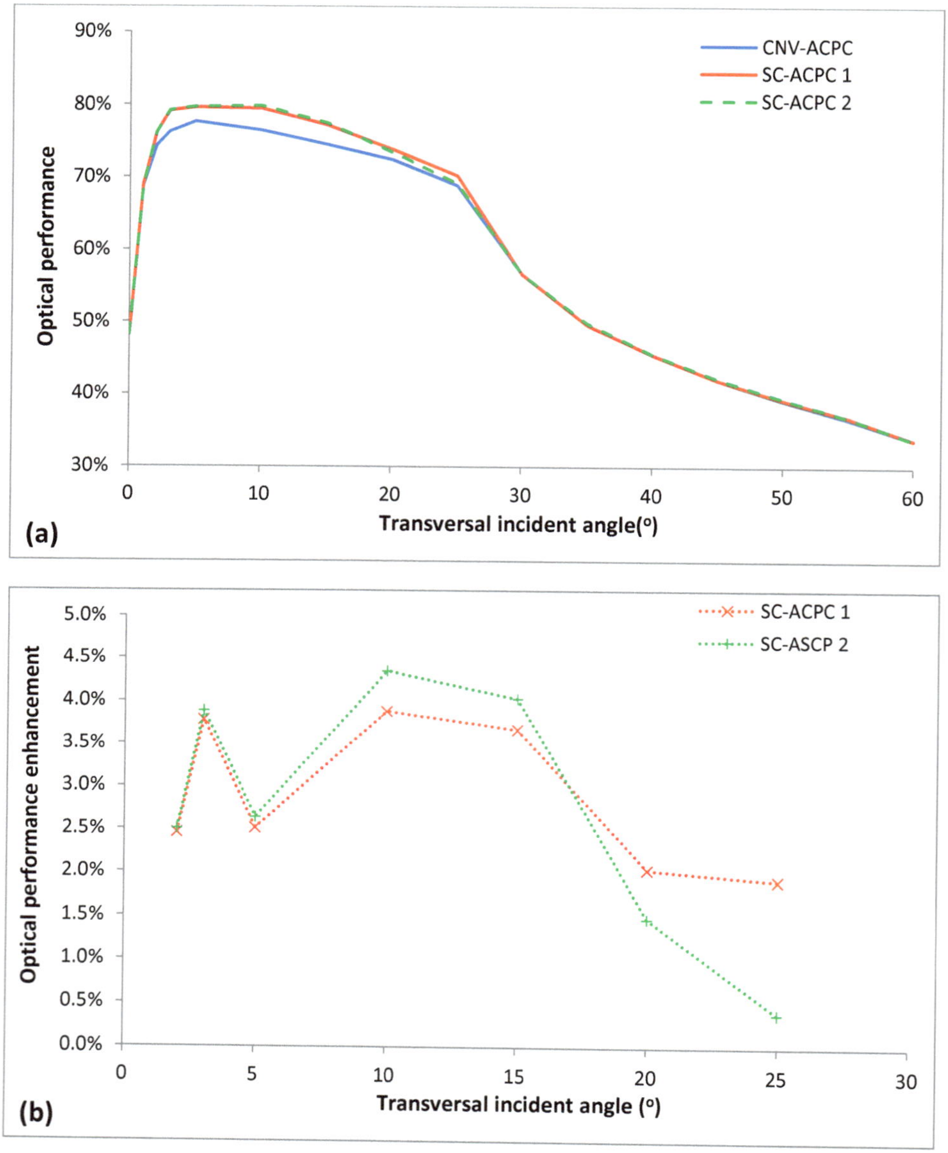

Figure 8. Optical performance comparison (**a**) Optical performance and (**b**) Enhancement.

3.3. Thermal Efficiency and Losses

In the present section, the thermal performance results will be presented. The thermal efficiency is depicted in Figure 9 for each one of the examined models. According to Figure 9a, the two proposed designs seem to exceed in thermal performance of the CNV-ACPC geometry. Particularly, the cavity configurations SC-ACPC 1 and SC-ACPC 2 appear sufficiently greater thermal efficiency in the whole operating range, with SC-ACPC 2 being the leader of this prevalence. The mean enhancement for the SC-ACPC 1 is equal to 2.92% with a maximum value of 3.70%, while in the case of the SC-ACPC 2 the maximum enhancement reaches 4.00% and the mean one is 3.20%. It is remarkable to mention that the minimum enhancement remains greater than 2.20% in both cases and appears in T_i = 80 °C. As regards the thermal losses in Figure 9b, there is a significant difference between the SC-ACPC geometries and the CNV-ACPC one. This difference exists due to the reason that the SC-ACPCs absorb more solar energy than the CNV-ACPC. Another reason for this deviation is that the flow tube in the SC-ACPCs is far from the edge points of the receiver compared to the CNV-ACPC since in the last one there are two straight parts of the U-pipe that are sharing the receiver's body in terms of heat transfer.

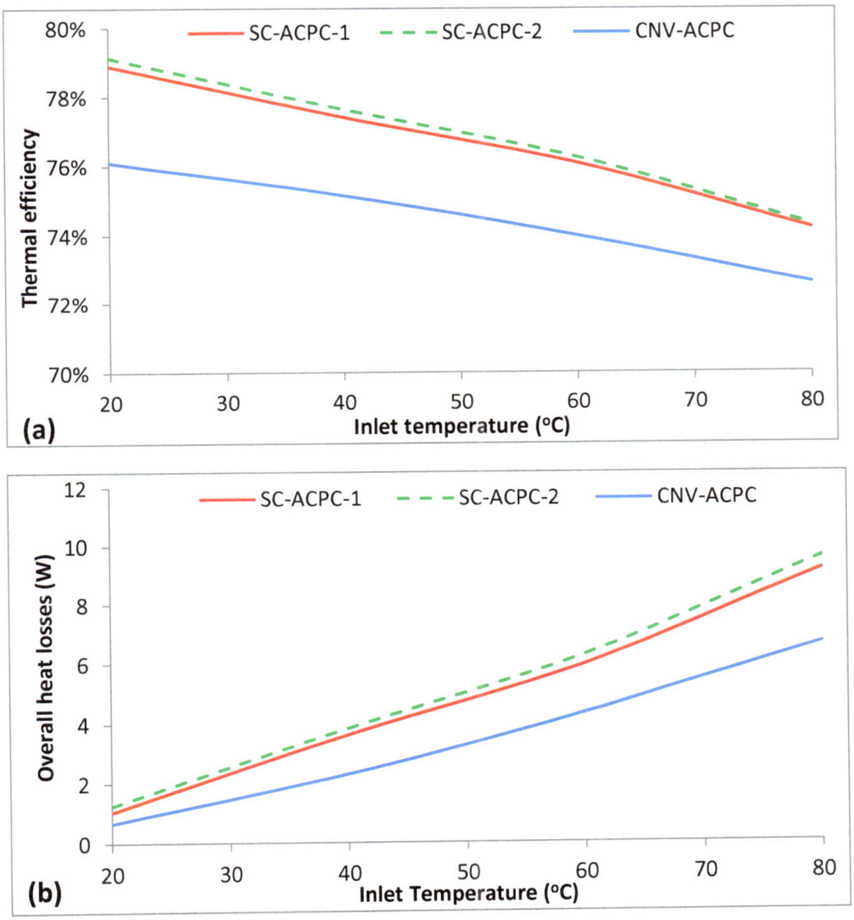

Figure 9. Thermal operation of the model (**a**) Thermal performance and (**b**) Thermal losses.

3.4. Allocations of the Temperature

In this section, various allocations of the SC-ACPC and the CNV-ACPC collectors for the simulation point with T_i = 60 °C are available for evaluation. Figure 10 gives the temperature distribution at a transversal section in the middle of each collector. The surface temperature distribution around the whole receiver in each collector is illustrated in Figure 11.

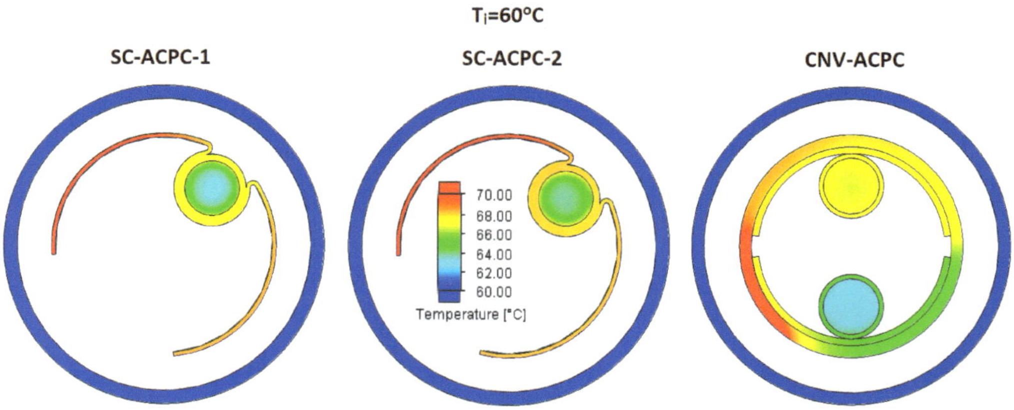

Figure 10. Transversal allocations in the middle of the ACPCs.

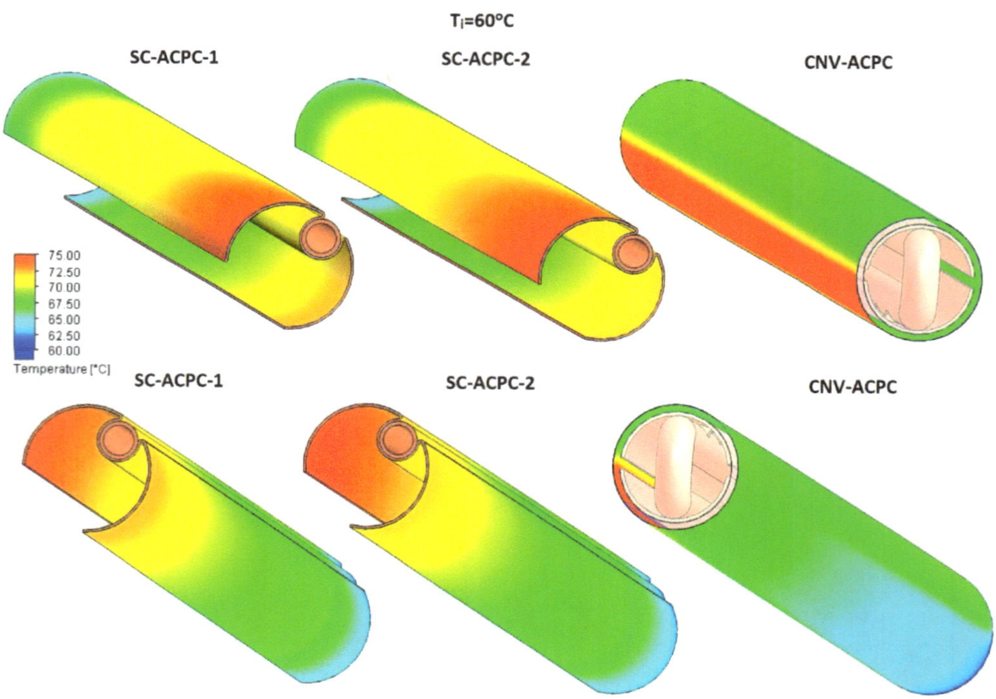

Figure 11. Receiver surface temperature distribution in each ACPC.

As is seen in Figure 10, the temperature regime differentiates among the APCs. In particular, the temperature fields seem to take their highest values in the SC-ACPC 2 case and the lowest ones in the CNV-ACPC geometry. This is reasonable since the SC-ACPC 2 absorbs the greatest amount of solar power in contradiction with the CNV-ACPC which absorbs the lowest one of all cases. It is, also, remarkable to notice the high-temperature areas in each collector. These areas represent the regions where most of the solar irradiation is being concentrated in each receiver. Figure 11 indicates one more solar irradiation concentration region and it declares that the SC-ACPC 2 configuration appears the highest temperatures on average compared to the other two collectors. It is, also, important to see that in all cases the temperature is reasonably becoming reduced going from the outlet to the inlet.

4. Conclusions

In this work, the integration of a linear cavity receiver in an ACPC, developed by the authors in a previous study, was investigated. This is the first time in literature that such integration is being proposed. Two different cavity receiver designs (SC-ACPC 1 and SC-ACPC 2) were proposed and compared with the previous study's design (CNV-ACPC). Next, the most important concluding remarks are listed.

- The numerical results were verified through an analytical solution developed by the authors with a sufficient agreement to be achieved (deviations less than 0.5% in thermal performance and 5.0% in thermal losses).
- Two new geometries were revealed from the optical optimization process.
- The proposed designs came out from a detailed optical performance optimization conducted through ray tracing.
- The cavity configurations lead to a significant enhancement of the optical performance of the CNV-ACPC geometry up to 4.4% for θ_T range of 23°.
- SC-ACPC 1 seems to be more suitable for wide incident angle range applications since it ensures a greater mean enhancement for θ_T in the range of 2° to 25° in contrast to SC-ACPC 2 which is better for θ_T in the range of 2° to 15°.
- SC-ACPC 1 was found to be the best solution between the two suggestions according to an evaluation process with an Efficiency Index (EI), having as the main criterion the balance between the maximization of the mean optical efficiency enhancement and the wider possible incident angle range.
- The mean enhancement with the use of SC-ACPC 2 for θ_T in the range of 2° to 20° is slightly greater than the respective of SC-ACPC 1 considering the same θ_T range (3.15% against 3.05%).
- SC-ACPC 2 ensures 3.2% mean thermal efficiency enhancement against CNV-ACPC. The maximum possible thermal efficiency enhancement, in this case, reaches 4.0% and it appears for T_i = 20 °C.
- SC-ACPC 1 configuration appears slightly lower mean and maximum enhancements than the SC-ACPC 2 geometry (2.92% and 3.70% respectively).

In general, the proposed designs ensure sufficient enhancements both in thermal and optical performance. These enhancements could be even greater when low-absorbing coatings with an absorptance of around 80% are applied, as in various studies [23,48].

Future work could be conducted for a deeper investigation of the proposed systems. It would be useful for the proposed configurations to be tested in terms of the convective regime inside the flow pipe and for possible enhancements in the field using flow inserts, nanofluids, or even a combination of them. Another interesting aspect would be the investigation of non-cylindrical cavities applied to the receiver geometry and the comparison of them with the cylindrical ones. This is recommended, to determine how the cavity's design affects the collector's performance.

Author Contributions: D.N.K.: Conceptualization, methodology, software, validation, formal analysis, investigation, data curation, writing—original draft preparation, writing—review and editing; E.B.: Methodology, software, formal analysis, investigation, writing—original draft preparation, writing—review and editing; C.T.: Conceptualization, software, writing—original draft preparation, supervision. All authors have read and agreed to the published version of the manuscript.

Funding: This research was funded by Bodossaki Foundation.

Data Availability Statement: Data available after request.

Acknowledgments: The first author would like to thank Bodossaki Foundation for its financial support in his post-doctoral research.

Conflicts of Interest: The authors declare no conflict of interest.

Nomenclature

Parameters

A	Area, m^2
C_p	Specific heat capacity, kJ/(kg K)
D	Diameter, m
G	Solar irradiation intensity, W/m2
h	Heat transfer coefficient, W/(m2K)
k	Thermal conductivity, W/(mK)
L	Length, m
$\dot{V}$	Volumetric flow rate, kg/s
p	Pressure level, N/m^2
Q	Heat rate, W
q	Distance between "F" and "G" points, m
T	Temperature level, °C
u	Fluid speed, m/s

Abbreviations

ACPC	Asymmetric CPC
CNV-ACPC	Conventional ACPC
CPC	Compound Parabolic Collector
CSTC	Concentrated Solar Thermal Collector
PTC	Parabolic Trough Collector
SC-ACPC	Single Cavity ACPC

Greek symbols

α	Absorbance
ε	Emittance
η	Efficiency
θ_T	Transversal incident angle
λ	Darcy friction factor
ν	Viscosity, m^2/s
ρ	Density, kg/m^3 Reflectivity
σ	Boltzmann constant, W/(m^{2}K^4)
τ	Transmittance
φ	Angular aperture of the cavity, °
ω	Rotation angle of the cavity, °

Dimensionless numbers

Nu	Nusselt number
Pr	Prandtl number
Re	Reynolds number

Subscripts

α	Ambient
abs	Absorbed
cr	Critical
eff	Effective
f	Working fluid
g	Glass
i	Inlet "for T" Inner "for D"
L	Thermal losses
lam	Laminar
m	Mean
o	Outlet "for T" Outer "for A"
opt	Optical
p	Absorber
s	Solar Wall of the tube
t	Flow tube
th	Thermal
u	Useful
w	wind

References

1. Pranesh, V.; Velraj, R.; Christopher, S.; Kumaresan, V. A 50 year review of basic and applied research in compound parabolic concentrating solar thermal collector for domestic and industrial applications. *Sol. Energy* **2019**, *187*, 293–340. [CrossRef]
2. Gomaa, M.R.; Mustafa, R.J.; Rezk, H. An experimental implementation and testing of a concentrated hybrid photovoltaic/thermal system with monocrystalline solar cells using linear Fresnel reflected mirrors. *Energy Res.* **2019**, *43*, 8660–8673. [CrossRef]
3. Korres, D.N.; Tzivanidis, C.; Koronaki, I.P.; Nitsas, M.T. Experimental, numerical and analytical investigation of a U-type evacuated tube collectors' array. *Renew. Energy* **2019**, *135*, 218–231. [CrossRef]
4. MGomaa, R.; Mustafa, R.J.; Rezk, H.; Al-Dhaifallah, M.; Al-Salaymeh, A. Sizing Methodology of a Multi-Mirror Solar Concentrated Hybrid PV/Thermal System. *Energies* **2018**, *11*, 3276.
5. Osorio, T.; Horta, P.; Collares-Pereira, M. Method for customized design of a quasi-stationary CPC-type solar collector to minimize the energy cost. *Renew. Energy* **2019**, *133*, 1086–1098. [CrossRef]
6. Korres, D.; Bellos, E.; Tzivanidis, C. Investigation of a nanofluid-based compound parabolic trough solar collector under laminar flow conditions. *Appl. Therm. Eng.* **2019**, *149*, 366–376. [CrossRef]
7. Rehan, M.A.; Ali, M.; Sheikh, N.A.; Khalil, M.S. Experimental performance analysis of low concentration ratio solar parabolic trough collectors with nanofluids in winter conditions. *Renew. Energy* **2018**, *118*, 742–751. [CrossRef]
8. Alnaqi, A.A.; Alsarraf, J.; Al-Rashed, A. Hydrothermal effects of using two twisted tape inserts in a parabolic trough solar collector filled with MgO-MWCNT/thermal oil hybrid nanofluid. *Sustain. Energy Technol. Assess.* **2021**, *47*, 101331. [CrossRef]

9. Kasaeian, A.; Daneshazarian, R.; Mahian, O.; Kolsi, L.; Chamkha, A.J.; Wongwises, S.; Pop, I. Nanofluid flow and heat transfer in porous media: A review of the latest developments. *Int. J. Heat Mass Transf.* **2017**, *107*, 778–791. [CrossRef]
10. Said, Z.; Arora, S.; Bellos, E. A review on performance and environmental effects of conventional and nanofluid-based thermal photovoltaics. *Renew. Sustain. Energy Rev.* **2018**, *94*, 302–316. [CrossRef]
11. Liu, Z.H.; Hu, R.-L.; Lu, L.; Zhao, F.; Xiao, H.-S. Thermal performance of an open thermosyphon using nanofluid for evacuated tubular high temperature air solar collector. *Energy Convers. Manag.* **2013**, *73*, 135–143. [CrossRef]
12. Bellos, E.; Tzivanidis, C.; Antonopoulos, K.A.; Gkinis, G. Thermal enhancement of solar parabolic trough collectors by using nanofluids and converging-diverging absorber tube. *Renew. Energy* **2016**, *94*, 213–222. [CrossRef]
13. Lu, L.; Liu, Z.-H.; Xiao, H.-S. Thermal performance of an open thermosyphon using nanofluids for high-temperature evacuated tubular solar collectors: Part 1: Indoor experiment. *Sol. Energy* **2011**, *85*, 379–387. [CrossRef]
14. Mwesigye, A.; Bello-Ochende, T.; Meyer, J.P. Multi-objective and thermodynamic optimisation of a parabolic trough receiver with perforated plate inserts. *Appl. Therm. Eng.* **2015**, *77*, 42–56. [CrossRef]
15. Liu, Y.; Chen, Q.; Hu, K.; Hao, J.-H. Flow field optimization for the solar parabolic trough receivers in direct steam generation systems by the variational principle. *Int. J. Heat Mass Transf.* **2016**, *102*, 1073–1081. [CrossRef]
16. Zhu, X.; Zhu, L.; Zhao, J. Wavy-tape insert designed for managing highly concentrated solar energy on absorber tube of parabolic trough receiver. *Energy* **2017**, *141*, 1146–1155. [CrossRef]
17. Bellos, E.; Tzivanidis, C. Investigation of a star flow insert in a parabolic trough solar collector. *Appl. Energy* **2018**, *224*, 86–102. [CrossRef]
18. Abad, M.T.J.; Saedodin, S.; Aminy, M. Experimental investigation on a solar parabolic trough collector for absorber tube filled with porous media. *Renew. Energy* **2017**, *107*, 156–163. [CrossRef]
19. Bellos, E.; Korres, D.; Tzivanidis, C.; Antonopoulos, K.A. Design, simulation and optimization of a compound parabolic collector. *Sustain. Energy Technol. Assess.* **2016**, *16*, 53–63. [CrossRef]
20. Arora, S.; Singha, H.P.; Sahotab, L.; Arorab, M.K.; Aryaa, R.; Singha, S.; Jaina, A.; Singha, A. Performance and cost analysis of photovoltaic thermal (PVT)-compound parabolic concentrator (CPC) collector integrated solar still using CNT-water based nanofluids. *Desalination* **2020**, *495*, 114595. [CrossRef]
21. Korres, D.N.; Tzivanidis, C. Development of two new semi-empirical formulas for estimation of solar absorptance in circular cavity receivers. *Therm. Sci. Eng. Prog.* **2019**, *10*, 147–153. [CrossRef]
22. Korres, D.N.; Tzivanidis, C. Investigation of a novel small-sized bifacial cavity PTC and comparison with conventional configurations. *Therm. Sci. Eng. Prog.* **2020**, *17*, 100355. [CrossRef]
23. Korres, D.N.; Tzivanidis, C. An innovative small-sized double cavity PTC under investigation and comparison with a conventional PTC. *Sustain. Energy Technol. Assess.* **2022**, *53A*, 102462.
24. Avargani, V.M.; Norton, B.; Rahimi, A. An open-aperture partially-evacuated receiver for more uniform reflected solar flux in circular-trough reflectors: Comparative performance in air heating applications. *Renew. Energy* **2021**, *176*, 11–24. [CrossRef]
25. Korres, D.N.; Tzivanidis, C. A Novel Asymmetric Compound Parabolic Collector Under Experimental and Numerical Investigation. *Renew. Energy* **2022**, *199*, 1580–1592. [CrossRef]
26. Korres, D.; Tzivanidis, C. A symmetric and an asymmetric mini Compound Parabolic Collector under Optical Investigation. In *The Role of Exergy in Energy and the Environment*; Green Energy and Technology; Springer: Cham, Switzerland, 2018; Chapter 46.
27. Karwa, N.; Jiang, L.; Winston, R.; Rosengarten, G. Receiver shape optimization for maximizing medium temperature CPC collector efficiency. *Sol. Energy* **2015**, *122*, 529–546. [CrossRef]
28. Waghmare, S.A.; Gulhane, N.P. Optimization of receiver height in compound parabolic collector by optical analysis and experimental method. *Optik* **2018**, *157*, 1331–1341. [CrossRef]
29. Loni, R.; Ghobadian, B.; Kasaeian, A.B.; Akhlaghi, M.M.; Bellos, E.; Najafi, G. Sensitivity analysis of parabolic trough concentrator using rectangular cavity receiver. *Appl. Therm. Eng.* **2020**, *169*, 114948. [CrossRef]
30. Ardeh, E.A.A.; Loni, R.; Najafi, G.; Ghobadian, B.; Bellos, E.; Wen, D. Exergy and economic assessments of solar organic Rankine cycle system with linear V-Shape cavity. *Energy Convers. Manag.* **2019**, *199*, 111997. [CrossRef]
31. Zhang, L.; Fang, J.; Wei, J.; Yang, G. Numerical investigation on the thermal performance of molten salt cavity receivers with different structures. *Appl. Energy* **2017**, *204*, 966–978. [CrossRef]
32. Lin, M.; Sumathy, K.; Dai, Y.J.; Wang, R.Z.; Chen, Y. Experimental and theoretical analysis on a linear Fresnel reflector solar collector prototype with V-shaped cavity receiver. *Appl. Therm. Eng.* **2013**, *51*, 963–972. [CrossRef]
33. Xiao, X.; Zhang, P.; Shao, D.D.; Li, M. Experimental and numerical heat transfer analysis of a V-cavity absorber for linear parabolic trough solar collector. *Energy Convers. Manag.* **2014**, *86*, 49–59. [CrossRef]
34. Lin, M.; Sumathy, K.; Dai, Y.J.; Zhao, X.K. Performance investigation on a linear Fresnel lens solar collector using cavity receiver. *Sol. Energy* **2014**, *107*, 50–62. [CrossRef]
35. Chen, F.; Li, M.; Zhang, P.; Luo, X. Thermal performance of a novel linear cavity absorber for parabolic trough solar concentrator. *Energy Convers. Manag.* **2015**, *90*, 292–299. [CrossRef]
36. Al-Nimr, M.A.; Al-Darawsheh, I.A.; AL-Khalayleh, L.A. A novel hybrid cavity solar thermal collector. *Renew. Energy* **2018**, *115*, 299–307. [CrossRef]
37. Liang, H.; Fan, M.; You, S.; Xia, J.; Zhang, H.; Wang, Y. An analysis of the heat loss and overheating protection of a cavity receiver with a novel movable cover for parabolic trough solar collectors. *Energy* **2018**, *158*, 719–729. [CrossRef]

38. Dassault Systemes. *Technical reference Solidworks Flow Simulation 2015*; Dassault Systems: Vélizy-Villacoublay, France, 2015.
39. Korres, D.N.; Tzivanidis, C. Numerical investigation and optimization of an experimentally analyzed solar CPC. *Energy* **2019**, *172*, 57–67. [CrossRef]
40. Korres, D.; Tzivanidis, C. A new mini-CPC under thermal and optical investigation. *Renew. Energy* **2018**, *128*, 529–540. [CrossRef]
41. Korres, D.N.; Tzivanidis, C. Thermal Analysis of a Serpentine Flat Plate Collector and Investigation of the flow and convection regime. *Therm. Sci.* **2019**, *23*, 47–59. [CrossRef]
42. Tzivanidis, C.; Bellos, E.; Korres, D.; Antonopoulos, K.A.; Mitsopoulos, G. Thermal and optical efficiency investigation of a parabolic trough collector. *Case Stud. Therm. Eng.* **2015**, *6*, 226–237. [CrossRef]
43. Ataee, S.; Ameri, M. Energy and exergy analysis of all-glass evacuated solar collector tubes with coaxial fluid conduit. *Solar Energy* **2015**, *118*, 575–591. [CrossRef]
44. Taler, D. Determining velocity and friction factor for turbulent flow in smooth tubes. *Int. J. Therm. Sci.* **2016**, *105*, 109–122. [CrossRef]
45. Bergman, T.L.; Lavine, A.S.; Incropera, F.P.; Dewitt, D.P. *Fundamentals of Heat and Mass Transfer*, 7th ed.; John Wiley & Sons: New York, NY, USA, 2011.
46. Fang, X.; Xu, Y.; Zhou, Z. New correlations of single-phase friction factor for turbulent pipe flow and evaluation of existing single-phase friction factor correlations. *Nucl. Eng. Design.* **2011**, *241*, 897–902. [CrossRef]
47. Rattner, A.; Bohren, J. *Heat and Mass Correlations*; University of Pennsylvania: Philadelphia, PA, USA, 2008; Available online: https://www.stwing.upenn.edu/~{}salexa/Documents/Correlations.pdf (accessed on 3 October 2022).
48. Katumba, G.; Lu, J.; Olumekor, L.; Westin, G.; Wäckelgard, E. Low Cost Selective Solar Coatings: Characteristics of Carbon-In-Silica Synthesized with Sol-Gel Thechnique. *J. Sol-Gel Sci. Technol.* **2005**, *36*, 33–43. [CrossRef]

energies

Article

The Self-Actuating Droplet That Can Turn: A Molecular Dynamics Simulation

Yalong Kong [1], Zhigang Liu [1], Lin Guo [1,*] and Yu Qiu [2,*]

[1] Energy Research Institute, Qilu University of Technology, Jinan 250014, China
[2] School of Energy Science and Engineering, Central South University, Changsha 410083, China
* Correspondence: linguo@sderi.cn (L.G.); yu.qiu@csu.edu.cn (Y.Q.)

Abstract: Water collection remains a fundamental challenge to stable and efficient operation of the solar desalination system. Functional surfaces that can realize self-actuation of droplets have shown great potential in improving droplet dynamics without external energy. Therefore, a surface that can make a droplet move spontaneously along a curve was designed for smart droplet manipulation, and the mechanism of the droplet motion was revealed through molecular dynamics simulations. Influences of the wettability difference between the curved track and the background, the width of curved track, and the temperature were evaluated via simulations. The results show that the surface on which the curved track and the background are both hydrophobic enables a faster actuating velocity of the droplet than the hydrophilic-hydrophobic surface and the hydrophilic-hydrophilic surface. The width of the curved track also affects the actuating velocity of the droplet and increasing the TRACK width can increase the actuating velocity of the droplet. However, actuation of the droplet slows down if the width of the curved track is too large. Overall, the mechanism driving the motion of the droplet along the curve was investigated, which opens new opportunities for the application and manufacturing of water collection in solar desalination.

Keywords: solar desalination; droplet smart manipulation; molecular dynamics simulation

1. Introduction

Solar desalination that utilizes abundantly available solar energy to facilitate seawater evaporation has gained growing attention as promising solutions to the ever-increasing scarcity of freshwater. Water collection, however, remains a fundamental challenge to stable and efficient operation of the solar desalination system. Dropwise condensation is the necessary process of water collection, and the departure of efficient condensates remains a critical bottleneck. Recently, functional surfaces that can realize self-actuation of droplets have been proposed, showing great potential in improving droplet dynamics without external energy. Through long-term evolution and development, animals and plants in nature have inspired and enlightened the design of such functional surfaces. For example, surfaces of rice leaves have self-cleaning properties [1]; cactus thorns [2,3] and cobwebs [4,5] collect water based on the Laplace pressure gradient and the surface energy gradient formed by their surface structures; the micro/nano-composite structures of nepenthes [6,7] optimize and enhance the capillary force in the transportation direction and allow directional motion of water on the surface of nepenthes; Namibian desert beetles [8,9] collect droplets using special hydrophilic spikes on their backs and transfer droplets to their mouths via the hydrophobic areas, which enable these beetles to survive in arid environments.

Scholars also achieved self-propelled motion of droplets by constructing irregular structures [10,11], curvature gradients [12–14], and wettability gradients [15–17] on surfaces, which have been reported in applications such as water collection [18,19], condensation heat transfer [20], liquid pumping [21,22], oil-water separation [23,24], and droplet-based microfluidics [25–27], etc. For example,, Tang et al. [28] prepared a taper copper (Cu) needle with a wettability gradient that combines the structure gradient and the chemical gradient, which realizes the directional collection of droplets. Based on the integrated physical conical structure and the wettability chemo-gradient, Feng et al. [29] achieved a high-efficiency water collection system. Lin et al. [30] devised an inclined column array, which coordinates with the surface wettability to achieve robust directional transmission of droplets via coordination of multiple gradients and multiple structures on the surface. The above research provides effective methods and strategies for fulfilling the self-actuation of droplets, although it is challenging to design and prepare these one-dimensional (1D) or three-dimensional (3D) structures on surfaces. On the contrary, two-dimensional (2D) surfaces are compatible with modern manufacturing technologies and have been widely applied in various cases [31]. For example, Hou et al. [25] employed the one-step electrodeposition method to impart a wettability gradient to the Cu surface. When placing a droplet on the surface, the droplet can be self-actuated under the wettability gradient. Additionally, patterned hydrophilic-hydrophobic surfaces have also been investigated. Hydrophilic-hydrophobic wedge-shaped patterned surfaces have been prepared in many experimental studies [32–40], and droplets can be self-actuated to move from the narrow to the wide conical end on such surfaces. Zheng et al. [38] and Guan et al. [39] applied lubricating oils on wedge-shaped patterned surfaces to reduce resistance of surfaces to droplets and accelerate the droplet motion. Apart from a single patterned surface, wedge-shaped patterns can also be permutated and combined. Bai et al. [41] designed star-patterned surfaces; Wang et al. [31] constructed wedge-shaped tree structures to realize continuous self-actuation of a large quantity of droplets.

Droplets in the research are always actuated along a linear direction, while how to control self-actuation of droplets along a curve on surfaces has not been reported. The non-linear self-actuation of droplets along a curve is important in lab-on-a-chip, microfluidic devices, and analysis of devices involving biological droplets. The non-linear self-actuation is highly flexible when performing complex, high-precision manipulations of microfluids, which enhances the level of integration of microsystems. The development of computer technology and the maturation of commercial numerical simulation software allows researchers to conduct numerical research into droplets on different surfaces through simulations [42–50]. Through molecular dynamics simulations, Xu et al. [43] studied droplet motion on wedge-shaped structural surfaces combining multiple gradients and explored the influences of different wettability gradients and cone angles thereon. Hao et al. [44] simulated the droplet motion in a wedge groove and found that the different initial locations of droplets in the wedge groove determine the direction of the motion of droplets. Guo et al. [45] conducted numerical research and theoretical analysis on the dynamic behaviors of droplets on a tapering fiber with different cross-sections. Mo et al. [46] simulated the directional motion of nanodroplets on an axially symmetric surface with a curvature gradient. Structural systems with different complexities can be established via molecular dynamics simulations. Experimental research from the microscopic perspective provides theoretical and mechanism support for macroscopic experiments, which reflects the superiority of molecular dynamics simulations.

We are aiming to examine a novel concept of surface design. A functional surface that drives droplet motion along a curve without external energy was designed and explored using a molecular dynamic simulations method. The droplet was driven by virtue of the wettability difference between hydrophilic and hydrophobic areas by molecular dynamics simulation. Influences of different factors including the wettability gradient, the width of curved tracks and the temperatures in relation to the motion characteristics of droplets were estimated. It is expected that this novel functional surface extends droplet manipulation strategies and provides water collection solutions in a solar desalination system.

2. Simulation Method

2.1. Molecular Dynamics Simulation System

Simulation System

To generate the curve track, two rectangular regions were built at first and shown in Figure 1. The *x*, *y*, *z*-coordinates of Block 1 were (−58.8, 98), (−39.2, 196), and (0, 3.92) Å, respectively. The *x*, *y*, *z*-coordinates of Block 2 were (0, 98), (0, 196), and (0, 3.92) Å, respectively. Two cylindrical regions were constructed. The centers of Cylinder 1 and Cylinder 2 were O_1 (0, 78.4) Å and O_2 (0, 74.48) Å, and the radii of Cylinder 1 and Cylinder 2 were 78.4 and 70.56 Å. Then the intersecting region of Block 2 and Cylinder 1 was set as Region 1, and the intersecting region of Block 2 and Cylinder 2 was set as Region 2. Steps to generate the curve track were shown as follows. Copper atoms were built in Block 1and set as type 1. Atoms of Region 1 were deleted. Copper atoms were rebuilt in Region 1, but set as type 2. Atoms in Region 2 were deleted, and copper atoms were created with type 1. Finally, the curve track was obtained and marked in blue in Figure 1. In addition, the curve track and the rest of the background were built with different kinds of atoms, which can be set with different wettability and discussed in the following section. A water cubic box with 50 × 50 × 50 $Å^3$ was generated at the narrow end of the curve and the centroid thereof was placed above the origin of the coordinates. In this way, the initial configuration of the simulation system was established and shown as Figure 1b.

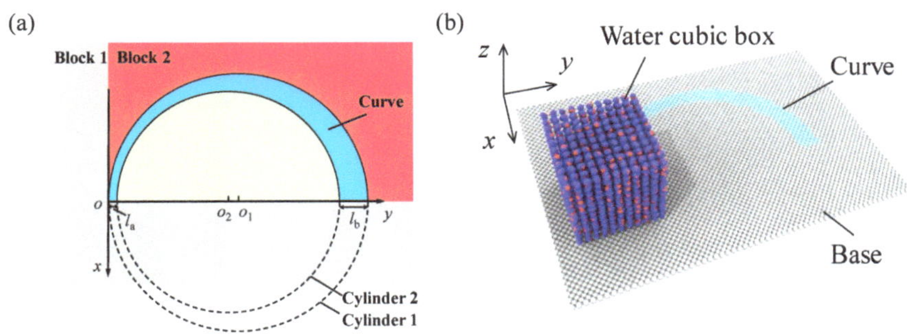

Figure 1. (**a**) Generation of the curve track. (**b**) Initial configuration of the simulation system.

2.2. Simulation Method and Procedure

2.2.1. Simulation Method

Molecular dynamics simulations were adopted to study the self-actuation of droplets along a curve and all simulations were performed using the software package for the large-scale atomic/molecular massively parallel simulator (LAMMPS). The simulation model included water molecules and the copper base. The simulation box measured 180 × 240 × 160 Å and periodic boundary conditions were applied to the *x*, *y*, and *z*-directions. In the whole simulation system, the mixed potential was used as the interaction potential of atomic pairs. The TIP4P model, which was a four-point model, was applied to the water molecule. The quantities of electric charges of oxygen atom and hydrogen atom were −1.0484 e and 0.5242 e, respectively. The bond angle was 104.52°, and the lj/cut/tip4p/long potential model was used for the interaction between atomic pairs in the water molecule.

In the current study, the interactions between copper and copper (Cu-Cu), copper and oxygen (Cu-O), and copper and hydrogen (Cu-H) were described by the Lennard-Jones potential function, shown as follows:

$$E = 4\varepsilon\left[\left(\frac{\sigma}{r}\right)^{12} - \left(\frac{\sigma}{r}\right)^{6}\right] \qquad r < r_c \tag{1}$$

where *r* represented the interatomic spacing; r_c denoted the truncation radius; ε and σ were energy and length parameters, respectively. The energy parameter ε and length parameter σ of copper-oxygen and copper-hydrogen interactions were calculated based on the Lorentz-Berthelot combining rule. The copper-oxygen interaction strength can be applied to regulate the interfacial wettability, which was determined by the energy coefficient $\varepsilon_{O\text{-}Cu}$ in the current study. The potential energy and relevant parameters used in the simulations were displayed in Table 1.

Table 1. Setting of potential energy parameter.

Inter-Particle Interaction	Type of Potential Energy	Potential Energy Parameter	
		ε/eV	σ/Å
O-O	lj/cut/tip4p/long	0.00802	3.1589
O-H	lj/cut/tip4p/long	0	0
H-H	lj/cut/tip4p/long	0	0
Cu-Cu	lj/cut	0.1656	2.471
H-Cu	lj/cut	0	0
O-Cu	lj/cut	Regulation variable	2.815

2.2.2. Simulation Procedure

The canonical ensemble (NVT) was used in the whole simulation during which the initial system temperature was set to 300 K. At first, the simulation system was relaxed for 10 ps, during which all molecules vibrated freely, and the cubic water molecule model took on the shape of a water-drop. The momentum of droplets in the x and y-directions was eliminated and the droplets were only allowed to move along the z-axis, which guaranteed the same initial location of the droplet centroid on the track and complete contact between the droplet and the solid surface. Then, the momentum constraint of the droplet was removed, and the droplet was able to move without additional force. The data acquisition stage was conducted in the microcanonicalensemble (NVE) for 5 ns to study the self-actuation of the droplet along the curve. The simulation data were output every 1000 timesteps, and the droplet trajectories were visualized by the open visualization software OVITO. The droplet velocity was averaged and exported every 5 ps.

2.3. *Surface Wettability Simulations*

The surface wettability can be defined by the contact angle θ of droplets. The surface was hydrophilic if θ was smaller than 90° and was hydrophobic if θ was greater than 90°. The energy coefficient $\varepsilon_{O\text{-}Cu}$ was applied in the present study to regulate the surface wettability, and the cylindrical droplet method was used to explore the relationship of $\varepsilon_{O\text{-}Cu}$ with the contact angle. Simulation details were shown as follows.

The computation box was $120 \times 120 \times 80$ Å^3 in length. A periodic boundary condition was imposed in the three directions of the computational domain. At first, a water box with 506 water molecules was placed at a prudent distance on top of a copper surface, and the system was allowed to equilibrate. Energy minimization of the system was conducted to remove any excess potential energy from the initial configuration. Then the system was equilibrated using a Nose/Hoover thermostat at 300 K with a time constant of 10 fs for 0.5 ns. After that, equilibration in the microcanonical ensemble with a time step of 10 fs for 0.5 ns was conducted, and snapshots of the system were collected for the next 3 ns.

To extract the profile of the contact line, cylindrical liquid slabs were equilibrated over flat surfaces. Then the microscopic contact angle was like its macroscopic counterpart as predicted by the modified Young equation. The contact angle calculation procedure was as follows: (a) The computational box was discretized into bins of square cross sections in the x–z plane, and the entire length of the computational box in the y direction was the depth of the bins. The resolution of the bins was set as 0.8×0.8 Å^2. The position of each atom of the tested group was stored in the appropriate bin for every snapshot, and the mass density was calculated as the average over time of the bins count per unit volume. (b) The averaged mass density of bins was plotted as contour maps, and the droplet interface was defined by the contour line in which $\rho(x, y) = \rho_l/2$, where ρ_l was the liquid density. (c) The coordinates of the surface position and the droplet interface are obtained and fitted using the equation of a circle. (d) Finally, the fitted equation was plotted, and the contact angle was measured.

3. Results and Discussion

3.1. Effects of Liquid-Solid Interactions on the Contact Angle

The relationship between the potential energy parameters $\varepsilon_{O\text{-}Cu}$ and the θ obtained by simulations was revealed and illustrated in Figure 2. The figure shows that the contact angle decreases as the energy parameter increases, and the surface wettability changes from hydrophobicity to hydrophilicity. Therefore, various contact angle combinations can be set for surfaces with the curve track, which are elaborated in the following investigations.

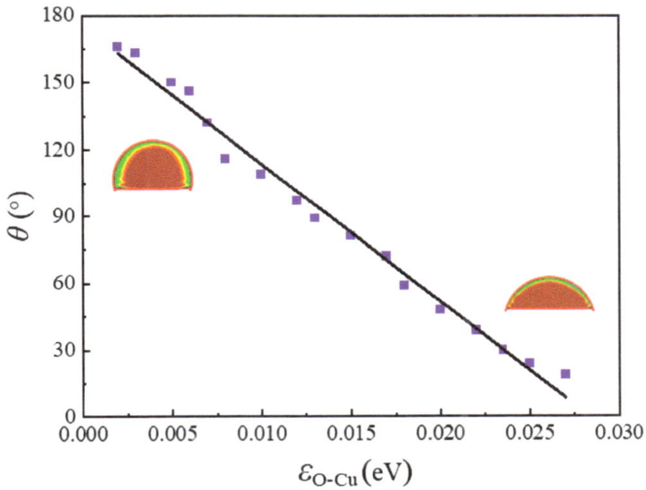

Figure 2. Relationship between θ and εO-Cu. Two density contours are given as examples of the hydrophobic and hydrophilic cases. The simulation results are shown as purple box, and the trend line is black line.

3.2. Simulation Validation

The underlying principle of the self-driving droplet on a conical curve track is like the self-driving droplet on a straight conical track, which all come from the Laplace pressure gradient [28]. Therefore, the present simulation results can be validated from two aspects. Case 100–160 was chosen as examples, as shown in Figure 3a. A straight conical track was constructed and shown as Figure 3b. A curve track with the same width was constructed and shown as Figure 3c. All the simulation details are kept the same. Snapshots of simulation results are exported and shown in Figure 3. Droplets can both move on tracks with width gradient, see cases a and b. In contrast, the droplet stays stationary on the curve track without width gradient, see case c. Therefore, the self-driving behavior of droplets on the conical curve track in the present study was verified.

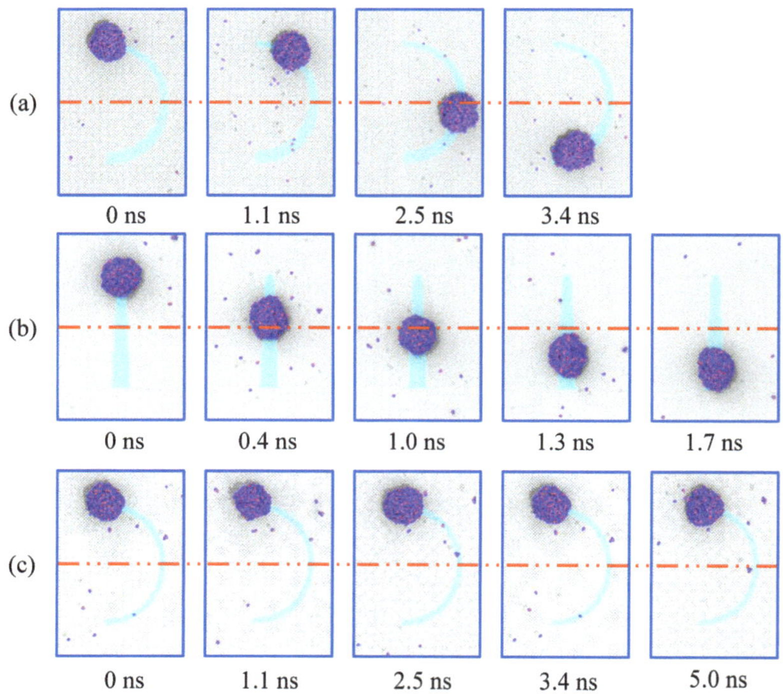

Figure 3. Snapshots of droplets on surfaces with patterns. (**a**) Curve track with width gradient. (**b**) Straight track with width gradient. (**c**) Curve track without width gradient. Red center line is marked to assist in observing droplet movement.

3.3. Theoretical Analysis

To understand the motion principle of a droplet on a curved track and its influencing factors, the model was theoretically analyzed. First, it is assumed that the droplet centroid moves on the curved track following an ideal trajectory, that is, the centerline of the curved track, as shown in Figure 4. The force equilibrium of the droplet is briefly expressed as follows:

$$F = F_L + F_W - F_H \quad (2)$$

where F_L is the Laplace pressure gradient caused by different menisci in the front and rear of the droplet along the track direction due to asymmetry of the curved track; F_W denotes the wettability gradient force generated due to the difference of the two sides of the curved track with the background in terms of wettability; F_H represents the force that resists motion of the droplet on surfaces of different wettabilities.

The droplet on the curved track is divided into four parts, in which a and b are the menisci of the droplet in the front and rear ends along the direction of motion; c and d are the droplet parts in contact with the two sides of the track, as shown in Figure 4.

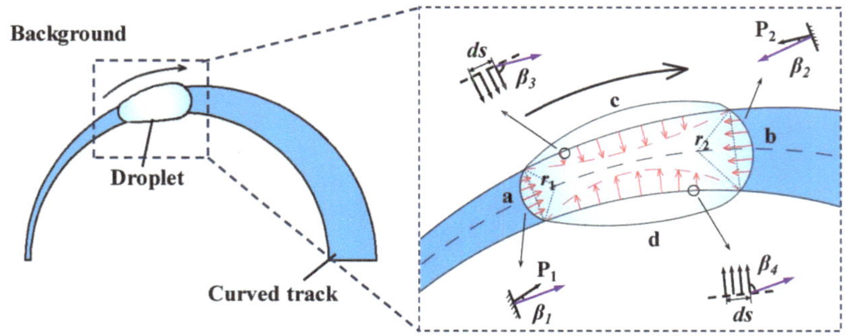

Figure 4. Theoretical model for motion of a droplet on the functional surface constituted by a gradually widening hydrophilic curved track and a hydrophobic surface.

After placing the droplet on the curved track, menisci with different radii of curvature are formed in the front and rear ends of the droplet along the direction of the curved track due to asymmetry of the hydrophilic track. Therefore, the Laplace pressure is generated, which can be expressed as follows:

$$P_1 \sim \gamma\frac{1}{r_1},\ P_2 \sim \gamma\frac{1}{r_2} \tag{3}$$

where r_1 and r_2 separately represent the local radii of the three-phase contact line in two sides of the droplet along the track direction; γ denotes the surface tension of water.

The Laplace pressure difference acts on the droplet to drive motion of the droplet. However, the directions of P_1 and P_2 have certain angles (β_1 and β_2) with the tangential direction of the motion trajectory of the droplet due to the special shape of the curved track. The values of β_1 and β_2 are similar at any given time, so they are both replaced with β for the convenience of computation. Therefore, the resultant force along the direction of motion produced by the Laplace pressure is expressed as follows:

$$F_1 = P_1 cos\beta_1 - P_2 cos\beta_2 = 2\gamma cos\beta\left(\frac{1}{r_1} - \frac{1}{r_2}\right) \tag{4}$$

Additionally, a force normal to the direction of motion is generated by the Laplace pressure due to the presence of an angle between the force and the direction of motion.

$$F_2 = P_1 sin\beta_1 + P_2 sin\beta_2 = 2\gamma sin\beta\left(\frac{1}{r_1} + \frac{1}{r_2}\right) \tag{5}$$

The direction of action of F_2 is from side d to side c.

The driving force produced by the wettability gradient is obtained by contact areas of the droplet in the two boundary regions with different wettability. The proportional relationship between the driving force F and the contact area σ is expressed as [43]:

$$F = A \times \sigma \tag{6}$$

where A is a constant. Similarly, the wettability gradient force also has certain angles with the direction of motion of the droplet and its included angles with c and d sides are separately β_3 and β_4. Therefore, a component force vertical to the direction of motion is also generated. F_3 and F_4 that are separately vertical to the tangential direction of the motion trajectory are:

$$F_3 = \int_c A\sigma cos\beta_3 ds + \int_d A\sigma cos\beta_4\, ds \tag{7}$$

$$F_4 = \int_c A\sigma sin\beta_3 ds - \int_d A\sigma sin\beta_4\, ds \tag{8}$$

where the direction of action of F_4 is from c side to d side. The arc shape of the track determines that the contact area of the droplet at the boundary on the c side is larger than that on the d side. The angle difference between β_3 and β_4 is almost negligible.

Combined with the above analysis, the resultant force on the droplet in two directions is expressed as follows:

The resultant force in the direction of motion is

$$F_n = F_1 + F_3 - F_H = 2\gamma cos\beta\left(\frac{1}{r_1} - \frac{1}{r_2}\right) + \int_c A\sigma cos\beta_3 ds + \int_d A\sigma cos\beta_4\, ds - F_H \quad (9)$$

The resultant force vertical to the direction of motion is

$$F_z = F_2 - F_4 = 2\gamma sin\beta\left(\frac{1}{r_1} + \frac{1}{r_2}\right) - \int_c A\sigma sin\beta_3 ds + \int_d A\sigma sin\beta_4\, ds \quad (10)$$

Driven by F_n, the droplet can move along the track; while F_z allows lateral motion of the droplet vertical to the direction of motion, which to some extent hinders the droplet motion. However, because the directions of F_2 and F_4 are opposite, the lateral motion of the droplet maintains dynamic equilibrium. Together with the adsorption of the hydrophilic track for the droplet, the droplet is always on the track. In addition, the force helps adjust the direction of motion of the droplet along the curve in the process of achieving dynamic equilibrium.

3.4. Influences of the Wettability Difference

Copper atoms were filled in the curved track region and the background region. After adding a droplet, the potential energy parameter $\varepsilon_{O\text{-}Cu}$ between oxygen atoms in water molecules and copper atoms in the base was set, which controls the base to show different water wettability. In this way, a controllable wettability difference between the curve and the background by the droplet can be achieved. The wettability difference of 60 degrees was applied in the present study to save simulation resources and set different contact angle combinations.

The wettability of the curve track and background by the droplet were separately adjusted to observe and assess influences of different wetting degrees and the wettability difference on the droplet motion along the curve. The self-actuation of the droplet on three groups of surfaces with different wettability was simulated under the same wettability gradient on the track and background (the track and background have a difference of 60° in terms of the CA θ). These groups include: (a) the hydrophobic surface (CAs θ in the track and background separately being 100° and 160°); (b) the hydrophilic-hydrophobic surface (CAs θ in the track and background separately being 60° and 120°); and (c) the hydrophilic surface (CAs θ in the track and background separately being 20° and 80°); (d) the hydrophobic-hydrophilic surface (CAs θ in the track and background separately being 120° and 60°).

Figure 5 instantaneously illustrates droplet motion along the curve track in four groups of molecular dynamics simulations. As shown, the droplet remains quasi-ellipsoidal in the motion process in group a and group b, which is in sharp contrast to the shape of the droplet in group c. As time goes on, the droplet spreads out along the track on the hydrophilic surface and the droplet is still in the early half of the curved track. This is because when the track with stronger hydrophily is driving the droplet forward, it needs to overcome the strong pinning effect of hydrophilicity of the background on the droplet. As for case d, the droplet tends to separate from both sides of the track. The hydrophilic background surface has a larger interfacial interaction, which makes the droplets tend to spread rather than move forward, and an effective droplet driving along the curved track cannot form.

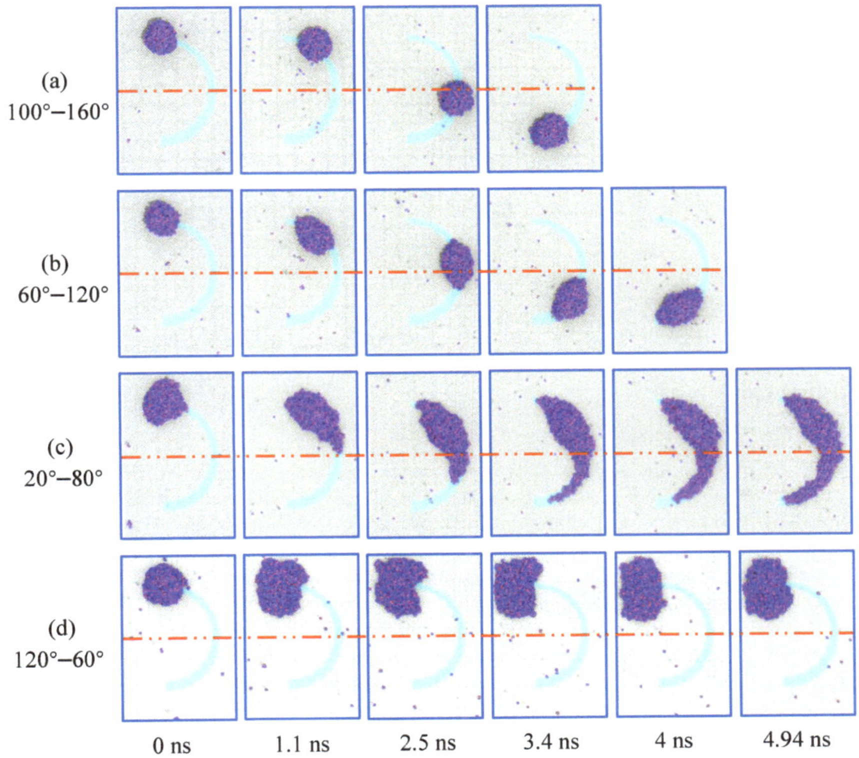

Figure 5. Instantaneous pictures for droplet motion on the curved track. (**a**) hydrophilic surface (CAs θ in the track and background separately being 100° and 160°); (**b**) hydrophilic–hydrophobic surface (CAs θ in the track and background separately being 60° and 120°); and (**c**) hydrophobic surface (CAs θ in the track and background separately being 20° and 80°); (**d**) hydrophobic–hydrophilic surface (CAs θ in the track and background separately being 120° and 60°). Red center line is marked to assist in observing droplet movement.

Data analysis indicates that the average motion velocity of the droplet in group a and group b separately reach 6.69 and 5.69 m/s, as shown in Figure 6a, while the droplet in group b moves much faster than that in group a, see Figure 6b. This is because the track and background in group a are both hydrophobic surfaces, which show weak adsorption for the droplet. In addition, the track is narrow in the initial section, which exerts slight influences on the actuating velocity of the droplet. After 1 ns, the motion velocity in group a exceeds that in group b. By observing the complete motion courses of the droplet over time in Figure 6b, the following result is found: the droplet does not always move forward on the curved track but also stagnates or, at times, retreats slightly. This is because the heading direction of the droplet constantly changes during motion on the curved track. Therefore, the forces applied by the surfaces on the droplet are also adjusted continuously in the motion process with varying directions, despite the forward motion along the track. Evidently, the self-actuation of the droplet can be realized on the functional surface and with a certain wettability difference, in which case the track and background are both hydrophobic and more conducive to the self-actuation of the droplet.

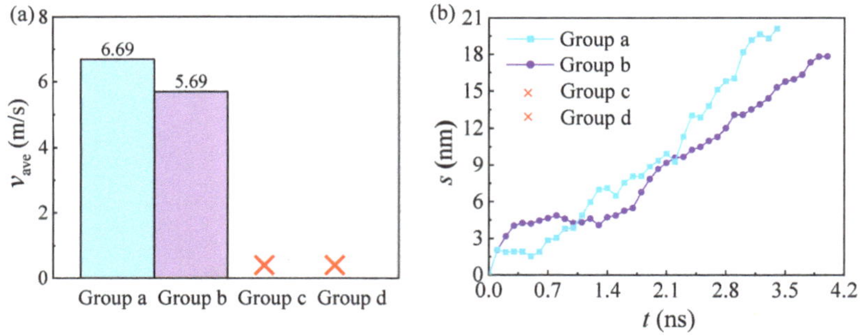

Figure 6. Influences of the wettability difference. (**a**) Average motion velocities of the droplet at varying wettability differences. (**b**) Motion courses of the droplet over time at varying wettability differences.

3.5. Influences of the Track Width

The hydrophobic surface (CAs θ in the track and background separately being 100° and 160°) was selected. The self-actuation performance of the droplet on the surface was simulated and observed by changing the track width. The width of the narrow end of the curved track and the radius of the cylinder 1 were kept at l_a = 3.92 Å and R_{O1} = 78.4 Å, while the radius R_{O2} of cylinder 2 was changed to 70.56, 66.64, and 62.72 Å (Figure 1), thus three groups of surfaces with different widths of the curved track.

The self-actuation of the droplet along the curve under three track widths was simulated. The instantaneous pictures for the motion are displayed in Figure 7; the curved tracks in group a, group b, and group c are narrowed in succession and the droplet separately takes 2.1, 1.82, and 2.22 ns to move from the start to the endpoint. The average motion velocities of the droplet in the three groups are 10.19, 11.14, and 8.83 m/s, respectively, see Figure 8a. After changing the track width, it can be found that, for a given initial width, the wider the track, the larger the rate of change of the track width in the direction of motion, which is more beneficial for the rapid motion of the droplet. Comparison of droplet motion in group a and group b shows that the droplet motion follows the trend; however, the droplet motion takes the longest time in group c with the widest track, corresponding to the lowest average motion velocity. Combined with analysis of Figure 8b, the motion of the droplet decelerates in the second half of the curved track. This is because the area of the most contact between the droplet and the surface is within the track as the track width increases, so that the contact area between the droplet and the more hydrophilic track enlarges, which affects the motion of the droplet. Meanwhile, the contact area between the droplet and the boundary shrinks; therefore, the driving force on the droplet is weakened, thus decelerating the motion of the droplet. The conclusion is, therefore, that an increase in the track width is conducive to the self-actuation of the droplet, while the self-actuation of the droplet is affected if the track is further widened, thus decreasing its velocity. Once the track width is larger than the spread-width of the droplet, that is, the droplet is completely within the track, the driving force on the droplet disappears, so the droplet stagnates on the track.

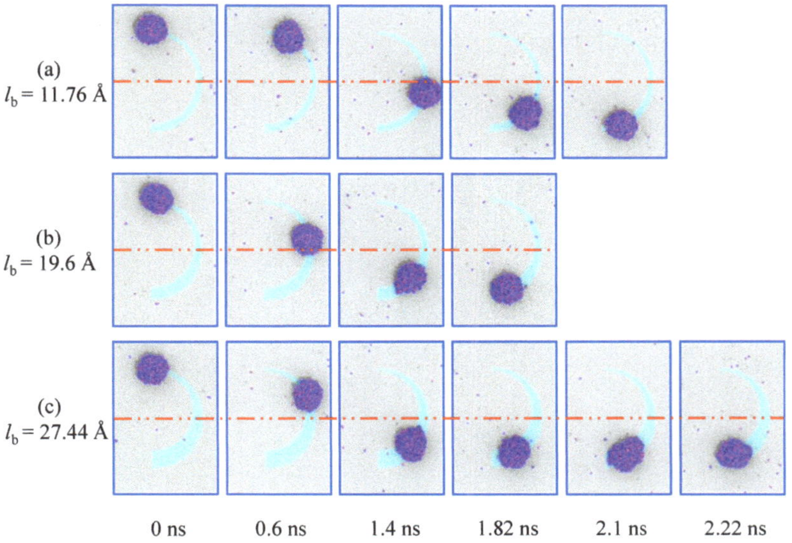

Figure 7. Instantaneous pictures showing the droplet motion on the surfaces of tracks with different widths. (**a**) The track witdh l_b is 11.76 Å; (**b**) The track witdh l_b is 19.6 Å; (**c**) The track witdh l_b is 27.44 Å. Red center line is marked to assist in observing droplet movement.

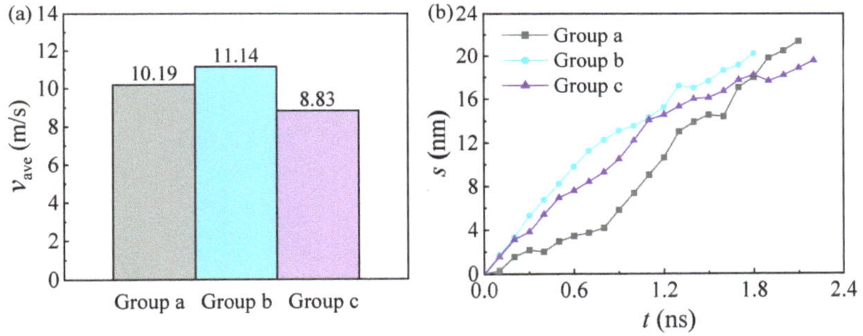

Figure 8. Influences of the track width: (**a**) Average motion velocities of the droplet correspond to different track widths. (**b**) Motion courses of the droplet over time correspond to different track widths.

3.6. Influences of the Droplet Temperature

The temperature parameter was adjusted to explore the droplet motion on the surface at a temperature of 280, 300, and 320 K in the simulation system. The instantaneous pictures of the droplet motion along the curve at different temperatures are shown in Figure 9. The results show that the droplets move increasingly fast along the curve with rising temperature. The motion velocity of the droplet is 8.12, 11.14, and 13.29 m/s with increasing temperature, as shown in Figure 10a. The movement process of the droplet over time is shown in Figure 10b. The simulations imply that in the simulation system under different temperatures, the same potential energy parameter $\varepsilon_{O\text{-}Cu}$ shows the same wettability of the droplet, that is, the contact angle between the droplet and the surface does not change due to different temperatures. Therefore, the analysis indicates that the motion velocity of the droplet changes mainly because the temperature change alters the intermolecular kinetic energy. As the temperature increases, molecular motion becomes more violent, thus increasing the interaction frequency of the droplet and surface and enabling faster motion of the droplet. In practical application, the motion velocity of the droplet can be adjusted by controlling the temperature.

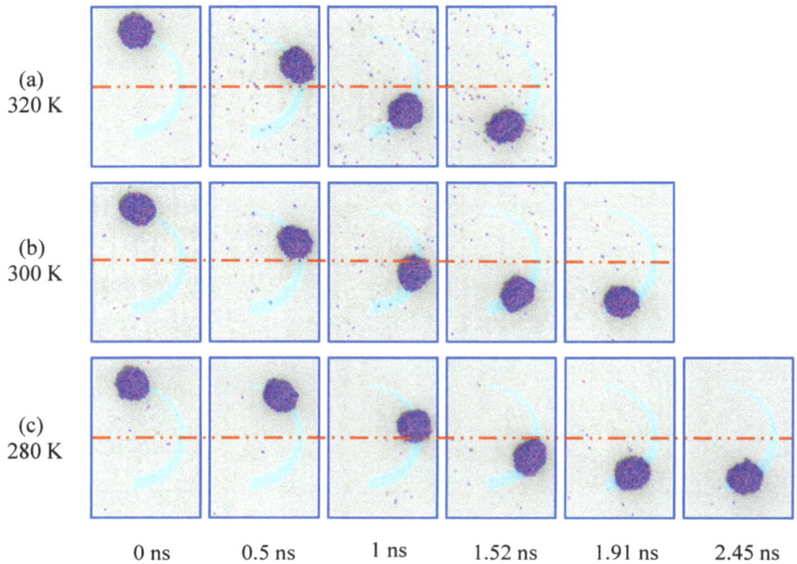

Figure 9. Instantaneous pictures of droplet motion at different temperatures. (**a**) 320 K; (**b**) 300 K; (**c**) 280 K. Red center line is marked to assist in observing droplet movement.

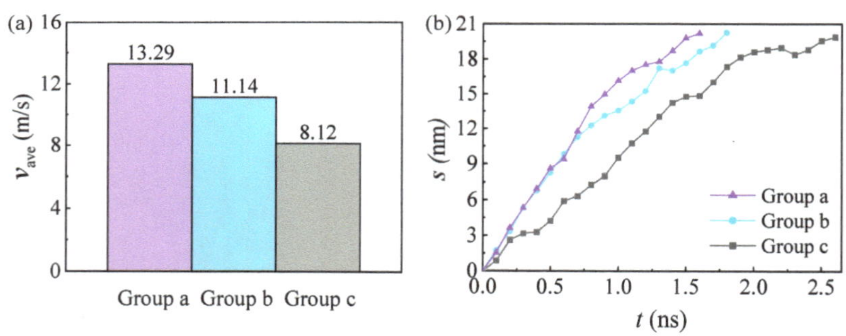

Figure 10. Influences of the droplet temperature: (**a**) Average motion velocity of the droplets at different temperatures; (**b**) Motion courses of the droplet over time at different temperatures.

4. Conclusions

Self-actuation of the droplet along the curve was designed and realized through molecular dynamics simulations. The motion mechanism and characteristics of the droplet on the surface were investigated. Moreover, influences of the wettability difference, the width of curved track, and the temperature on droplet motion were evaluated. The results show that:

(1) Droplet motion is achieved according to the joint action of the Laplace pressure and the wettability gradient force. The component force in the direction of motion promotes forward motion of the droplet, while the component force vertical to the direction of motion induces the droplet to turn in the direction of dynamic equilibrium. This realises droplet motion along the curved track.

(2) Under the same wettability difference, the hydrophobic-hydrophobic combination of the curved track and the background provides a larger driving force on the droplet than the hydrophilic-hydrophobic surface and the hydrophilic-hydrophilic surface.
(3) Increasing the width of the curved track leads to an increasingly fast motion of the droplet. However, a too large track width shrinks the contact area between the droplet and the boundary, thus decelerating the motion of the droplet and even causing the motion to stagnate. On the premise of keeping the droplet in a liquid state, the temperature rise accelerates the motion velocity of the droplet along the curved track. This is because the temperature rise increases the internal energy of molecules, leading to more violent motion, and improving the contact frequency between the droplet molecules and the surface.

The research investigates the self-actuation of the droplet along the curve from the microscopic perspective, which provides a basis for macroscopic experiments. It also reveals new opportunities for development in fields, including the lab-on-a-chip, microfluidic devices, and analytical devices using biological droplets.

Author Contributions: Methodology, Zhigang Liu; Writing—original draft, Yalong Kong; Writing—review & editing, Lin Guo and Yu Qiu. All authors have read and agreed to the published version of the manuscript.

Funding: This work was financially supported by the National Natural Science Foundation of China under contract Nos. 52206093 and 52076113, the Natural Science Foundation of Shandong Province (No. ZR2021QE078), the Collaborative Innovation Project of Colleges in Jinan (No. 2021GXRC045), and the International Cooperation Project of Science, Education, Industry Integration in Qilu University of Technology (No. 2022GH021).

Data Availability Statement: Not applicable.

Conflicts of Interest: The authors declare that they have no known competing financial interests or personal relationships that could appear to influence the work reported in this paper.

Nomenclature

E	The lj/cut styles compute the standard 12/6 Lennard-Jones potential
ε	Energy units
σ	Distance units
r	Interatomic spacing
r_c	Cutoff
θ	Contact angle between droplet and copper base
O_1	The center of the outer circle of the curved track
O_2	The center of the inside circle of the curved track
R_{O1}	The radius of the outer circle of the curved track
R_{O2}	The radius of the inside circle of the curve track
l_a	Narrow end distance of curved track
l_b	Wide end distance of curved track
F	The resultant force exerted by the curved track on the droplet
F_L	Laplace pressure gradient
F_W	Wetting gradient force
F_H	The force of surface with different wettability hindering the movement of droplets
P_1	Laplace pressure generated by the curved liquid surface at the rear end of the droplet along the track direction
P_2	Laplace pressure generated by the curved liquid surface at the front end of the droplet along the track direction
F_1	The force generated by Laplace pressure in the direction of droplet movement
F_2	The force generated by Laplace pressure perpendicular to the direction of droplet movement

F_3	The force generated by the wettability gradient in the direction of droplet movement
F_4	The force generated by the wettability gradient perpendicular to the direction of droplet motion
F_n	The resultant force of the droplet in the direction of the curve tracks
F_z	The resultant force of the droplet perpendicular to the direction of motion
γ	Surface tension of water
r_1	Local radius of the three-phase contact lines behind the droplet along the direction of the track
r_2	Local radius of the three-phase contact line in front of the droplet along the track direction
$\beta = \beta_1 = \beta_2$	
β_1	The angle formed by force P_1 and the direction of droplet movement
β_2	The angle formed by force P_2 and the direction of droplet movement
β_3	The angle between the c side wetting gradient force and the direction of droplet movement.
β_4	The angle between the d side wetting gradient force and the direction of droplet movement.
S	Contact area of droplets on both sides of curved track

References

1. Bixler, G.D.; Bhushan, B. Fluid drag reduction and efficient self-cleaning with rice leaf and butterfly wing bioinspired surfaces. *Nanoscale* **2013**, *5*, 7685–7710. [CrossRef]
2. Ju, J.; Bai, H.; Zheng, Y.M.; Zhao, T.Y.; Fang, R.C.; Lei, J. A multi-structural and multi-functional integrated fog collection system in cactus. *Nat. Commun.* **2012**, *3*, 1247. [CrossRef]
3. Guo, L.; Tang, G.H. Experimental study on directional motion of a single droplet on cactus spines. *Int. J. Heat Mass Transf.* **2015**, *84*, 198–202. [CrossRef]
4. Bai, H.; Ju, J.; Zheng, Y.M.; Jiang, L. Functional fibers with unique wettability inspired by spider silks. *Adv. Mater.* **2012**, *24*, 2786–2791. [CrossRef]
5. Ju, J.; Zheng, Y.M.; Jiang, L. Bioinspired one-dimensional materials for directional liquid transport. *Acc. Chem. Res.* **2014**, *47*, 2342–2352. [CrossRef] [PubMed]
6. Chen, H.W.; Zhang, P.F.; Zhang, L.W.; Liu, H.L.; Jiang, Y.; Zhang, D.Y.; Han, Z.W. Continuous directional water transport on the peristome surface of Nepenthes alata. *Nature* **2016**, *532*, 85–89. [CrossRef] [PubMed]
7. Jiao, Y.L.; Lv, X.D.; Zhang, Y.Y.; Li, C.Z.; Li, J.W.; Wu, H.; Xiao, Y.; Wu, S.Z.; Hu, Y.L.; Wu, D.; et al. Pitcher plant-bioinspired bubble slippery surface fabricated by femtosecond laser for buoyancy-driven bubble self-transport and efficient gas capture. *Nanoscale* **2019**, *11*, 1370–1378. [CrossRef] [PubMed]
8. Hamilton, W.J.; Seely, M.K. Fog basking by the Namib Desert beetle, Onymacris Unguicu-laris. *Nature* **1976**, *262*, 284–285. [CrossRef]
9. Parker, A.R.; Lawrence, C.R. Water capture by a desert beetle. *Nature* **2001**, *414*, 33–34. [CrossRef]
10. Cieplak, M.; Koplik, J.; Banavar, J.R. Nanoscale fluid flows in the vicinity of patterned surfaces. *Phys. Rev. Lett.* **2006**, *96*, 114502. [CrossRef] [PubMed]
11. Paradisanos, I.; Fotakis, C.; Anastasiadis, S.H.; Stratakis, E. Gradient induced liquid motion on laser structured black Si surfaces. *Appl. Phys. Lett.* **2015**, *107*, 111603. [CrossRef]
12. Lv, C.J.; Chen, C.; Chuang, Y.C.; Tseng, F.G.; Yin, Y.J.; Grey, F.; Zheng, Q.S. Substrate curvature gradient drives rapid droplet motion. *Phys. Rev. Lett.* **2014**, *113*, 26101. [CrossRef] [PubMed]
13. Bai, H.; Tian, X.L.; Zheng, Y.M.; Ju, J.; Zhao, Y.; Jiang, L. Direction controlled driving of tiny water drops on bioinspired artificial spider silks. *Adv. Mater.* **2010**, *22*, 5521–5525. [CrossRef] [PubMed]
14. Du, M.; Zhao, Y.; Tian, Y.; Li, K.; Jiang, L. Electrospun multiscale structured membrane for efficient water collection and directional transport. *Small* **2016**, *12*, 1000–1005. [CrossRef]
15. Hou, Y.P.; Feng, S.L.; Dai, L.M.; Zheng, Y.M. Droplet manipulation on wettable gradient surfaces with micro-/nano-hierarchical structure. *Chem. Mater.* **2016**, *28*, 3625–3629. [CrossRef]
16. Kou, J.L.; Mei, M.F.; Lu, H.J.; Wu, F.M.; Fan, J.T. Unidirectional motion of a water nanodroplet subjected to a surface energy gradient. *Phys. Rev. E Stat. Nonlin. Soft. Matter Phys.* **2010**, *85*, 988–1000. [CrossRef]
17. Wang, Z.K.; Zheng, H.Y.; Xia, H.M. Femtosecond laser-induced modification of surface wettability of PMMA for fluid separation in microchannels. *Microfluid. Nanofluid.* **2011**, *10*, 225–229. [CrossRef]
18. Gurera, D.; Bhushan, B. Multistep wettability gradient on bioinspired conical surf-aces for water collection from fog. *Langmuir* **2019**, *35*, 16944–16947. [CrossRef]

19. Chen, H.W.; Tong, R.; Yang, G.; Zhou, J.J.; Zhang, Y.; Zhang, L.W.; Zhang, D.Y.; Jiang, L. Ultrafast water harvesting and transport in hierarchical microchannels. *Nat. Mater.* **2018**, *17*, 935–942. [CrossRef]
20. Park, K.C.; Kim, P.; Grinthal, A.; He, N.; Fox, D.; Weaver, J.C.; Aizenberg, J. Condensation on slippery asymmetric bumps. *Nature* **2016**, *531*, 78–82. [CrossRef]
21. Meng, Q.A.; Xu, B.J.; He, M.J.; Bian, R.X.; Meng, L.L.; Wang, P.W.; Jiang, L.; Liu, H. Bioinspired controllable liquid manipulation by fibrous array driven by elasticity. *ACS Appl. Mater.* **2018**, *10*, 26819–26824. [CrossRef]
22. Liu, W.J.; Fan, P.X.; Cai, M.Y.; Luo, X.; Chen, C.H.; Pan, R.; Zhang, H.J.; Zhong, M.L. An integrative bioinspired venation network with ultra-contrasting wettability for large-scale strongly self-driven and efficient water collection. *Nanoscale* **2019**, *11*, 8940–8949. [CrossRef] [PubMed]
23. Li, K.; Ju, J.; Xue, Z.X.; Ma, J.; Feng, L.; Gao, S.; Jiang, L. Structured cone arrays for continuous and effective collection of micron-sized oil droplets from water. *Nat. Commun.* **2013**, *4*, 2276. [CrossRef] [PubMed]
24. Seo, K.; Kim, M.; Kim, D.H. Candle-based process for creating a stable superhydrophobic surface. *Carbon* **2014**, *68*, 583–596. [CrossRef]
25. Pei, Y.Y.; Song, Q.; Li, P. Research progress of biomimetic micro/nano-structured antibacterial surfaces. *Surf. Technol.* **2019**, *48*, 200–210. [CrossRef]
26. Lan, F.; Haliburton, J.R.; Yuan, A.; Abate, A.R. Droplet barcoding for massively parallel single-molecule deep sequencing. *Nat. Commun.* **2016**, *7*, 11784. [CrossRef] [PubMed]
27. Parashar, V.K.; Wacker, J.B.; Gijs, M.A. Spherical superstructures of oxide nanoparticles for catalytic reactions in microchemical reactors. *Mater. Lett.* **2015**, *139*, 182–186. [CrossRef]
28. Tang, X.; Huang, J.X.; Guo, Z.G.; Liu, W.M. A combined structural and wettability gradient surface for directional droplet transport and efficient fog collection. *J. Colloid Interface Sci.* **2021**, *604*, 526–536. [CrossRef] [PubMed]
29. Feng, S.L.; Wang, Q.Q.; Xing, Y.; Hou, Y.P.; Zheng, Y.M. Continuous directional water transport on integrating tapered surfaces. *Adv. Mater. Interfaces* **2020**, *7*, 2000081. [CrossRef]
30. Lin, Y.C.; Hu, Z.Y.; Gao, C.L.; Guo, Z.Y.; Li, C.; Zheng, Y.M. Directional droplet spreading transport controlled on tilt-angle pillar arrays. *Adv. Mater. Interfaces* **2018**, *5*, 1800962. [CrossRef]
31. Wang, M.; Liu, Q.; Zhang, H.R.; Wang, C.; Wang, L.; Xiang, B.X.; Fan, Y.T.; Guo, C.F.; Ruan, S.C. Laser direct writing of tree-shaped hierarchical cones on a superhydrophobic film for high-efficiency water collection. *ACS Appl. Mater.* **2017**, *9*, 29248–29254. [CrossRef] [PubMed]
32. Khoo, H.S.; Tseng, F.G. Spontaneous high-speed transport of subnanoliter water droplet on gradient nanotextured surfaces. *Appl. Phys. Lett.* **2009**, *95*, 063108. [CrossRef]
33. Zhang, J.L.; Han, Y.C. Shape-gradient composite surfaces: Water droplets move uphill. *Langmuir* **2007**, *23*, 6136–6141. [CrossRef]
34. Huang, D.J.; Leu, T.S. Fabrication of a wettability-gradient surface on copper by screen-printing techniques. *J. Micromech. Microeng.* **2015**, *25*, 1325–1336. [CrossRef]
35. Ody, T.J.; Panth, M.; Sommers, A.D.; Eid, K.F. Controlling the motion of ferrofluid droplets using surface tension gradients and magnetoviscous pinning. *Langmuir* **2016**, *32*, 6967–6976. [CrossRef] [PubMed]
36. Deng, S.Y.; Shang, W.F.; Feng, S.L.; Zhu, S.P.; Xing, Y.; Li, D.; Hou, Y.P.; Zheng, Y.M. Controlled droplet transport to target on a high adhesion surface with multi-gradients. *Sci. Rep.* **2017**, *7*, 45687. [CrossRef] [PubMed]
37. Liu, M.; Yao, Y.; Li, J.J.; Peng, Z.L.; Chen, S.H. Directional sliding behavior of a water droplet on a wedge-shape patterned functional surface. *J. Phys. Chem. B* **2020**, *124*, 6905–6912. [CrossRef] [PubMed]
38. Zheng, Y.F.; Chen, J.; Zhou, C.L.; Xing, H.T.; Wen, X.F.; Pi, P.H.; Xu, S.P. Dro-plet motion on a shape gradient surface. *Langmuir* **2017**, *33*, 4172–4177. [CrossRef] [PubMed]
39. Guan, J.H.; Élfego, R.G.; Xu, B.B.; Wood, D.; Mchale, G.; Ledesma-Aguilar, R.; Wells, G.G. Drop transport and positioning on lubricant-impregnated surfaces. *Soft. Matter* **2017**, *13*, 3404–3410. [CrossRef] [PubMed]
40. Zhang, C.H.; Zhang, B.; Ma, H.Y.; Li, Z.; Xiao, X.; Zhang, Y.H.; Cui, X.Y.; Yu, C.M.; Cao, M.Y.; Jiang, L. Bioinspired pressure-tolerant asymmetric slippery surface for continuous self-transport of gas bubbles in aqueous environment. *ACS Nano* **2018**, *12*, 2048–2055. [CrossRef]
41. Bai, H.; Wang, L.; Ju, J.; Sun, R.Z.; Zheng, Y.M.; Jiang, L. Efficient water collection on integrative bioinspired surfaces with star-shaped wettability patterns. *Adv. Mater.* **2014**, *26*, 5025–5030. [CrossRef] [PubMed]
42. Wang, S.; Wang, C.; Peng, Z.L.; Chen, S.H. Moving behavior of nanodroplets on wedge-shaped functional surfaces. *J. Phys. Chem. C* **2019**, *123*, 1798–1805. [CrossRef]
43. Xu, B.; Chen, Z.Q. Droplet movement on a composite wedge-shaped surface with multi-gradients and different gravitational field by molecular dynamics. *Microgravity Sci. Tec.* **2018**, *30*, 571–579. [CrossRef]
44. Hao, S.Q.; Xie, Z.; Li, Z.; Kou, J.L.; Wu, F.M. Initial-position-driven opposite directional transport of water droplet on wedge-shaped groove. *Nanoscale* **2021**, *13*, 15963–15972. [CrossRef]
45. Guo, L.; Kumar, S.; Yang, M.Y.; Tang, G.H.; Liu, Z.G. Role of the microridges on cactus spines. *Nanoscale* **2022**, *14*, 525–533. [CrossRef]
46. Mo, J.W.; Wang, C.; Zeng, J.Y.; Sha, J.J.; Li, Z.G.; Chen, Y.F. Directional passive transport of nanodroplets on general axisymmetric surfaces. *Phys. Chem. Chem. Phys.* **2022**, *24*, 9727–9734. [CrossRef] [PubMed]

47. Tan, X.H.; Zhu, Y.Y.; Shi, T.L.; Tang, Z.R.; Liao, G.L. Patterned gradient surface for spontaneous droplet transportation and water collection: Simulation and experiment. *J. Micromech. Microeng.* **2016**, *26*, 115009. [CrossRef]
48. Halverson, J.D.; Maldarelli, C.; Couzis, A.; Koplik, J. A molecular dynamics study of the motion of a nanodroplet of pure liquid on a wetting gradient. *J. Chem. Phy.* **2008**, *129*, 827. [CrossRef] [PubMed]
49. Huang, Q.J.; Zhang, Z.Q.; Liu, Z.; Zhang, F.J.; Cheng, G.G.; Ding, J.N. Pinning effect in droplet self-driving and its reduction mechanism by monolayer graphene. *Appl. Surf. Sci.* **2021**, *542*, 148666. [CrossRef]
50. Zhang, Z.Q.; Guo, X.F.; Tang, H.Y.; Ding, J.N.; Zheng, Y.G.; Li, S.F. Unidirectional self-driving liquid droplet transport on a monolayer graphene-covered textured substrate. *ACS Appl. Mater.* **2019**, *11*, 28562–28570. [CrossRef] [PubMed]

energies

MDPI

Review

Review of Concentrated Solar Power Technology Applications in Photocatalytic Water Purification and Energy Conversion: Overview, Challenges and Future Directions

Cheng Zhang [1], Na Li [1] and Guangqi An [2,*]

[1] National Energy Key Laboratory for New Hydrogen-Ammonia Energy Technologies, Foshan Xianhu Laboratory, Foshan 528200, China; amdain@sina.cn (C.Z.); lina@xhlab.cn (N.L.)

[2] Graduate School of Life and Environmental Science, University of Tsukuba, 1-1-1 Tennodai, Tsukuba 305-8572, Ibaraki, Japan

* Correspondence: s2236023@u.tsukuba.ac.jp; Tel./Fax: +81-070-3145-8576

Abstract: Photocatalysis, a promising semiconductor-based technology activated by free and eternal solar energy, has great potential for addressing environmental remediation and energy conversion challenges. Concentrated solar power (CSP) technologies, namely parabolic trough reflectors, solar power towers, parabolic dish reflectors and linear Fresnel reflectors, exhibited excellent feasibility for boosting solar-driven photocatalytic processes. Based on the structural characteristics of CSP technologies, the CSP-based photocatalytic reactors could be divided into concentrated types and non/low-concentrated types. This academic review comprehensively investigated the integration of CSP technology in photocatalysis, emphasizing the feasibility of sunlight as an ideal energy source. Additionally, considering the optimal light irradiance and reaction temperature demands for achieving efficient photocatalytic processes, the significance of introducing CSP into solar light-driven photocatalytic reactions was highlighted. Moreover, the current challenges that exist in CSP-based photoreactors were identified, and potential solutions were proposed accordingly. This work hopes to provide some references for the future study of CSP-based photocatalytic reactors under the theme of sustainable development.

Keywords: concentrated solar power (CSP) technology; photocatalysis; solar energy; photoreactor; sustainable development

1. Introduction

The sustainable development of the global economy and society is being disrupted by environmental deterioration, energy crises and other aspects [1]. The traditional way of production and life requires a large amount of non-renewable energy to drive, which not only further aggravates the problem of insufficient energy supply but also continues to destroy the ecological environment. Faced with this dilemma, the United Nations put forward the Sustainable Development Goals (SDGs), which emphasize the utilization of renewable energy and new production development modes [2]. Motivated by this, significant effort has been put into searching for new process strategies for energy conservation and emission reduction and studying their utilization patterns.

In the last few decades, photocatalysis has been intensively studied [3–5]. Under the illumination of incident light, the photocatalyst, as a semiconductor, could be excited to release electrons and holes, thereby leading to the subsequent generation of free radicals (such as superoxide radicals, hydroxyl radicals, etc.) with strong oxidation–reducing properties [6]. These free radicals could act on the structure of organic molecules and destroy their molecular bonds, resulting in the destruction and degradation of organic particles with no change occurring on the photocatalyst itself. Owing to the non-selectivity, cost-effectiveness, non-toxicity and eco-friendliness, photocatalytic technology exhibited impressive application potential in environment remediation (e.g., wastewater treatment [7,8], abatement of noxious gases [9], sterilization ([3,10], etc.), energy conversion (e.g., water splitting for hydrogen generation [11,12], hydrogen formation via ammonia decomposition [13,14], microalgae biorefinery [1], etc.) and so on. In order to boost this promising technology further into practical applications, the sources of energy used to drive photocatalytic reactions need to be carefully selected. Solar energy, as a natural source of light, has the following exciting characteristics: (1) permanent supply of energy, (2) wide distribution on earth, (3) easy utilization availability and (4) wide spectral band of wavelengths. Although the proportion of high-energy UV photons is only 3% to 5% of the total solar energy, it does not affect solar light to be the ideal driving force to motivate the photocatalytic processes [1,3,4,7].

As a bridge between sources of driving forces (solar energy) and users of driving forces (photocatalysis), photocatalytic reactors have been studied tirelessly in recent decades. The simplest photoreactor was a beaker placed under simulated solar light irradiation, which was filled with a photocatalyst and reactant [15,16]. Despite initiating the photocatalytic reaction, it was obviously far from practical application. Going one step further, some primitive but systematic prototype photoreactors have been developed and studied, including the classic inclined plate reactor (IPR) [7]. The structure of an IPR is characterized by an inclined plane facing the direction of solar light incidence, and the angle of inclination can be adjusted according to the local latitude. The photocatalyst is immobilized or flows down with wastewater in powder form, and the photocatalytic process is implemented upon the inclined plate. This simple and inexpensive design propelled IPR into widespread use, and an IPR-based wastewater treatment plant was even manufactured for purifying industrial wastewater from textile mills [17]. Nevertheless, one obvious disadvantage of IPR is that it leads to insufficient photocatalytic activity because of the low solar energy collecting efficiency brought on by passive acceptance of solar light [7]. Aiming to promote photocatalytic efficiency under actual sunlight, new photoreactor concepts are required. When it comes to effectively harvesting sunlight, it is natural to think of concentrated solar power (CSP) technology. CSP technology refers to the renewable energy project that concentrates, collects and converts solar energy into heat flux through a photo-thermal conversion process [18]. It has attracted great attention, owing to its unique advantages such as its superior capability for light and thermal collection, wide feasibility for multiple utilization purposes and high technical maturity. Generally, CSP technology can be divided into parabolic trough reflectors (PTRs), solar power towers (SPTs), parabolic dish reflectors (PDRs) and linear Fresnel reflectors (LFRs) according to their system configurations [19]. Since the effective solar energy collection capability of CSP technology matches the demand for solar energy-driven photocatalytic processes, it is not surprising that CSP-based photoreactors are widely developed and studied.

Based on the above background, this academic review is dedicated to demonstrating the integration and application of CSP technology in photocatalytic processes, especially water purification and energy conversion. Preferentially, the principle of the action between solar energy and photocatalysis is explained, and the legitimacy and rationality of sunlight as an ideal light source for photocatalysis is highlighted. Then, the reasons why CSP technology is meaningful and necessary to be introduced into solar energy-driven photocatalytic reactions is explained from the perspective of energy demand for efficient photocatalysis. Subsequently, four typical CSP configurations (including PTR, SPT, PDR and LFR), as well as their traditional applications, are described. In addition, as the highlight of this review, the original idea, development process, current status and future research trends of each CSP-based photocatalytic reactor are emphatically introduced. In this section, CSP-based photocatalytic photoreactors are classified as solar-concentrated types and solar non/low-concentrated types according to their solar concentration ratio, which will be introduced in order to present the technological changes. Moreover, some of the state-of-art photocatalytic reactor technologies, such as solar energy control strategies and 24 h all-weather systems, will also be demonstrated. Furthermore, the current challenges faced by CSP-based photocatalytic reactors are also pointed out, and some potential ideas for solving these problems are provided.

2. Feasibility of Solar Energy for Photocatalytic Applications

The formation of electrons and holes in photocatalysts requires photon excitation with sufficient energy. In order to excite electrons from the valence band (VB) of the photocatalyst to the conduction band (CB), the photon energy carried by the incident light needs to overcome the band gap between the VB and CB (as shown in Figure 1a). After decades of development, a large family of photocatalysts has been formed. According to the element composition, synthesis conditions and defect location, different photocatalysts have different band gaps. Figure 1b shows the light wavelength required for activating some typical photocatalysts (data from [8,20–37]). It can be seen that although the sun emits light over a wide range of wavelengths, traditional photocatalysts such as TiO_2 can only be excited by UV light with wavelengths less than 380 nm [28]. Since UV energy only accounts for a small proportion of solar energy, this results in low UV radiation intensity and low photocatalytic efficiency. In recent years, newly developed photocatalysts such as $P/Ag/Ag_2O/Ag_3PO_4/TiO_2$ composites [31] have expanded the utilizable solar radiation band, enabling it to be excited by both visible light and UV, significantly improving the photocatalytic efficiency compared to TiO_2 alone. Since solar energy is an infinite and widely distributed light source possessing a broad spectrum wavelength ranging from 300 to 2500 nm [7], it could excite the photocatalyst with any band gap, thus implementing the photocatalytic reaction process.

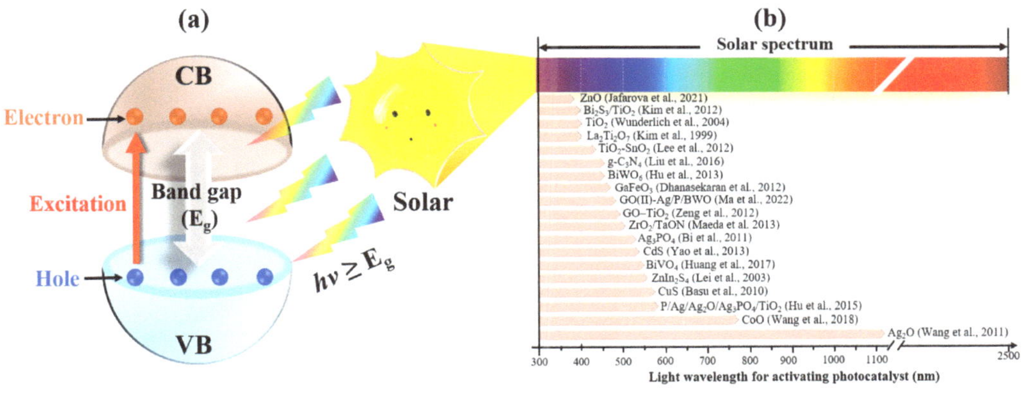

Figure 1. Activation of photocatalysts under solar light. (**a**) Generation of electron and hole; (**b**) light wavelength required for activating different photocatalysts [8,20–37].

In addition, from a practical point of view, to make photocatalytic technology applicable, efficient photocatalytic processes need to be pursued. Therefore, suitable photocatalytic reaction conditions need to be provided and maintained. Since solar light energy mainly provides light irradiance and the reaction temperature, whether natural sunlight is qualified to meet the appropriate reaction conditions for efficient photocatalysis needs to be investigated. Figure 2a illustrates the optimal light irradiance required for various photocatalytic reactions. It can be observed that for a wide range of photocatalytic reactions, including the removal of organic dyes (rhodamine b, tartrazine, reactive black 5, etc.) [7,38,39], the degradation of harmful antibiotics (metronidazole, tetracycline, cephalexin, etc.) [40,41], the inactivation of pathogenic bacteria (*Escherichia coli*, etc.) [42] and the oxidation of organic matters (aerobic oxidation, etc.) [43], the light irradiance in the range of 600–1600 W/m^2 is conducive to the implementation of efficient photocatalytic reactions (Figure 2a). However, under real environmental conditions, the solar light irradiance (data from [7,44–49]) is much lower than the appropriate light irradiance required for efficient photocatalysis

(Figure 2a). For example, Basem et al. reported that the maximum sunlight irradiance on a typical sunny day (during December) in Iraq was only 500 W/m^2 [49]. Additionally, Cao et al. measured the annual solar irradiance in Northwest China and found that the average value was 641 W/m^2 [48]. Moreover, Zhang et al. investigated the whole-year solar light irradiance in Japan, which ranged from 0–767 W/m^2 for Tsukuba (36.1° N) and 0–633 W/m^2 for Sapporo (43.4° N) [7]. Since Tsukuba-Sapporo is located in Japan's latitude zone, which is the latitude range with the most densely populated areas in the world [7], the natural solar irradiance of most of the world can also be inferred to be within 800 W/m^2. Therefore, natural solar light irradiance needs to be enhanced to meet the needs of efficient photocatalytic reactions.

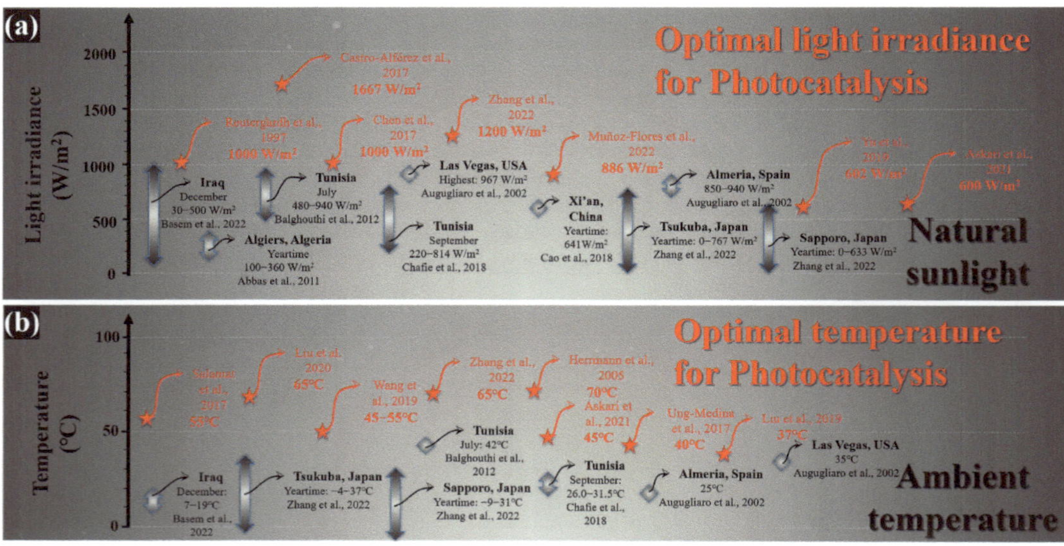

Figure 2. Comparison of optimum photocatalytic reaction conditions with natural conditions. (**a**) Light irradiance [7,38–49]; (**b**) temperature [7,10,40,45–47,49–54].

On the other hand, in order to further investigate whether natural solar energy is qualified to support efficient photocatalytic reactions, a comparison between ambient temperature and the favorable reaction temperature was carried out. As shown in Figure 2b, a wide range of photocatalytic reactions, including the treatment of steel mill wastewater [51], organic wastewater treatment (containing acid blue 9 or rhodamine b) [7,54], sterilization (*Escherichia coli*, etc.) [10], treatment of harmful antibiotics (cephalexin, metronidazole, tetracycline, etc.) [40,53] and oxidation of acetone [52], require a suitable reaction temperature ranging from 37 to 70 °C for maximizing their efficiency (Figure 2b). In addition, the influence of the reaction temperature on the kinetics of photocatalysis was also studied, which indicated that 70 °C might be a superior temperature for maximizing photocatalytic efficiency [50]. However, many researchers have reported their local ambient temperature, confirming that ambient temperatures directly converted by natural sunlight may not be conducive to efficient photocatalytic reactions (Figure 2b). For example, November temperatures in Iraq ranged from 7.0–19.0 °C [49], September temperatures in Tunisia ranged from 26.0–31.5 °C [47], a typical sunny day in Las Vegas in the US was approximately 35.0 °C [46] and year-round temperatures in Tsukuba and Sapporo in Japan ranged from −4.0 to 37.0 °C and −9.0 to 31.0 °C, respectively [7]. Although solar energy could effectively initiate photocatalytic reactions, its inherent low energy density makes it difficult to achieve optimal photocatalytic efficiency when using sunlight directly. Hence, to boost the photocatalytic reaction efficiency by adopting free and eternal solar energy, sunlight needs to be concentrated. CSP technology, an effective approach for solar energy convergence, could have significant advantages in meeting this demand.

3. Brief Overview of Concentrated Solar Power Technologies

The history of CSP probably dates back to the beginning of attempts to harness photo-thermal energy. As technology continues to advance, CSP design and manufacturing techniques continue to improve. With the development of materials science and manufacturing processes, people began to use more durable and more reflective materials to create CSP. In the past decades, with the increasingly prominent energy crisis and environmental problems, solar energy technology has become more widely used and researched. During this period, CSP design and manufacturing techniques were further improved. For example, computer-aided design (CAD) and optical techniques are being used to optimize the structure and reflection properties of solar technologies, and applications of novel materials and manufacturing processes are being explored. Currently, according to the configuration characteristics, CSP can be classified as PTR, PDR, SPT and LFR.

3.1. *Parabolic Trough Reflectors*

The PTR might be the most common and widely deployed CSP technology. As shown in Figure 3a, the most important technical feature of the PTR is that it is equipped with a parabolic-shaped concentrator mirror and a receiver tube located at the focal line of the parabola [45]. Owing to the optical properties of the paraboloid, it can efficiently focus incident parallel sunlight onto its focal line. This capability allows the receiver to harness a substantial amount of light energy, subsequently converting it into a concentrated heat flux. In order for the parabolic mirror to accurately converge parallel solar rays, the aperture of the parabola must always face the incidence direction of sunlight accurately. Therefore, parabolic mirrors are usually equipped with a controlling system to track solar motion. Due to its simpler structure compared with other CSP technologies, PTR has been successfully applied in various application scenarios. The most interesting application of PTR might be in the field of solar thermal power generation [55]. By concentrating a large amount of light and heat energy, the thermal medium (such as thermal oil and molten salts) in the receiver tube is heated, thus heating the water into high-temperature steam, which drives the turbine to generate electricity. Additionally, the solar cooking system also consists of an aluminum reflector with a vacuum tube, which provides a suitable cooking temperature of over 200 °C under actual solar conditions [56]. Moreover, the systems employed for desalination were also established by combining the PTR with an evacuated tube [57]. Furthermore, a PTR-based solar adsorption refrigeration system was also investigated, which used olive waste as the adsorbent and methanol as the adsorbate [58]. Other applications of PTR include but are not limited to photo-thermal direct sterilization systems [59], photoelectrochemical reactors for contaminant removal [60], photo-Fenton reactors [61] and industrial process heat [62]. These applications demonstrate the wide applicability and flexible deployability of PTR technology.

3.2. *Parabolic Dish Reflectors*

The main components of a PDR system (shown in Figure 3b) include a parabolic dish concentrator mirror and a thermal receiver mounted at the focus point of the parabolic dish [63]. Similar to PTR technology, PDR also requires the incident direction of the parallel sunlight to be perpendicular to the aperture of the parabolic dish, thus enabling effective solar energy harvesting. Therefore, a two-axis sun tracking system is also indispensable for PDR. Generally, PDR systems are characterized by a high efficiency, autonomous operation and excellent modularity [63]. Many PDR-based power plants have been manufactured worldwide, which exhibit high solar-to-electric conversion efficiency and have the potential to become one of the most effective renewable energy utilization approaches [44,49,63]. Additionally, owing to their inherent simplicity, portability and cost-effectiveness, solar cooker systems are also regarded as one of the most significant application fields of

PDR [64–66]. Moreover, the solar desalination system based on PDR was also experimentally tested, in which a triple basin glass solar still was fabricated and located upon the focus point of the parabola dish for adsorbing solar energy [67]. Therefore, PDR systems also share merits for wide deployment in different fields to achieve multiple goals.

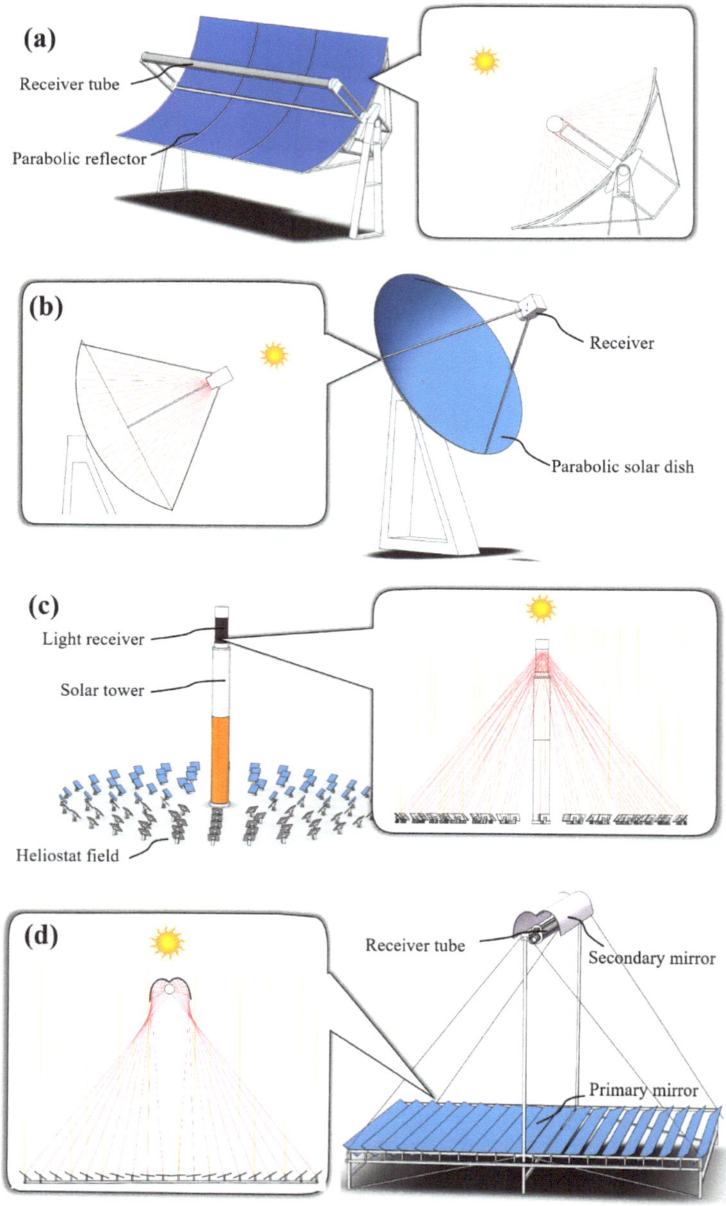

Figure 3. Schematic diagram of major concentrated solar power (CSP) technologies. (**a**) Parabolic trough reflector (PTR); (**b**) parabolic dish reflector (PDR); (**c**) solar power tower (SPT) and (**d**) linear Fresnel reflector (LFR).

3.3. *Solar Power Towers*

In recent decades, as a representative of high-temperature concentration engineering, SPT has been a popular topic in CSP engineering applications [68]. As shown in Figure 3c, in the SPT system, a large number of heliostats are arranged in an oval shape around the heat collecting tower. The heliostat could automatically track the movement of the sun and focus sunlight to the receiver located above the tower, thereby converting light energy into heat flux. The solar concentration ratio of a SPT can reach the equivalent of hundreds to thousands of suns, thereby achieving ultra-high operating temperatures. In recent years, to achieve the goal of saving energy and reducing emissions, SPT-based photo-thermal power plants employing the Rankine or Brayton cycle have been intensively studied and deployed [69]. The above deployment and operation of SPT power plants demonstrates their excellent availability and realizability in solar energy collection.

3.4. *Linear Fresnel Reflectors*

In the past few decades, LFRs have also been studied and developed, and they have been widely used in photo-thermal power generation projects [70,71]. As shown in Figure 3d, the main structure of an LFR includes segmented flat mirrors which are mounted in parallel (as a primary mirror) and a secondary mirror located above the primary mirror array. The light receiver is mounted at the focus position of the secondary mirror. Each row of primary mirrors has a rotation axis, and all the primary mirrors track the position of the sun in the sky under the control of the driving mechanism, focusing the sunlight on the secondary mirror [7]. The shape of the secondary mirror is a compound parabolic collector (CPC) that can capture and reflect incoming sunlight in all directions to its focal point [72]. CPC, as a supporting CSP technology for LFR, plays a crucial role in improving the efficiency and heat flow uniformity of LFR systems. CPC is usually formed by a variety of shape combinations (such as involute, parabola, etc.), thus endowing it with the unique capability to collect beam radiation and diffuse radiation within a specific acceptable half-angle and without the utilization of a complex solar tracking system [73]. Owing to the cooperation of the primary mirror and secondary mirror (CPC), the heat flux could be distributed uniformly over the receiver's surface, thus enabling the achievement of a high thermal efficiency in LFR systems. Usually, LFR can reach a solar concentrating ratio equivalent to tens to hundreds of suns, thereby achieving an operating temperature of 500 °C [29,71]. Due to the low cost of flat primary mirrors and single-axis control systems, LFRs are generally considered to have superior economic efficiency for solar energy collection [7].

4. CSP Technologies Applied in Photocatalysis

To effectively provide solar energy for photocatalytic reactions, a large amount of photocatalytic reactors have been designed and developed. Due to the excellent performance of CSP technology in harvesting solar light energy and heat energy conversion, it is naturally applied as the design basis of photocatalytic reactors. As shown in Table 1, according to the structure of the solar concentrator, these photoreactors could be classified into PTR, PDR, LFR, FL, V-groove and CPC, which are organized in a timeline. Reviewing the past photocatalytic reactors based on CSP technology, they can be further divided into solar concentrated types and solar non/low-concentrated types, which are depicted below.

Table 1. Summary of photocatalytic reactors employing CSP technology.

Thematic Axis of Reactor	Dimension of Sunlight Reflector	Reactant Volume	Temperature	Irradiance	Deployed Location	Photocatalyst	Treatment Target	Timeline	Ref.
PTR	-	-	-	-	U.S. (latitude: 38.6° N)	TiO_2	Groundwater	1991	[74]
	Width: 1.8 m; Length: 4.5 m	-	-	-	Spain (latitude: 37.4° N)	TiO_2	Pentachlorophenol	1993	[75]
	Width: 1.8 m; Length: 4.5 m	220 L	-	-	Spain (latitude: 37.4° N)	TiO_2	Atrazine	1996	[75]
	29.1 m^2	260 L	-	-	Spain (latitude: 37.4° N)	TiO_2	2,4-Dichlorophenol	1997	[76]
	-	80 L	-	-	Germany (latitude: 50.6° N)	TiO_2	$C_2H_5NH_2$ and $(C_2H_5)_2NH$	2000	[77]
	Width: 1 m; Length: 1 m	-	200–250 °C	-	Japan (latitude: 36° N)	Pt-TiO_2	Toluene, acetaldehyde	2004	[78]
	0.72 m^2	10 L	-	-	Mexico (latitude: 19° N)	TiO_2	Oxalic acid	2004	[79]
	0.042 m^2	1 L	<38 °C	-	Spain (latitude: 37° N)	TiO_2	*Escherichia coli*	2004	[80]
	Width: 0.5 m; Length: 1.4 m	-	-	-	Iran (latitude: 22.4° N)	g-C_3N_4/TiO_2	Methylene blue and rhodamine b	2021	[81]
PDR	0.785 m^2	3 L	30–58 °C	10–35 W/m^2 (UV)	Japan (latitude: 35.4° N)	TiO_2	Detergent	2004	[82]
LFR	Width: 0.055 m; Length: 0.23 m	1 L	≤52 °C	24–40 W/m^2 (UV)	Japan (latitude: 36.1° N)	TiO_2	Rhodamine b, amoxicillin, *Escherichia coli*	2022	[7]
	0.077 m^2	1 L	40 °C	23.6 W/m^2 (UV)	Japan (latitude: 36.1° N)	TiO_2	*Chlorella vulgaris*	2023	[1]
FL	-	0.05 L	93 °C	15,000 W/m^2	China (latitude: 34.5° N)	Cu-TiO_{2-x}	Urea, urine	2022	[83]
V-groove	0.72 m^2	10 L	-	-	Mexico (latitude: 19° N)	TiO_2	Oxalic acid	2004	[79]
	0.057 m^2	1 L	<38 °C	-	Spain (latitude: 37° N)	TiO_2	*Escherichia coli*	2004	[80]
CPC	8.9 m^2	247 L	-	-	Spain (latitude: 37.4° N)	TiO_2	2,4-Dichlorophenol	1997	[76]
	4.5 m^2	25 L	<40 °C	15–25 W/m^2 (UV) 20–35 W/m^2 (UV)	Spain (latitude: 43° N) Morocco (latitude: 34° N)	TiO_2	*Escherichia coli, Enterococcus faecalis*	1999	[84]
	3.08 m^2	39 L	-	-	Spain (latitude: 37° N)	TiO_2	Azo-dyes	2002	[46]
	0.72 m^2	10 L	-	-	Mexico (latitude: 19° N)	TiO_2	Oxalic acid	2004	[79]
	0.057 m^2	1 L	<38 °C	-	Spain (latitude: 37° N)	TiO_2	*Escherichia coli*	2004	[80]
	3.08 m^2	35 L	<45 °C	-	Spain (latitude: 37° N)	TiO_2	Peptone, meat extract, urea, etc.	2004	[85]
	0.25 m^2	11 L	15.8–36.5 °C	-	Spain (latitude: 37.4° N)	TiO_2	*Escherichia coli*	2005	[86]
	3.08 m^2	39 L	-	-	Spain (latitude: 37° N)	TiO_2	Lincomycin	2005	[87]
	0.8 m^2	17.5 L	25–48 °C	400 ± 25 W/m^2	Spain (latitude: 40° N)	Ruthenium (II) tris–chelate complex	*Escherichia coli, Escherichia faecalis*	2006	[88]
	0.4 m^2	14 L	<30 °C	16.5 W/m^2 (UV)	Spain (latitude: 37.09° N)	TiO_2	*Escherichia coli, Fusarium solani, Fusarium anthophilum*	2007	[89]

Table 1. *Cont.*

Thematic Axis of Reactor	Dimension of Sunlight Reflector	Reactant Volume	Temperature	Irradiance	Deployed Location	Photocatalyst	Treatment Target	Timeline	Ref.
CPC	0.2 m^2	7 L	<32 °C	2.8–7.2 W/m^2 (UV)	Spain (latitude: 37° N)	TiO_2	*Escherichia coli*	2012	[90]
	0.25 m^2	5–7 L	-	35–46 W/m^2 (UV)	Spain (latitude: 38.5° N)	TiO_2	Acetaminophen, antipyrine, bisphenol A, caffeine, metoprolol and testosterone	2015	[91]
	Length: 1.2 m	40 L	-	30 W/m^2 (UV)	Colombia (latitude: 3.5° N)	TiO_2	17-β estradiol and nomegestrol acetate	2015	[92]
	Width: 0.207 m	-	-	-	Colombia (latitude: 3.4° N)	TiO_2		2017	[93]
	1 m^2	20 L	26.1–38.2 °C	15.7–33.1 W/m^2 (UV)	Spain (latitude: 37.09° N)	TiO_2	*Curvularia* sp.	2017	[94]
	0.228 m^2	5 L	37 °C	-	Spain (latitude: 41.4° N)	TiO_2	Valproic acid sodium salt	2017	[95]
	1 m^2	15 L	16.6–43.2 °C	40 W/m^2 (UV)	Spain (latitude: 37.84° N)	TiO_2	*Pseudomonas*, *Rheinheimera* and *Methylotenera*	2018	[96]
	0.228 m^2	5 L	30 ± 5 °C	12.45–49.78 W/m^2 (UV)	Spain (latitude: 41.4° N)	TiO_2	Diphenhydramine hydrochloride	2018	[97]
	-	100 L	-	-	Spain (latitude: 37.6° N)	TiO_2	Endocrine disruptors	2018	[98]
	2.56 m^2	11.4 L	-	≈2083 W/m^2	China (latitude: 34.5° N)	$Cd_{0.5}Zn_{0.5}S$	Water splitting	2018	[48]
	1.74 m^2	36 L	33 °C	1573 W/m^2	Colombia (latitude: 10.2° N)	TiO_2	Methylene blue, dichloroacetic acid,	2018	[73]
	1.74 m^2	65 L		1259 W/m^2			4-Chlorophenol, phenol		
	3.2 m^2	35 L	-	30 W/m^2 (UV)	Spain (latitude: 37.09° N)	TiO_2-rGO	Methomyl, Isoproturon, Alachlor	2020	[99]
	Width: 0.75 m Width: 3.6 m	-	-	-	U.S. (latitude: 38.5° N)	g-C_3N_4/chitosan	Atrazine, phenol, sulfamethoxazole, carbamazepine, *Escherichia coli*	2020	[100]
	1 m^2	20 L	12–20 °C	664–1018 W/m^2	Chile (latitude: 37.1° S)	Cu@C	Tartrazine	2022	[38]

4.1. Solar Concentrated Type

4.1.1. Parabolic Trough Reflector-Based Photoreactors

At present, solar-concentrated types of photoreactors mainly apply three technologies, including PTR, PDR and LFR. As a compact and efficient configuration for harvesting solar energy, it is not surprising that PTRs have been widely adopted as the basis for the construction of photocatalytic reactors [74–81,101]. As early as 1991, Anderson et al. reported a PTR-based photocatalytic reactor for groundwater remediation (shown in Figure 4a), which was widely considered to be the first on-site application project [101]. This project was built at the Lawrence Livermore Laboratory in the United States. The system structure consisted of rows of parabola troughs arranged in a field to form a trough array, which could reflect solar rays onto the reaction tube filled with photocatalyst particles. Subsequently, from 1993, a series of PTR-based photoreactors employing TiO_2 as a photocatalyst for the degradation of various organic pollutants was tested and evaluated at the Plataforma Solar de Almeria (PSA), Spain. For example, Minero et al. established a large solar plant for the photocatalytic degradation of pentachlorophenol, which achieved a water treatment capacity of cubic meters per hour (Figure 4b) [78]. The system consisted of a total of twelve heliostat modules that could track solar motion, and every heliostat module was equipped with four PTRs. The nominal aperture area of each heliostat module was 32 m^2, thus providing sufficient light energy for the reaction tube. They also carried out the photocatalytic degradation of atrazine by employing one of the heliostat modules in the above solar plant, which exhibited an atrazine removal ratio of 98% in 2 h. Other photocatalytic experiments using PTR-based photoreactors include the inactivation of *Escherichia coli* [75], the elimination of methylene blue and rhodamine b [80] and the removal of oxalic acid [81]. In addition, experiments for the removal of harmful gases (such as toluene, acetaldehyde, etc.) were also carried out [74]. The above applications of PTRs in the design and development of photocatalytic reactors demonstrate their availability in photocatalysis. Moreover, it is worth noting that in addition to the combination of the heliostat and PTR configuration described above, PTR-based photoreactors can also be constructed by integrating an inclined plate with PTR. For instance, Barzegar et al. reported a parabolic solar collector unit [80] which comprised several PTR components located upon a 30° titled inclined plate (shown in Figure 4c). The system was established facing south to maximize the incident solar rays, and the reaction tubes in each PTR component were connected in sequence. Compared to the traditional heliostat–PTR configuration, since there is no need for a complex two-axis control mechanism, the PTR system equipped with an inclined plate could be more cost-effective, while a decrease in system efficiency might be inevitable.

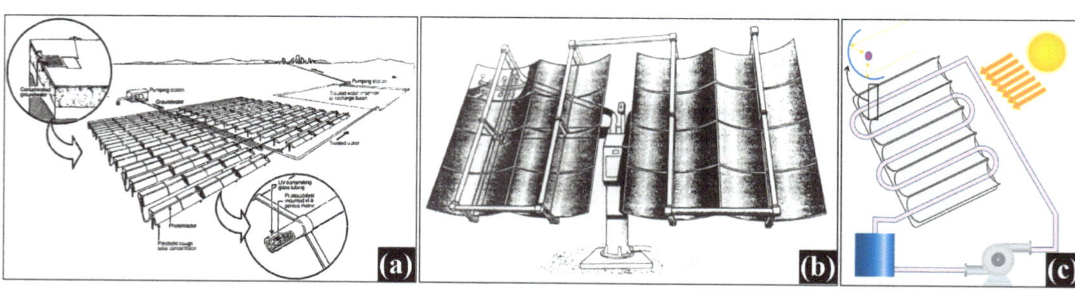

Figure 4. Configurations of parabolic trough collectors. (**a**) Parabolic trough-based photocatalytic system for treating contaminated groundwater (reused with permission from ref. [74]. Copyright 1991 Elsevier B.V.); (**b**) a heliostat-type parabolic trough-based photocatalytic system for degrading pentachlorophenol (reused with permission from ref. [75]. Copyright 1993 Elsevier B.V.); (**c**) a parabolic trough photoreactor installed on the inclined plate [81].

4.1.2. Parabolic Dish Reflector-Based Photoreactors

Due to the high efficiency and compactness of PDR technology in concentrating solar energy, unique PDR-based photoreactors have also been explored. Oyama et al. presented a batch-mode photoreactor system aiming to achieve the photocatalytic degradation of commercial detergent under real sunlight (as shown in Figure 5) [82]. The system consisted of a round concave mirror (parabolic dish reflector) with an aperture diameter of 1.0 m and a flask-type reaction vessel. The parabolic dish reflector employed could achieve a geometric concentration ratio of 70 suns. The reaction vessel was mounted at the focal point of the parabolic dish for adsorbing the incident solar energy. TiO_2, as the photocatalyst, was used in suspension and flowed inside the system circulation loop. The photocatalytic degradation of the refractory detergent driven by concentrated sunlight exhibited a much higher treatment efficiency than conventional biodegradation.

Figure 5. Parabolic dish collector for photocatalytic degradation of detergent (adapted with permission from ref. [82]. Copyright 2004 Elsevier B.V.).

4.1.3. Fresnel Condenser-Based Photoreactors

Linear Fresnel reflectors, due to their excellent solar energy harvesting capabilities and widely acknowledged low costs, have also shown strengths in the structural exploration of photocatalytic systems. Zhang et al. developed an LFR-based photoreactor for the photocatalytic treatment of various organic contaminants including organic dye, antibiotics and pathogenic microorganisms (Figure 6a) [7]. In this system, six flat primary mirrors were mounted on the pedestal of the reactor and driven by stepper motors. The system was located in the north–south direction, with the primary continuously tracking and reflecting solar rays to the reaction tubes. The quartz glass reaction tubes were located above the primary mirror array, which could be moved in a north–south direction to accommodate the solar altitude angle in different seasons. Different from previous studies employing TiO_2 particles, this LFR-based photoreactor utilized TiO_2-coated silicone beads as the photocatalyst, which filled the inside of the reaction tubes. Rhodamine b, amoxicillin and *Escherichia coli* were successfully eliminated under actual weather conditions including a sunny day, cloudy day and rainy day, which demonstrated the feasibility of the LFR configuration for providing solar energy to photocatalysis. In addition, Zhang et al. also reported the fabrication of an LFR-based microalgae biorefinery system [1]. Six flat mirrors were deployed as the primary mirror array for providing the solar flux to a transparent reaction box. The photocatalytic meshes, which were arranged in a stack, were installed inside the reaction box for receiving the reflected solar light. The microalgal solution could flow through the photocatalytic meshes and be destroyed by the photocatalytic effect, thereby achieving the release of carbohydrates, proteins and lipids. Moreover, Ma et al. carried out a hydrogen formation experiment through the photocatalytic conversion of aqueous urea and urine employing a Fresnel lens (FL) (shown in Figure 6b) [83], which provided ultra-high light irradiance (15,000 W/m^2) and a temperature close to the boiling point of water (93 °C). A cascade reaction of heat-triggered urea hydrolysis and thermally assisted ammonia photolysis was established, which resulted in a high hydrogen formation efficiency.

Figure 6. Fresnel condenser-based photoreactors. (**a**) A linear Fresnel collector for photocatalytic wastewater treatment (reused with permission from ref. [7]. Copyright 2023 Elsevier B.V.); (**b**) a Fresnel lens for photocatalytic conversion of aqueous urea and urine (adapted with permission from ref. [83]. Copyright 2022 Elsevier B.V.).

4.1.4. Problems in Solar-Concentrated Photoreactors

Although solar energy-concentrated photocatalytic reactors have demonstrated the capability of improving photocatalytic efficiency, their inherent disadvantages appeared with further research. First, solar-concentrated photoreactors may provide excessively high light irradiance and reaction temperatures for photocatalytic reactions [17,102]. Figure 7a shows the ability of some typical PTRs [45,62,103–105], PDRs [49,106–108], SPTs [69] and LFRs [70,72,109] to concentrate solar energy. Usually, these solar concentrators are used to concentrate solar energy efficiently to produce high temperatures which can be employed in thermal engineering processes such as in photo-thermal power plants. A solar concentration ratio equivalent to hundreds or thousands of suns was achieved, resulting in ultra-high light irradiance (Figure 7a). For example, Mussard et al. reported a PTR system that could achieve a light irradiance of 62,000 W/m^2 [104]. El Ouederni et al. carried out experimental tests and confirmed that the PDR could achieve a 185,000 W/m^2 illumination irradiance [108]. In addition, Nepveu even used PDR to concentrate a light irradiance of 1,068,000 W/m^2 [107]. LFR also achieved a light irradiance of more than 17,000 W/m^2 [70]. Moreover, SPT has the capacity to achieve a light energy concentration of 700,000 W/m^2 [69]. Due to a large amount of light energy converging, the photo-thermal conversion is also substantial, resulting in high temperatures.

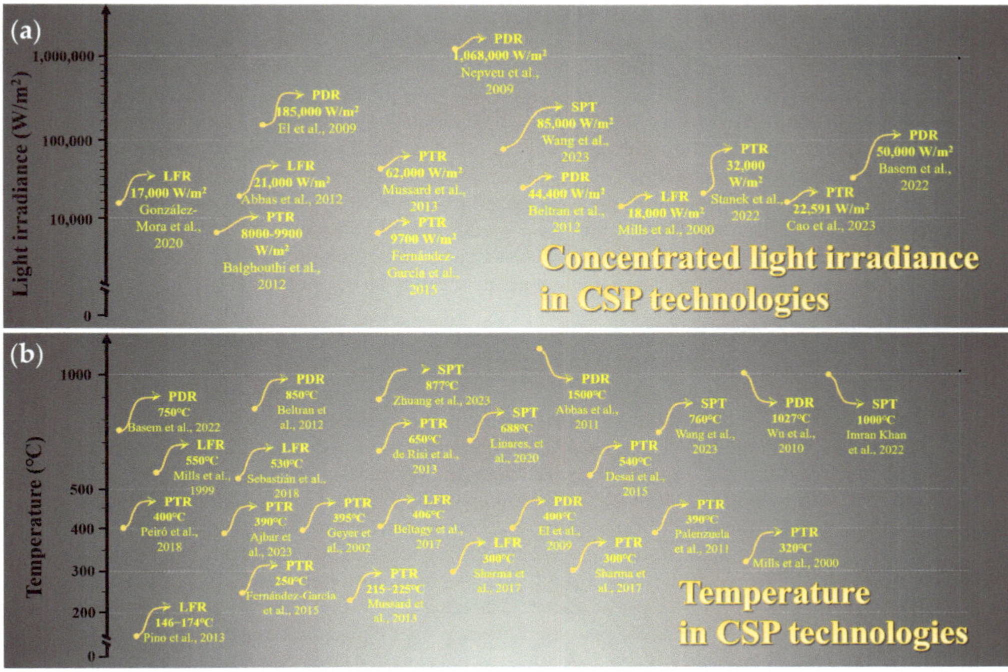

Figure 7. Solar thermal energy provided by typical CSP technologies. (**a**) Concentrated light irradiance [45,49,62,69,70,72,103–109]; (**b**) temperature [18,19,44,49,62,63,68,69,71,72,104,106,108,110–119].

Figure 7b illustrates the temperature in some typical CSP projects, including PTR [18,19,62,72,104,110–114], PDR [44,49,63,106,108], SPT [68,69,115,116] and LFR [71,111,117–119]. For example, de Risi et al. optimized a PTR-based solar collector, which achieved a nanofluid outlet temperature of 650 °C [114]. Khan et al. introduced a next-generation SPT system with a temperature of more than 1000 °C [68]. Additionally, a dish Stirling system was developed by Abbas et al., who demonstrated its ability to achieve a temperature in excess of 1500 °C [44]. Meanwhile, the LFR system proposed by Mills et al. also reached a temperature of 550 °C [119]. Although these extreme conditions are beneficial for some photocatalytic reactions (such as the degradation of volatile organic compounds carried out by Sano et al. [74] and the photo-thermal-induced conversion of aqueous urea by Ma et al. [83]), it is fatal for the vast majority of scenarios in which photocatalytic technology is utilized. A large number of studies have confirmed that due to the fast electron-hole recombination rate, excessive light irradiance is not helpful in improving photocatalytic efficiency. Additionally, the high temperatures caused by excessive light irradiance also have negative effects on the reaction. For example, Askari et al. reported that at temperatures above 35 °C, due to the faster recombination of charge carriers and a decrease in the dissolved oxygen concentration in the water, the photodegradation performance was hindered [40]. When the reaction temperature exceeded 65 °C, Zhang et al. found that due to the evaporation of water and the release of dissolved gas, the reaction volume was crowded out by a large number of bubbles. This not only resulted in a low mass transfer rate, but it also seriously hindered the contact efficiency between the photocatalyst and the treatment target [7]. Under the further increased reaction temperature conditions, such as those greater than 80 °C, Ung-Medina et al. found that the recombination of charge carries was promoted and the adsorption of organic contaminants on the catalyst surface was inhibited [54].

Secondly, although PTR, PDR, SPT and LFR can achieve effective solar energy convergence under sunny conditions with sufficient direct light, their photothermal harvesting capability will plummet on cloudy days, thus greatly limiting the efficiency of the photocatalytic reaction [1,7]. Thirdly, for these solar concentrator configurations, complex mechanisms for driving these structures are unavoidable due to the need to actively track the incidence angle of the solar rays [17]. This would increase the overall design difficulty, construction investment, operation and maintenance costs of photocatalytic reactors. Due to the above issues, traditional CSP-based photoreactors, mainly PTRs, are widely considered to be an outdated technology. In order to find photoreactors that can achieve appropriate enhancement of natural solar energy for achieving promoted photocatalytic efficiency, non/low-concentrated photoreactors began to be developed.

4.2. Non/Low-Concentrated Photoreactors

4.2.1. V-Groove-Based Photoreactors

The V-groove, as a variant of the PTR, simplifies the profile of the parabola into two flat mirrors, thereby dramatically reducing the solar concentrating ratio (as shown in Figure 8a). V-groove-based photoreactors were mainly studied in 2004 by McLoughlin et al. and Bandala et al. to explore the advantages and disadvantages of PTR, V-groove and CPC technologies in photocatalytic applications [75,81]. TiO_2 was adopted as the photocatalyst for the degradation of various organic targets including oxalic acid and *Escherichia coli* under real sunlight. Due to the low concentration of sunlight, the reaction temperature of the V-groove-based photoreactor never exceeded 38 °C. In comparison experiments, the CPC-based photoreactor clearly demonstrated a higher photocatalytic efficiency compared to the V-groove-based photoreactor. This may be due to the fact that the V-groove has no optical focus; therefore, the flat mirror cannot effectively provide the reflected light to the reaction tube, thereby resulting in its low efficiency. As a transitional photocatalytic reactor technology, the V-groove-based photoreactor was quickly eliminated by CPC-based photoreactors.

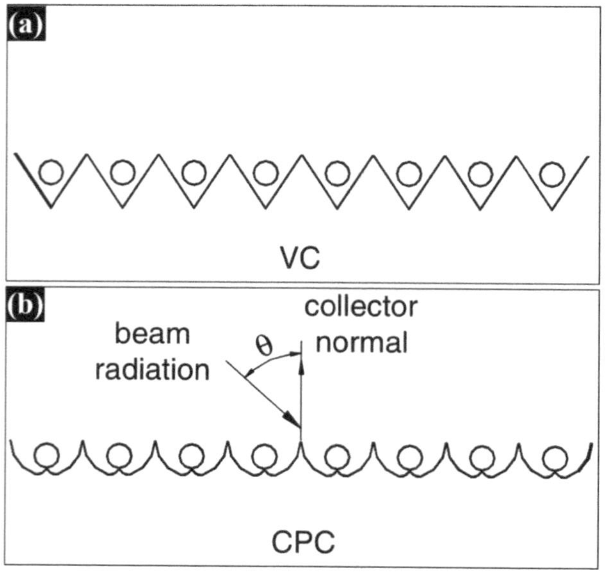

Figure 8. Cross sections of (**a**) V-groove and (**b**) compound parabolic collector-based photoreactors (adapted with permission from ref. [79]. Copyright 2004 Elsevier B.V.).

4.2.2. Compound Parabolic Collector-Based Photoreactors

As mentioned above, in traditional applications, CPC was used as a secondary mirror for LFR, which has the unique advantage of being able to focus the light incoming from any direction to its focus point [17]. Applying a similar principle, under sunny conditions, a fixed CPC could focus direct solar light with different incidence angles at its focal point or reflect scattered light at its focal point on cloudy days. Due to the excellent adaptability of CPC to actual weather conditions, currently, CPC-based photoreactors (shown in Figure 8b) have gradually become the mainstream configuration of photocatalytic reactors, and a large number of CPC photoreactor experiments have been conducted (Table 1) [38,46,48,73,75,79,81,84–100]. For example, Malato et al. established a pilot-scaled CPC photoreactor for the degradation of 2,4-Dichlorophenol [79]. This system was inclined at 37° and located facing south. It was compared with a traditional PTR-based photoreactor under different weather conditions. Under sunny conditions, although the PTR provided much higher light irradiance (6–7 suns) than the CPC photoreactor (only 1 sun), no obvious difference was identified in their photocatalytic efficiencies. This study demonstrated that concentrating too much solar energy has limited value in improving photocatalytic efficiency. Meanwhile, under cloudy weather conditions, a significant improvement in the photocatalytic efficiency was achieved in the CPC photoreactor compared with the PTR photoreactor. This might be because, even in overcast conditions with scattered light, the CPC could still concentrate this light in different directions onto the reaction tube, whereas the PTR could not. This study clearly indicated the advantages of CPC over traditional PTR-based photoreactors, founding a new generation of photoreactors based on CPC technology.

Afterwards, another CPC-based treatment plant was established by Vidal et al. for detoxification and disinfection of contaminated water [84]. Various typical organic pollutants such as pesticides (EPTC, butiphos and γ-lindane) and microorganisms (*Escherichia coli*, *Enterococcus faecalis*) were selected as the treatment targets. All the organic contaminants were nearly 100% removed after 30 min of sunlight exposure, further demonstrating the applicability of CPC photoreactors to different treatment targets. Furthermore, a variant of CPC photoreactors has also been explored to promote its photocatalytic reaction performance. Ochoa-Gutiérrez et al. developed a new CPC photoreactor prototype (named OMTP), adopting multiple reaction tubes for the degradation of several water contaminants (methylene blue, dichloroacetic acid, 4-chlorophenol and phenol) [73]. Compared to a traditional CPC photoreactor, which was installed with only one reaction tube, the OMTP demonstrated its improved photocatalytic treatment efficiency across all of the selected contaminants. This was due to the much-enhanced ratio of the solar irradiated area to the reactant volume in the OMTP, which benefited from the boosted activation of photocatalysts, thereby leading to a superior efficiency. The above early studies of CPC photoreactors have preliminarily confirmed their applicability in photocatalysis, especially highlighting their excellent adaptability to real weather.

In recent years, to further enhance the efficiency of CPC photoreactors in different application directions, photocatalysts with improved solar response capabilities were applied. Luna-Sanguino et al. investigated the degradation of harmful pesticides (methomyl, isoproturon and alachlor) by a CPC photoreactor [99]. In this study, a composite photocatalyst combining TiO_2 with reduced graphene oxide (TiO_2-rGO) was utilized, which possessed broader absorption in the solar visible spectrum region compared with TiO_2 alone. Cao et al. reported several CPC-based water splitting photoreactor systems employing a composite photocatalyst ($Cd_{0.5}Zn_{0.5}S$) and sacrificial agent (Na_2S/Na_2SO_3) [48]. Their CPC system was developed over three generations. The first generation was installed in a closed indoor condition, which utilized the Xe lamp as a light source. Then, a lab-scaled CPC hydrogen production unit was deployed under real sunlight conditions, which was reported to be the first reactor to achieve direct solar photocatalytic hydrogen production. A third-generation CPC photoreactor built to a pilot scale was also deployed under real sunlight, while the reaction medium was naturally circulated to reduce the overall operating energy consumption. Moreover, Muñoz-Flores et al. reported a pilot-scale CPC photoreactor adopting a carbon-containing Cu-based (Cu@C) photocatalyst for the degradation of tartrazine [38]. Compared to the traditional TiO_2, which has low photoactivity under visible light, the employed Cu@C, as a visible light-responsive composite photocatalyst, could be excited not only by UV, but also by the solar visible light, containing significant energy. Therefore, this CPC photoreactor showed a high photocatalytic efficiency.

Despite the above encouraging application of CPC photoreactors in photocatalysis and the continued efficiency improvements brought by numerous efforts, some shortcomings specific to the CPC structure still need to be noted. First of all, although CPC could facilitate light converge uniformly on the reaction tube to a certain extent, it would inevitably cause excessive light irradiance and overheating in some areas of the reaction tube [73]. Therefore, in order to achieve the ideal uniform distribution of light energy on the reaction tube, intensive optimization work and fine assembly of the CPC components are required. On the other hand, manufacturing CPC mirrors, as precision optical devices, poses an additional challenge beyond their complex design and optimization. Ensuring the mirror is produced according to the design blueprint may be also a challenge that needs to be faced. Additionally, the CPC mirrors used in actual engineering are usually manufactured through high-temperature hot bending processes; therefore, this relatively high-cost reflector may also raise some concerns for large-scale application of CPC photoreactors.

5. Relevant Problems in CSP-Based Photoreactors

Due to the variability of actual weather and the complexity of real industrial processes, CSP technology-based photoreactors may also face multiple challenges in the future real large-scale application scenarios. The following sections envisage some of the issues of interest, addressing which may advance the application of CSP technology in photocatalysis and ultimately contribute to the sustainable development of society.

5.1. *Instability of Real Weather*

Although solar energy is an ideal energy source that is widely distributed and could be utilized at a low cost, the actual weather inevitably causes instability in the natural solar energy reaching the ground [7]. For the photocatalytic process, appropriate light irradiance is required to maximize the excitation of the photocatalyst and favorable heat energy is required to promote the reaction rate [7]. Therefore, natural variations in solar energy could inevitably affect the photocatalytic reaction rate. It also needs to be pointed out that modern industrial processes pursue the high robustness of system operation, and uncontrolled decline in the production rate is usually unacceptable. Therefore, if solar-energy-driven photocatalysis technology is to be truly extended to large-scale industrial processes (such as wastewater treatment, hydrogen production, biomass conversion, etc.), searching for solutions that could mitigate or even avoid the effects of natural solar variability on photocatalytic rates needs to be prioritized.

At present, some efforts have been put into solving this problem by researchers. In the development of a photocatalytic reactor based on a linear Fresnel concentrator constructed by Zhang et al., a novel solar energy regulation strategy was employed [7]. Traditional linear Fresnel concentrators utilize multiple primary mirrors to track the sun's movement and reflect the sunlight to the focus to obtain ultra-high energy. Meanwhile, in their reactors, the innovation is that each primary mirror was controlled to rotate through its own independent control system. Under different weather conditions and at different times during the day, there were different numbers of primary mirrors used to focus sunlight. On sunny days when natural sunlight is strong, only two or three primary mirrors focus sunlight on the photocatalytic reaction tube, thus providing superior light irradiance and reaction temperatures for maximizing photocatalytic efficiency. Meanwhile, under cloudy, overcast and other weather conditions with weak natural sunlight, all the primary mirrors participate in the convergence of solar energy to provide optimized light irradiance and reaction temperatures to the photocatalytic reaction. Compared with the reactor that passively receives solar energy, the photocatalytic reaction rate of this solar concentration-controlled reactor was increased by more than 200% [7], thus demonstrating the great potential of this solar control strategy in photocatalytic applications. A similar solar optimization control system has also been successfully adopted by Zhang et al. in another reactor for microalgae biorefining, further demonstrating the broad applicability of this strategy [1]. The above studies provided empirical experience for realizing efficient and stable photocatalytic reactions driven by real sunlight and promoted the practical process of photocatalytic technology. However, for other CSP configurations such as PTR, SPT and PDR, how to realize the effective regulation and optimization of solar energy is still a topic worth exploring.

5.2. Nighttime Operation

Photocatalytic reactions require continuous input of light and heat energy to proceed smoothly. Considering the requirement of high robustness for actual industrial process, it is also a challenge to maintain efficient photocatalytic processes at night when there is no natural sunlight. The adoption of artificial light sources for nighttime photocatalytic lighting is straightforward in theory, but in the actual engineering project design, it is necessary to integrate the natural sunlight collector with the artificial light source. Portela et al. proposed a new solar/lamp-irradiated CPC photoreactor for air treatment [9]. The designed tubular photoreactor consisted of two concentric borosilicate glass tubes, and the space between them was filled with a photocatalyst. Natural solar light was employed as the radiation energy source under sunny conditions during the daytime, which was reflected onto the reaction tube. While during the nighttime and totally overcast conditions, since there was almost no natural light source, an artificial lamp was used as the complement of natural solar energy to activate the photocatalyst, thus allowing 24 h operation in any climate. An automated system was used to switch on or off the lamp depending on the natural light radiation. In the photocatalytic conversion of H_2S, this system demonstrated a sustained reaction capability, demonstrating its availability under real solar conditions. This study provided an idea of a compact reactor layout for designing photocatalytic reactions that could continuously carry out 24 h operation.

5.3. Configuration of Photocatalyst Immobilization Substrate

Most CSP-based photoreactors utilize suspended photocatalysts, which may lead to complex post-separation and increased operating costs. In order to facilitate the large-scale application of photocatalytic technology, the immobilized photocatalyst needs to be equipped in the photoreactors.

In the previous studies, Villén et al. tested the efficiency of two CPC prototypes with different substrate configurations of photocatalysts, namely coaxial- and a fin-type photocatalysts (as shown in Figure 9a) [88]. The results demonstrated that the fin-type photochemical reactor always showed higher inactivation of waterborne bacteria than the coaxial-type reactor. This phenomenon might be due to the lower moving rate of water flowing through the fin-type system, which may lead to a higher number of bacteria attached to the immobilized photocatalyst. Additionally, Alrousan et al. studied the photocatalytic disinfection efficiency in CPC photoreactors equipped with several kinds of photocatalyst-immobilized configurations (as shown in Figure 9b) [90]. The substrate structures included an uncoated single tube, coated single tube, coated double coaxial tube, coated external–uncoated internal coaxial tube, uncoated external–coated internal coaxial tube and uncoated double coaxial tube. Compared with the coated external–uncoated internal coaxial tube system, the uncoated external–coated internal coaxial tube exhibited a significant enhancement in its sterilization capability, which may be because the latter configuration could make use of all the available incident UV solar energy for photolytic inactivation. Meanwhile, the concentric coated tube arrangement showed the highest efficiency among these supporting structures owing to its superior light utilization capability. More recently, Zhang et al. integrated stacked photocatalytic meshes with an

LFR photoreactor [1]. These photocatalytic meshes were inclined at 36.1° to receive real solar energy vertically, thus ensuring the maximized acceptance of incident solar energy. Additionally, a spacing interval of 2 mm was kept between adjacent mesh, thereby not only allowing the reactant to flow smoothly, but also improving the mass transfer rate during the reaction. Moreover, all the photocatalytic meshes were arranged into a three-layer stacked configuration, which not only ensured the maximum utilization of light energy but also reduced the complexity of the system design.

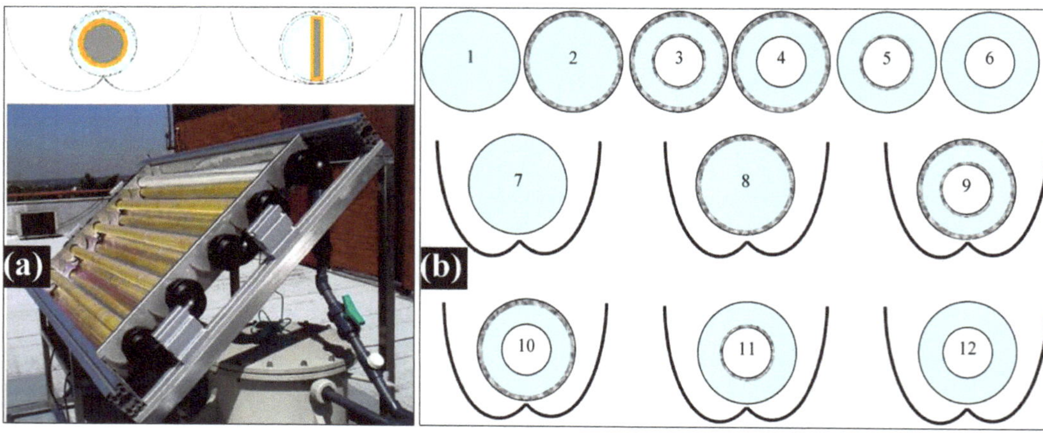

Figure 9. Configurations of photocatalyst immobilization substrates. (**a**) Comparison of coaxial-type and fin-type CPC photoreactors (reused with permission from ref. [88]. Copyright 2006 Elsevier B.V.); (**b**) CPC photoreactor equipped with several photocatalyst-immobilized configurations (reused with permission from ref. [90]. Copyright 2012 Elsevier B.V.).

It is worth noting that the above-mentioned development and optimization of the supporting configuration of the immobilized photocatalysts are all oriented towards specific application scenarios (e.g., sterilization and microalgae processing). The conclusions obtained may not be applicable to a wide range of application scenarios. For other application scenarios (such as gas-phase treatment, water treatment, water splitting, etc.) and different working conditions (such as reactant concentrations, flow rates, etc.), appropriate catalyst substrate configurations are needed. Generally, in a CSP-based photoreactor, the immobilization of photocatalyst may need to meet the following requirements: (1) achieve effective incident light energy utilization; (2) avoid obstructing the flow of the reactant and (3) load a large amount of photocatalyst. Based on the above requirements, an immobilization substrate and configuration that are suitable for CSP-based photocatalytic reactors need to be designed correspondingly.

5.4. Economic Analysis

An economic evaluation is one of the criteria used to judge whether a new technology can be applied to practical applications. Due to the absence of practical application examples of large-scale photocatalytic facilities, an economic analysis of photocatalytic reactors is required. In a study by Vidal et al., a large-scale CPC photocatalytic wastewater treatment plant with an occupied area of 500 m^2 was assumed [84]. The operation cost of this plant was estimated to be 0.7 USD for 1 m^3 water treatment (based on 1997 construction cost indices), which exhibited excellent cost-effectiveness compared with conventional technologies (around 1.0 dollars/m^3). Then, Vela et al. carried out an economic comparison of CPC photoreactors using different photocatalysts (ZnO and TiO_2) [98]. Assuming 365 sunny days in a year and 8 h of system operation per day, the cost per cubic meter of water treated by the TiO_2-adopted CPC photoreactor was 1.45-fold higher than that of the ZnO system.

The difference in process cost caused by the different catalysts was mainly due to the different times the catalysts required for pollutants degradation. Additionally, it was also found that different weather had a significant impact on the system operating cost. The cost of photocatalytic wastewater treatment under real whole-year weather condition was estimated to be 1.40-fold higher than that of ideal year-round sunny conditions, which was due to the inefficient photocatalytic rate on cloudy days. These results further indicated that the stable operation of a CSP-based photoreactor under different weather conditions might be an urgent problem to be solved, as mentioned in Section 5.1. Furthermore, Zhang et al. also implemented a global economic analysis of a LFR photoreactor, which indicated that photocatalytic reactors have significant application potential in every country [1]. However, due to latitude, altitude, climate and other actual factors, the performance of photocatalytic reactors in different countries would be significantly different, and generally, photocatalytic reactors deployed in low-latitude countries could be more effective than in high-latitude countries. Despite studies evaluating photocatalytic reactors from different perspectives that have been proposed, a more comprehensive and professional evaluation of the establishment of real large-scale plants needs to be conducted rigorously.

6. Prospects for Further Work

At present, photocatalytic reactors developed based on CSP technology have shown encouraging application potential in photocatalytic water purification and energy conversion, which has laid the foundation for the early realization of sustainable social development. However, most of the developed CSP-based photoreactors are still in the batch stage, lacking application capabilities in real situations. For industrial and large-scale application scenarios, the establishment of a flow reactor system is necessary, while it requires further technological maturity to promote the realization of this goal. In the past 5 years, some important research results have been obtained, which may set future trends for the future development of CSP-based photoreactors. For example, the solar energy controllable strategy proposed in 2021 could provide optimized light irradiance and temperature for photocatalytic reactions under unstable real solar conditions [7], which provides a reference for the design of efficient and stable photocatalytic reactors in the future. In addition, the installation of multiple reaction tubes in a photoreactor reported in 2018 could improve the ratio of solar irradiated area to reactant volume, which provides a new way for further improving the reaction efficiency [73]. Additionally, the multilayer stacked photocatalytic mesh system proposed in 2023 could achieve efficient reception of solar energy, a large amount of photocatalyst loading, smooth flow of reactants and long-term stability, providing a feasible technical solution to solve the problem of post-separation [1]. The economic evaluation conducted in 2018 clearly indicated that the use of high-performance catalysts in CSP-based photoreactors could significantly reduce reactor operating costs [98]. Visible light-activated photocatalysts, such as graphene oxide-modified $Ag/Ag_2O/BiPO_4/Bi_2WO_6$ developed in 2022 [20], showed excellent photocatalytic capabilities for the removal of organic pollutants and bacteria, sharing great potential to be utilized in future CSP-based photoreactors. Based on the above recent studies, future development of CSP-based photoreactors may focus on the innovation in the reactor structure and the application of high-performance immobilized photocatalysts. These advancements will facilitate the achievement of high-efficiency photocatalytic reactions and low system costs, finally realizing industrial applications.

7. Conclusions

This review illustrated the application of CSP technology in photocatalysis based on three dimensions: the development history of photoreactors, current technical challenges and future work. Efficient photocatalytic reactions require appropriate light irradiance and temperatures, and the low energy density of natural solar energy highlights the necessity of combining CSP technology with photocatalysis. Although existing CSP-based photoreactor technology has significantly improved the practical availability of photocatalysis, the challenges in unstable real weather, nighttime operation, post-separation and economic concern remain to be solved. Based on the latest reports, the adoption of a solar energy control strategy, employment of UV-visible responsive photocatalyst and immobilization substrate, innovations in reactor structure, economic evaluation of systems and establishment of reactor industry standards may be favorable for technological breakthroughs and engineering applications of future CSP-based photoreactors.

Author Contributions: Conceptualization, C.Z.; methodology, C.Z.; software, C.Z.; validation, C.Z.; formal analysis, C.Z. and N.L.; investigation, C.Z.; resources, G.A.; data curation, C.Z.; writing—original draft preparation, C.Z.; writing—review and editing, C.Z., G.A. and N.L.; visualization, C.Z.; supervision, G.A.; project administration, N.L. and G.A.; funding acquisition, C.Z. All authors have read and agreed to the published version of the manuscript.

Funding: This research was funded by the Science and Technology Innovation Project 2021 under grant number XHD2021-002 and the Science and Technology Innovation Project 2022 under grant number XHD2022-003 from the Foshan Xianhu Laboratory.

Conflicts of Interest: The authors declare no conflicts of interest.

References

1. Zhang, C.; Ming, J.; Sun, X.; Zhu, Y.; An, G.; Chen, G.; Yang, Y. Development of a Green and Efficient Photocatalytic Mesh Microalgae Biorefinery (PMMB) System for Sustainable Biomass Conversion under Real Solar Light. *Chem. Eng. J.* **2023**, *466*, 143260. [CrossRef]
2. Akpan, J.O.O. Sustainable Energy Development: History and Recent Advances. *Energies* **2023**, *16*, 7049. [CrossRef]
3. Ming, J.; Sun, X.; Ma, Q.; Liu, N.; Zhang, C.; Kawazoe, N.; Chen, G.; Yang, Y. Advanced Photocatalytic Sterilization for Recalcitrant Enterococcus Sp. Contaminated Water by Newly Developed Z-Scheme Bi_2WO_6 Based Composites under Solar Light. *Chemosphere* **2023**, *310*, 136912. [CrossRef] [PubMed]
4. Ma, Q.; Hu, X.; Liu, N.; Sharma, A.; Zhang, C.; Kawazoe, N.; Chen, G.; Yang, Y. Polyethylene Glycol (PEG)-Modified Ag/Ag2O/Ag3PO4/Bi2WO6 Photocatalyst Film with Enhanced Efficiency and Stability under Solar Light. *J. Colloid Interface Sci.* **2020**, *569*, 101–113. [CrossRef] [PubMed]
5. Zhu, Q.; Liu, N.; Ma, Q.; Sharma, A.; Nagai, D.; Sun, X.; Zhang, C.; Yang, Y. Sol-Gel/Hydrothermal Two-Step Synthesis Strategy for Promoting Ag Species–Modified TiO_2-Based Composite Activity toward H_2 Evolution under Solar Light. *Mater. Today Energy* **2021**, *20*, 100648. [CrossRef]
6. Hu, X.; Ma, Q.; Wang, X.; Yang, Y.; Liu, N.; Zhang, C.; Kawazoe, N.; Chen, G.; Yang, Y. Layered $Ag/Ag_2O/BiPO_4/Bi_2WO_6$ Heterostructures by Two-Step Method for Enhanced Photocatalysis. *J. Catal.* **2020**, *387*, 28–38. [CrossRef]
7. Zhang, C.; Liu, N.; Ming, J.; Sharma, A.; Ma, Q.; Liu, Z.; Chen, G.; Yang, Y. Development of a Novel Solar Energy Controllable Linear Fresnel Photoreactor (LFP) for High-Efficiency Photocatalytic Wastewater Treatment under Actual Weather. *Water Res.* **2022**, *208*, 117880. [CrossRef]
8. Bi, Y.; Ouyang, S.; Umezawa, N.; Cao, J.; Ye, J. Facet Effect of Single-Crystalline Ag_3PO_4 Sub-Microcrystals on Photocatalytic Properties. *J. Am. Chem. Soc.* **2011**, *4*, 6490–6492. [CrossRef]
9. Portela, R.; Suárez, S.; Tessinari, R.F.; Hernández-Alonso, M.D.; Canela, M.C.; Sánchez, B. Solar/Lamp-Irradiated Tubular Photoreactor for Air Treatment with Transparent Supported Photocatalysts. *Appl. Catal. B Environ.* **2011**, *105*, 95–102. [CrossRef]
10. Liu, N.; Zhu, Q.; Zhang, N.; Zhang, C.; Kawazoe, N.; Chen, G.; Negishi, N.; Yang, Y. Superior Disinfection Effect of Escherichia Coli by Hydrothermal Synthesized TiO_2-Based Composite Photocatalyst under LED Irradiation: Influence of Environmental Factors and Disinfection Mechanism. *Environ. Pollut.* **2019**, *247*, 847–856. [CrossRef] [PubMed]
11. Zhang, G.; Lan, Z.A.; Lin, L.; Lin, S.; Wang, X. Overall Water Splitting by Pt/g-C_3N_4 Photocatalysts without Using Sacrificial Agents. *Chem. Sci.* **2016**, *7*, 3062–3066. [CrossRef]
12. Liu, J.; Liu, Y.; Liu, N.; Han, Y.; Zhang, X.; Huang, H.; Lifshitz, Y.; Lee, S.T.; Zhong, J.; Kang, Z. Metal-Free Efficient Photocatalyst for Stable Visible Water Splitting via a Two-Electron Pathway. *Science* **2015**, *347*, 970–974. [CrossRef]
13. Reli, M.; Edelmannová, M.; Šihor, M.; Praus, P.; Svoboda, L.; Mamulová, K.K.; Otoupalíková, H.; Čapek, L.; Hospodková, A.; Obalová, L.; et al. Photocatalytic H_2 Generation from Aqueous Ammonia Solution Using ZnO Photocatalysts Prepared by Different Methods. *Int. J. Hydrogen Energy* **2015**, *40*, 8530–8538. [CrossRef]
14. Kominami, H.; Nishimune, H.; Ohta, Y.; Arakawa, Y.; Inaba, T. Photocatalytic Hydrogen Formation from Ammonia and Methyl Amine in an Aqueous Suspension of Metal-Loaded Titanium(IV) Oxide Particles. *Appl. Catal. B Environ.* **2012**, *111–112*, 297–302. [CrossRef]
15. Vidyasagar, D.; Ghugal, S.G.; Kulkarni, A.; Mishra, P.; Shende, A.G.; Jagannath; Umare, S.S.; Sasikala, R. Silver/Silver(II) Oxide (Ag/AgO) Loaded Graphitic Carbon Nitride Microspheres: An Effective Visible Light Active Photocatalyst for Degradation of Acidic Dyes and Bacterial Inactivation. *Appl. Catal. B Environ.* **2018**, *221*, 339–348. [CrossRef]
16. Deng, J.; Liang, J.; Li, M.; Tong, M. Enhanced Visible-Light-Driven Photocatalytic Bacteria Disinfection by g-C3N4-AgBr. *Colloids Surf. B Biointerfaces* **2017**, *152*, 49–57. [CrossRef]

17. Braham, R.J.; Harris, A.T. Review of Major Design and Scale-up Considerations for Solar Photocatalytic Reactors. *Ind. Eng. Chem. Res.* **2009**, *48*, 8890–8905. [CrossRef]
18. Ajbar, W.; Hernández, J.A.; Parrales, A.; Torres, L. Thermal Efficiency Improvement of Parabolic Trough Solar Collector Using Different Kinds of Hybrid Nanofluids. *Case Stud. Therm. Eng.* **2023**, *42*, 102759. [CrossRef]
19. Desai, N.B.; Bandyopadhyay, S. Optimization of Concentrating Solar Thermal Power Plant Based on Parabolic Trough Collector. *J. Clean. Prod.* **2015**, *89*, 262–271. [CrossRef]
20. Ma, Q.; Ming, J.; Sun, X.; Liu, N.; Chen, G.; Yang, Y. Visible Light Active Graphene Oxide Modified $Ag/Ag_2O/BiPO_4/Bi_2WO_6$ for Photocatalytic Removal of Organic Pollutants and Bacteria in Wastewater. *Chemosphere* **2022**, *306*, 135512. [CrossRef]
21. Jafarova, V.N.; Orudzhev, G.S. Structural and Electronic Properties of ZnO: A First-Principles Density-Functional Theory Study within LDA(GGA) and LDA(GGA)+U Methods. *Solid State Commun.* **2021**, *325*, 114166. [CrossRef]
22. Huang, C.K.; Wu, T.; Huang, C.W.; Lai, C.Y.; Wu, M.Y.; Lin, Y.W. Enhanced Photocatalytic Performance of $BiVO_4$ in Aqueous $AgNO_3$ Solution under Visible Light Irradiation. *Appl. Surf. Sci.* **2017**, *399*, 10–19. [CrossRef]
23. Lee, S.S.; Bai, H.; Liu, Z.; Sun, D.D. Electrospun TiO_2/SnO_2 Nanofibers with Innovative Structure and Chemical Properties for Highly Efficient Photocatalytic H_2 Generation. *Int. J. Hydrogen Energy* **2012**, *37*, 10575–10584. [CrossRef]
24. Yao, W.; Song, X.; Huang, C.; Xu, Q.; Wu, Q. Enhancing Solar Hydrogen Production via Modified Photochemical Treatment of Pt/CdS Photocatalyst. *Catal. Today* **2013**, *199*, 42–47. [CrossRef]
25. Zeng, P.; Zhang, Q.; Zhang, X.; Peng, T. Graphite Oxide-TiO_2 Nanocomposite and Its Efficient Visible-Light-Driven Photocatalytic Hydrogen Production. *J. Alloys Compd.* **2012**, *516*, 85–90. [CrossRef]
26. Kim, J.; Kang, M. High Photocatalytic Hydrogen Production over the Band Gap-Tuned Urchin-like Bi_2S_3-Loaded TiO_2 Composites System. *Int. J. Hydrogen Energy* **2012**, *37*, 8249–8256. [CrossRef]
27. Liu, J.; Cheng, B.; Yu, J. A New Understanding of the Photocatalytic Mechanism of the Direct Z-Scheme g-C_3N_4/TiO_2 Heterostructure. *Phys. Chem. Chem. Phys.* **2016**, *18*, 31175–31183. [CrossRef]
28. Wunderlich, W.; Oekermann, T.; Miao, L.; Hue, N.T.; Tanemura, S.; Tanemura, M. Electronic Properties of Nano-Porous TiO_2- and ZnO-Thin Films-Comparison of Simulations and Experiments. *J. Ceram. Process. Res.* **2004**, *5*, 343–354.
29. Lei, Z.; You, W.; Liu, M.; Zhou, G.; Takata, T.; Hara, M.; Domen, K.; Li, C. Photocatalytic Water Reduction under Visible Light on a Novel $ZnIn_2S_4$ Catalyst Synthesized by Hydrothermal Method. *Chem. Commun.* **2003**, *3*, 2142–2143. [CrossRef]
30. Wang, X.; Li, S.; Yu, H.; Yu, J.; Liu, S. Ag_2O as a New Visible-Light Photocatalyst: Self-Stability and High Photocatalytic Activity. *Chem. A Eur. J.* **2011**, *17*, 7777–7780. [CrossRef] [PubMed]
31. Hu, X.; Zhu, Q.; Wang, X.; Kawazoe, N.; Yang, Y. Nonmetal-Metal-Semiconductor-Promoted $P/Ag/Ag_2O/Ag_3PO_4/TiO_2$ Photocatalyst with Superior Photocatalytic Activity and Stability. *J. Mater. Chem. A* **2015**, *3*, 17858–17865. [CrossRef]
32. Maeda, K.; Lu, D.; Domen, K. Direct Water Splitting into Hydrogen and Oxygen under Visible Light by Using Modified Taon Photocatalysts with D0 Electronic Configuration. *Chem. A Eur. J.* **2013**, *19*, 4986–4991. [CrossRef]
33. Kim, H.G.; Hwang, D.W.; Kim, J.; Kim, Y.G.; Lee, J.S. Highly Donor-Doped (110) Layered Perovskite Materials as Novel Photocatalysts for Overall Water Splitting. *Chem. Commun.* **1999**, *2*, 1077–1078. [CrossRef]
34. Dhanasekaran, P.; Gupta, N.M. Factors Affecting the Production of H_2 by Water Splitting over a Novel Visible-Light-Driven Photocatalyst $GaFeO_3$. *Int. J. Hydrogen Energy* **2012**, *37*, 4897–4907. [CrossRef]
35. Wang, Y.; Ge, H.X.; Chen, Y.P.; Meng, X.Y.; Ghanbaja, J.; Horwat, D.; Pierson, J.F. Wurtzite CoO: A Direct Band Gap Oxide Suitable for a Photovoltaic Absorber. *Chem. Commun.* **2018**, *54*, 13949–13952. [CrossRef]
36. Hu, S.P.; Xu, C.Y.; Zhen, L. Solvothermal Synthesis of Bi_2WO_6 Hollow Structures with Excellent Visible-Light Photocatalytic Properties. *Mater. Lett.* **2013**, *95*, 117–120. [CrossRef]
37. Basu, M.; Sinha, A.K.; Pradhan, M.; Sarkar, S.; Negishi, Y.; Govind; Pal, T. Evolution of Hierarchical Hexagonal Stacked Plates of CuS from Liquid—Liquid Interface and Its Photocatalytic Application for Oxidative Degradation of Different Dyes under Indoor Lighting. *Environ. Sci. Technol.* **2010**, *44*, 6313–6318. [CrossRef] [PubMed]
38. Muñoz-Flores, P.; Poon, P.S.; Ania, C.O.; Matos, J. Performance of a C-Containing Cu-Based Photocatalyst for the Degradation of Tartrazine: Comparison of Performance in a Slurry and CPC Photoreactor under Artificial and Natural Solar Light. *J. Colloid Interface Sci.* **2022**, *623*, 646–659. [CrossRef]
39. Reutergårdh, L.B.; Iangphasuk, M. Photocatalytic Decolourization of Reactive Azo Dye: A Comparison between TiO_2 and CdS Photocatalysis. *Chemosphere* **1997**, *35*, 585–596. [CrossRef]
40. Askari, N.; Beheshti, M.; Mowla, D.; Farhadian, M. Facile Construction of Novel Z-Scheme MnWO4/Bi_2S_3 Heterojunction with Enhanced Photocatalytic Degradation of Antibiotics. *Mater. Sci. Semicond. Process.* **2021**, *127*, 105723. [CrossRef]
41. Yu, J.; Kiwi, J.; Zivkovic, I.; Rønnow, H.M.; Wang, T.; Rtimi, S. Quantification of the Local Magnetized Nanotube Domains Accelerating the Photocatalytic Removal of the Emerging Pollutant Tetracycline. *Appl. Catal. B Environ.* **2019**, *248*, 450–458. [CrossRef]
42. Castro-Alférez, M.; Polo-López, M.I.; Marugán, J.; Fernández-Ibáñez, P. Mechanistic Model of the Escherichia Coli Inactivation by Solar Disinfection Based on the Photo-Generation of Internal ROS and the Photo-Inactivation of Enzymes: CAT and SOD. *Chem. Eng. J.* **2017**, *318*, 214–223. [CrossRef]
43. Chen, Y.; Wang, Y.; Li, W.; Yang, Q.; Hou, Q.; Wei, L.; Liu, L.; Huang, F.; Ju, M. Enhancement of Photocatalytic Performance with the Use of Noble-Metal-Decorated TiO_2 Nanocrystals as Highly Active Catalysts for Aerobic Oxidation under Visible-Light Irradiation. *Appl. Catal. B Environ.* **2017**, *210*, 352–367. [CrossRef]

44. Abbas, M.; Boumeddane, B.; Said, N.; Chikouche, A. Dish Stirling Technology: A 100 MW Solar Power Plant Using Hydrogen for Algeria. *Int. J. Hydrogen Energy* **2011**, *36*, 4305–4314. [CrossRef]
45. Balghouthi, M.; Chahbani, M.H.; Guizani, A. Investigation of a Solar Cooling Installation in Tunisia. *Appl. Energy* **2012**, *98*, 138–148. [CrossRef]
46. Augugliaro, V.; Baiocchi, C.; Prevot, A.B.; García-López, E.; Loddo, V.; Malato, S.; Marcí, G.; Palmisano, L.; Pazzi, M.; Pramauro, E. Azo-Dyes Photocatalytic Degradation in Aqueous Suspension of TiO_2 under Solar Irradiation. *Chemosphere* **2002**, *49*, 1223–1230. [CrossRef] [PubMed]
47. Chafie, M.; Ben Aissa, M.F.; Guizani, A. Energetic End Exergetic Performance of a Parabolic Trough Collector Receiver: An Experimental Study. *J. Clean. Prod.* **2018**, *171*, 285–296. [CrossRef]
48. Cao, F.; Wei, Q.; Liu, H.; Lu, N.; Zhao, L.; Guo, L. Development of the Direct Solar Photocatalytic Water Splitting System for Hydrogen Production in Northwest China: Design and Evaluation of Photoreactor. *Renew. Energy* **2018**, *121*, 153–163. [CrossRef]
49. Basem, A.; Moawed, M.; Abbood, M.H.; El-Maghlany, W.M. The Energy and Exergy Analysis of a Combined Parabolic Solar Dish—Steam Power Plant. *Renew. Energy Focus* **2022**, *41*, 55–68. [CrossRef]
50. Salamat, S.; Younesi, H.; Bahramifar, N. Synthesis of Magnetic Core-Shell Fe_3O_4@TiO_2 Nanoparticles from Electric Arc Furnace Dust for Photocatalytic Degradation of Steel Mill Wastewater. *RSC Adv.* **2017**, *7*, 19391–19405. [CrossRef]
51. Ung-Medina, F.; Caudillo-Flores, U.; Correa-González, J.C.; Maya-Yescas, R.; Chávez-Parga, M.D.C.; Cortés, J.A. Use of an Annular Non-Sleeve Photoreactor for Photocatalytic Dye Degradation: Study of Temperature and Light Intensity Effects. *Environ. Prog. Sustain. Energy* **2017**, *36*, 1083–1088. [CrossRef]
52. Wang, H.; Sun, Y.; Wu, Y.; Tu, W.; Wu, S.; Yuan, X.; Zeng, G.; Xu, Z.J.; Li, S.; Chew, J.W. Electrical Promotion of Spatially Photoinduced Charge Separation via Interfacial-Built-in Quasi-Alloying Effect in Hierarchical $Zn_2In_2S_5/Ti_3C_2$(O, OH)x Hybrids toward Efficient Photocatalytic Hydrogen Evolution and Environmental Remediation. *Appl. Catal. B Environ.* **2019**, *245*, 290–301. [CrossRef]
53. Liu, B.; Wu, H.; Parkin, I.P. New Insights into the Fundamental Principle of Semiconductor Photocatalysis. *ACS Omega* **2020**, *5*, 14847–14856. [CrossRef]
54. Herrmann, J.M. Heterogeneous Photocatalysis: State of the Art and Present Applications In honor of Pr. R.L. Burwell Jr. (1912–2003), Former Head of Ipatieff Laboratories, Northwestern University, Evanston (Ill). *Top. Catal.* **2005**, *34*, 49–65. [CrossRef]
55. Al-Soud, M.S.; Hrayshat, E.S. A 50 MW Concentrating Solar Power Plant for Jordan. *J. Clean. Prod.* **2009**, *17*, 625–635. [CrossRef]
56. Balzar, A.; Stumpf, P.; Eckhoff, S.; Ackermann, H.; Grupp, M. A Solar Cooker Using Vacuum-Tube Collectors with Integrated Heat Pipes. *Sol. Energy* **1996**, *58*, 63–68. [CrossRef]
57. Jafari Mosleh, H.; Mamouri, S.J.; Shafii, M.B.; Hakim Sima, A. A New Desalination System Using a Combination of Heat Pipe, Evacuated Tube and Parabolic through Collector. *Energy Convers. Manag.* **2015**, *99*, 141–150. [CrossRef]
58. Abu-Hamdeh, N.H.; Alnefaie, K.A.; Almitani, K.H. Design and Performance Characteristics of Solar Adsorption Refrigeration System Using Parabolic Trough Collector: Experimental and Statistical Optimization Technique. *Energy Convers. Manag.* **2013**, *74*, 162–170. [CrossRef]
59. Bigoni, R.; Kötzsch, S.; Sorlini, S.; Egli, T. Solar Water Disinfection by a Parabolic Trough Concentrator (PTC): Flow-Cytometric Analysis of Bacterial Inactivation. *J. Clean. Prod.* **2014**, *67*, 62–71. [CrossRef]
60. Fernandez-Ibañez, P.; Malato, S.; Enea, O. Photoelectrochemical Reactors for the Solar Decontamination of Water. *Catal. Today* **1999**, *54*, 329–339. [CrossRef]
61. García-Montaño, J.; Pérez-Estrada, L.; Oller, I.; Maldonado, M.I.; Torrades, F.; Peral, J. Pilot Plant Scale Reactive Dyes Degradation by Solar Photo-Fenton and Biological Processes. *J. Photochem. Photobiol. A Chem.* **2008**, *195*, 205–214. [CrossRef]
62. Fernández-García, A.; Rojas, E.; Pérez, M.; Silva, R.; Hernández-Escobedo, Q.; Manzano-Agugliaro, F. A Parabolic-Trough Collector for Cleaner Industrial Process Heat. *J. Clean. Prod.* **2015**, *89*, 272–285. [CrossRef]
63. Wu, S.Y.; Xiao, L.; Cao, Y.; Li, Y.R. A Parabolic Dish/AMTEC Solar Thermal Power System and Its Performance Evaluation. *Appl. Energy* **2010**, *87*, 452–462. [CrossRef]
64. El-Kassaby, M.M. New Solar Cooker of Parabolic Square Dish: Design and Simulation. *Renew. Energy* **1991**, *1*, 59–65. [CrossRef]
65. Grupp, M.; Balmer, M.; Beall, B.; Bergler, H.; Cieslok, J.; Hancock, D.; Schröder, G. On-Line Recording of Solar Cooker Use Rate by a Novel Metering Device: Prototype Description and Experimental Verification of Output Data. *Sol. Energy* **2009**, *83*, 276–279. [CrossRef]
66. Badran, A.A.; Yousef, I.A.; Joudeh, N.K.; Hamad, R.; Al Halawa, H.; Hassouneh, H.K. Portable Solar Cooker and Water Heater. *Energy Convers. Manag.* **2010**, *51*, 1605–1609. [CrossRef]
67. Srithar, K.; Rajaseenivasan, T.; Karthik, N.; Periyannan, M.; Gowtham, M. Stand Alone Triple Basin Solar Desalination System with Cover Cooling and Parabolic Dish Concentrator. *Renew. Energy* **2016**, *90*, 157–165. [CrossRef]
68. Imran Khan, M.; Asfand, F.; Al-Ghamdi, S.G. Progress in Technology Advancements for next Generation Concentrated Solar Power Using Solid Particle Receivers. *Sustain. Energy Technol. Assess.* **2022**, *54*, 102813. [CrossRef]
69. Wang, Q.; Yao, Y.; Shen, Z.; Hu, M.; Yang, H. Concentrated Solar Power Tower Systems Coupled Locally with Spectrally Selective Coatings for Enhancement of Solar-Thermal Conversion and Economic Performance. *Green Energy Resour.* **2023**, *1*, 100001. [CrossRef]
70. González-Mora, E.; Dolores Durán García, M. Methodology for an Opto-Geometric Optimization of a Linear Fresnel Reflector for Direct Steam Generation. *Energies* **2020**, *13*, 355. [CrossRef]

71. Sebastián, A.; Abbas, R.; Valdés, M.; Casanova, J. Innovative Thermal Storage Strategies for Fresnel-Based Concentrating Solar Plants with East-West Orientation. *Appl. Energy* **2018**, *230*, 983–995. [CrossRef]
72. Mills, D.R.; Morrison, G.L. Compact Linear Fresnel Reflector Solar Thermal Powerplants. *Sol. Energy* **2000**, *68*, 263–283. [CrossRef]
73. Ochoa-Gutiérrez, K.S.; Tabares-Aguilar, E.; Mueses, M.Á.; Machuca-Martínez, F.; Li Puma, G. A Novel Prototype Offset Multi Tubular Photoreactor (OMTP) for Solar Photocatalytic Degradation of Water Contaminants. *Chem. Eng. J.* **2018**, *341*, 628–638. [CrossRef]
74. Anderson, J.V.; Link, H.; Bohn, M.; Gupta, B. Development of Solar Detoxification Technology in the USA—An Introduction. *Sol. Energy Mater.* **1991**, *24*, 538–549. [CrossRef]
75. Minero, C.; Pelizzetti, E.; Malato, S.; Blanco, J. Large Solar Plant Photocatalytic Water Decontamination: Degradation of Pentachlorophenol. *Chemosphere* **1993**, *26*, 2103–2119. [CrossRef]
76. Malato, S.; Blanco, J.; Richter, C.; Curco, D.; Gimenez, J. Low-Concentrating CPC Collectors for Photocatalytic Water Detoxification: Comparison with a Medium Concentrating Solar Collector. *Water Sci. Technol.* **1997**, *35*, 157–164. [CrossRef]
77. Klare, M.; Scheen, J.; Vogelsang, K.; Jacobs, H.; Broekaert, J.A.C. Degradation of Short-Chain Alkyl- and Alkanolamines by TiO_2- and Pt/TiO_2-Assisted Photocatalysis. *Chemosphere* **2000**, *41*, 353–362. [CrossRef] [PubMed]
78. Sano, T.; Negishi, N.; Takeuchi, K.; Matsuzawa, S. Degradation of Toluene and Acetaldehyde with Pt-Loaded TiO_2 Catalyst and Parabolic Trough Concentrator. *Sol. Energy* **2004**, *77*, 543–552. [CrossRef]
79. Bandala, E.R.; Arancibia-Bulnes, C.A.; Orozco, S.L.; Estrada, C.A. Solar Photoreactors Comparison Based on Oxalic Acid Photocatalytic Degradation. *Sol. Energy* **2004**, *77*, 503–512. [CrossRef]
80. McLoughlin, O.A.; Kehoe, S.C.; McGuigan, K.G.; Duffy, E.F.; Al Touati, F.; Gernjak, W.; Oller Alberola, I.; Malato Rodríguez, S.; Gill, L.W. Solar Disinfection of Contaminated Water: A Comparison of Three Small-Scale Reactors. *Sol. Energy* **2004**, *77*, 657–664. [CrossRef]
81. Barzegar, M.H.; Sabzehmeidani, M.M.; Ghaedi, M.; Avargani, V.M.; Moradi, Z.; Roy, V.A.L.; Heidari, H. S-Scheme Heterojunction g-C3N4/TiO_2 with Enhanced Photocatalytic Activity for Degradation of a Binary Mixture of Cationic Dyes Using Solar Parabolic Trough Reactor. *Chem. Eng. Res. Des.* **2021**, *174*, 307–318. [CrossRef]
82. Oyama, T.; Aoshima, A.; Horikoshi, S.; Hidaka, H.; Zhao, J.; Serpone, N. Solar Photocatalysis, Photodegradation of a Commercial Detergent in Aqueous TiO_2 Dispersions under Sunlight Irradiation. *Sol. Energy* **2004**, *77*, 525–532. [CrossRef]
83. Ma, R.; Su, H.; Sun, J.; Li, D.; Zhang, Z.; Wei, J. Concentrating Photo-Thermo-Organized Single-Atom and 2D-Raft Cu Catalyst for Full-Spectrum Solar Harmonic Conversion of Aqueous Urea and Urine into Hydrogen. *Appl. Catal. B Environ.* **2022**, *315*, 121493. [CrossRef]
84. Vidal, A.; Díaz, A.I.; El Hraiki, A.; Romero, M.; Muguruza, I.; Senhaji, F.; González, J. Solar Photocatalysis for Detoxification and Disinfection of Contaminated Water: Pilot Plant Studies. *Catal. Today* **1999**, *54*, 283–290. [CrossRef]
85. Kositzi, M.; Poulios, I.; Malato, S.; Caceres, J.; Campos, A. Solar Photocatalytic Treatment of Synthetic Municipal Wastewater. *Water Res.* **2004**, *38*, 1147–1154. [CrossRef] [PubMed]
86. Fernández, P.; Blanco, J.; Sichel, C.; Malato, S. Water Disinfection by Solar Photocatalysis Using Compound Parabolic Collectors. *Catal. Today* **2005**, *101*, 345–352. [CrossRef]
87. Augugliaro, V.; García-López, E.; Loddo, V.; Malato-Rodríguez, S.; Maldonado, I.; Marcì, G.; Molinari, R.; Palmisano, L. Degradation of Lincomycin in Aqueous Medium: Coupling of Solar Photocatalysis and Membrane Separation. *Sol. Energy* **2005**, *79*, 402–408. [CrossRef]
88. Villén, L.; Manjón, F.; García-Fresnadillo, D.; Orellana, G. Solar Water Disinfection by Photocatalytic Singlet Oxygen Production in Heterogeneous Medium. *Appl. Catal. B Environ.* **2006**, *69*, 1–9. [CrossRef]
89. Sichel, C.; Tello, J.; de Cara, M.; Fernández-Ibáñez, P. Effect of UV Solar Intensity and Dose on the Photocatalytic Disinfection of Bacteria and Fungi. *Catal. Today* **2007**, *129*, 152–160. [CrossRef]
90. Alrousan, D.M.A.; Polo-López, M.I.; Dunlop, P.S.M.; Fernández-Ibáñez, P.; Byrne, J.A. Solar Photocatalytic Disinfection of Water with Immobilised Titanium Dioxide in Re-Circulating Flow CPC Reactors. *Appl. Catal. B Environ.* **2012**, *128*, 126–134. [CrossRef]
91. Quiñones, D.H.; Álvarez, P.M.; Rey, A.; Beltrán, F.J. Removal of Emerging Contaminants from Municipal WWTP Secondary Effluents by Solar Photocatalytic Ozonation. A Pilot-Scale Study. *Sep. Purif. Technol.* **2015**, *149*, 132–139. [CrossRef]
92. Colina-Márquez, J.; Machuca-Martínez, F.; Puma, G.L. Modeling the Photocatalytic Mineralization in Water of Commercial Formulation of Estrogens 17-β Estradiol (E2) and Nomegestrol Acetate in Contraceptive Pills in a Solar Powered Compound Parabolic Collector. *Molecules* **2015**, *20*, 13354–13373. [CrossRef]
93. Otálvaro-Marín, H.L.; Mueses, M.A.; Crittenden, J.C.; Machuca-Martinez, F. Solar Photoreactor Design by the Photon Path Length and Optimization of the Radiant Field in a TiO_2-Based CPC Reactor. *Chem. Eng. J.* **2017**, *315*, 283–295. [CrossRef]
94. Aguas, Y.; Hincapie, M.; Fernández-Ibáñez, P.; Polo-López, M.I. Solar Photocatalytic Disinfection of Agricultural Pathogenic Fungi (Curvularia Sp.) in Real Urban Wastewater. *Sci. Total Environ.* **2017**, *607–608*, 1213–1224. [CrossRef] [PubMed]
95. Haranaka-Funai, D.; Didier, F.; Giménez, J.; Marco, P.; Esplugas, S.; Machulek-Junior, A. Photocatalytic Treatment of Valproic Acid Sodium Salt with TiO_2 in Different Experimental Devices: An Economic and Energetic Comparison. *Chem. Eng. J.* **2017**, *327*, 656–665. [CrossRef]

96. Moreira, N.F.F.; Narciso-da-Rocha, C.; Polo-López, M.I.; Pastrana-Martínez, L.M.; Faria, J.L.; Manaia, C.M.; Fernández-Ibáñez, P.; Nunes, O.C.; Silva, A.M.T. Solar Treatment (H_2O_2, TiO_2-P_{25} and GO-TiO_2 Photocatalysis, Photo-Fenton) of Organic Micropollutants, Human Pathogen Indicators, Antibiotic Resistant Bacteria and Related Genes in Urban Wastewater. *Water Res.* **2018**, *135*, 195–206. [CrossRef]
97. López, N.; Marco, P.; Giménez, J.; Esplugas, S. Photocatalytic Diphenhydramine Degradation under Different Radiation Sources: Kinetic Studies and Energetic Comparison. *Appl. Catal. B Environ.* **2018**, *220*, 497–505. [CrossRef]
98. Vela, N.; Calín, M.; Yáñez-Gascón, M.J.; Garrido, I.; Pérez-Lucas, G.; Fenoll, J.; Navarro, S. Solar Reclamation of Wastewater Effluent Polluted with Bisphenols, Phthalates and Parabens by Photocatalytic Treatment with $TiO_2/Na_2S_2O_8$ at Pilot Plant Scale. *Chemosphere* **2018**, *212*, 95–104. [CrossRef]
99. Luna-Sanguino, G.; Ruíz-Delgado, A.; Tolosana-Moranchel, A.; Pascual, L.; Malato, S.; Bahamonde, A.; Faraldos, M. Solar Photocatalytic Degradation of Pesticides over TiO_2-RGO Nanocomposites at Pilot Plant Scale. *Sci. Total Environ.* **2020**, *737*, 140286. [CrossRef]
100. Zheng, Q.; Aiello, A.; Choi, Y.S.; Tarr, K.; Shen, H.; Durkin, D.P.; Shuai, D. 3D Printed Photoreactor with Immobilized Graphitic Carbon Nitride: A Sustainable Platform for Solar Water Purification. *J. Hazard. Mater.* **2020**, *399*, 123097. [CrossRef]
101. Minero, C.; Pelizzetti, E.; Malato, S.; Blanco, J. Large Solar Plant Photocatalytic Water Decontamination: Degradation of Atrazine. *Sol. Energy* **1996**, *56*, 411–419. [CrossRef]
102. Malato, S.; Blanco, J.; Vidal, A.; Richter, C. Photocatalysis with Solar Energy at a Pilot-Plant Scale: An Overview. *Appl. Catal. B Environ.* **2002**, *37*, 1–15. [CrossRef]
103. Cao, F.; Pang, J.; Gu, X.; Wang, M.; Shangguan, Y. Performance Simulation of Solar Trough Concentrators: Optical and Thermal Comparisons. *Energies* **2023**, *16*, 1673. [CrossRef]
104. Mussard, M.; Nydal, O.J. Charging of a Heat Storage Coupled with a Low-Cost Small-Scale Solar Parabolic Trough for Cooking Purposes. *Sol. Energy* **2013**, *95*, 144–154. [CrossRef]
105. Stanek, B.; Węcel, D.; Bartela, Ł.; Rulik, S. Solar Tracker Error Impact on Linear Absorbers Efficiency in Parabolic Trough Collector—Optical and Thermodynamic Study. *Renew. Energy* **2022**, *196*, 598–609. [CrossRef]
106. Beltran, R.; Velazquez, N.; Espericueta, A.C.; Sauceda, D.; Perez, G. Mathematical Model for the Study and Design of a Solar Dish Collector with Cavity Receiver for Its Application in Stirling Engines. *J. Mech. Sci. Technol.* **2012**, *26*, 3311–3321. [CrossRef]
107. Nepveu, F.; Ferriere, A.; Bataille, F. Thermal Model of a Dish/Stirling Systems. *Sol. Energy* **2009**, *83*, 81–89. [CrossRef]
108. El Ouederni, A.R.; Salah, M.B.; Askri, F.; Nasrallah, M.B.; Aloui, F. Experimental Study of a Parabolic Solar Concentrator. *J. Renew. Energ.* **2009**, *12*, 395–404.
109. Abbas, R.; Montes, M.J.; Piera, M.; Martínez-Val, J.M. Solar Radiation Concentration Features in Linear Fresnel Reflector Arrays. *Energy Convers. Manag.* **2012**, *54*, 133–144. [CrossRef]
110. Geyer, M.; Lüpfert, E.; Osuna, R.; Nava, P.; Langenkamp, J.; Mandelberg, E. EUROTROUGH-Parabolic Trough Collector Developed for Cost Efficient Solar Power Generation. In Proceedings of the 11th International Symposium on Concentrating Solar Power and Chemical Energy Technologies, Zurich, Switzerland, 4–6 September 2002; Volume 7.
111. Sharma, A.K.; Sharma, C.; Mullick, S.C.; Kandpal, T.C. GHG Mitigation Potential of Solar Industrial Process Heating in Producing Cotton Based Textiles in India. *J. Clean. Prod.* **2017**, *145*, 74–84. [CrossRef]
112. Palenzuela, P.; Zaragoza, G.; Alarcón-Padilla, D.C.; Guillén, E.; Ibarra, M.; Blanco, J. Assessment of Different Configurations for Combined Parabolic-Trough (PT) Solar Power and Desalination Plants in Arid Regions. *Energy* **2011**, *36*, 4950–4958. [CrossRef]
113. Peiró, G.; Prieto, C.; Gasia, J.; Jové, A.; Miró, L.; Cabeza, L.F. Two-Tank Molten Salts Thermal Energy Storage System for Solar Power Plants at Pilot Plant Scale: Lessons Learnt and Recommendations for Its Design, Start-up and Operation. *Renew. Energy* **2018**, *121*, 236–248. [CrossRef]
114. de Risi, A.; Milanese, M.; Laforgia, D. Modelling and Optimization of Transparent Parabolic Trough Collector Based on Gas-Phase Nanofluids. *Renew. Energy* **2013**, *58*, 134–139. [CrossRef]
115. Linares, J.I.; Montes, M.J.; Cantizano, A.; Sánchez, C. A Novel Supercritical CO_2 Recompression Brayton Power Cycle for Power Tower Concentrating Solar Plants. *Appl. Energy* **2020**, *263*, 114644. [CrossRef]
116. Zhuang, X.; Wang, H.; Lu, H.; Yang, Z.; Guo, H. Numerical Investigation of Heat Transfer and Flow Characteristics of Supercritical CO_2 in Solar Tower Microchannel Receivers at High Temperature. *Energies* **2023**, *16*, 6445. [CrossRef]
117. Pino, F.J.; Caro, R.; Rosa, F.; Guerra, J. Experimental Validation of an Optical and Thermal Model of a Linear Fresnel Collector System. *Appl. Therm. Eng.* **2013**, *50*, 1463–1471. [CrossRef]
118. Beltagy, H.; Semmar, D.; Lehaut, C.; Said, N. Theoretical and Experimental Performance Analysis of a Fresnel Type Solar Concentrator. *Renew. Energy* **2017**, *101*, 782–793. [CrossRef]
119. Mills, D.R.; Morrison, G.L. Modelling Study for Compact Fresnel Reflector Power Plant. *Le J. De Phys. IV* **1999**, *9*, Pr3-159–Pr3-165. [CrossRef]

MDPI AG
Grosspeteranlage 5
4052 Basel
Switzerland
Tel.: +41 61 683 77 34

Energies Editorial Office
E-mail: energies@mdpi.com
www.mdpi.com/journal/energies

www.ingramcontent.com/pod-product-compliance
Lightning Source LLC
Chambersburg PA
CBHW041349210926
51850CB00013B/3026
* 9 7 8 3 7 2 5 8 4 2 5 3 7 *